MATERIALS SELECTION

IN

MECHANICAL DESIGN

MATERIALS SELECTION

IN

MECHANICAL DESIGN

SECOND EDITION

MICHAEL F. ASHBY
Department of Engineering, Cambridge University, England

OXFORD AUCKLAND BOSTON JOHANNESBURG MELBOURNE NEW DELHI

Butterworth-Heinemann
Linacre House, Jordan Hill, Oxford OX2 8DP
225 Wildwood Avenue, Woburn, MA 01801-2041
A division of Reed Educational and Professional Publishing Ltd

A member of the Reed Elsevier plc group

First published by Pergamon Press Ltd 1992
Reprinted with corrections 1993
Reprinted 1995, 1996, 1997
Second edition 1999
Reprinted 2000 (twice)

British Library Cataloguing in Publication Data
A catalogue record for this book is available from the British Library

Library of Congress Cataloguing in Publication Data
A catalogue record for this book is available from the Library of Congress

ISBN 0 7506 4357 9

Typeset by Laser Words, Madras, India
Printed in Great Britain

FOR EVERY TITLE THAT WE PUBLISH, BUTTERWORTH-HEINEMANN
WILL PAY FOR BTCV TO PLANT AND CARE FOR A TREE.

Contents

Preface

'Materials, of themselves, affect us little; it is the way we use them which influences our lives'. Epictetus, AD 50–100, *Discourses* Book 2, Chapter 5.

New materials advanced engineering design in Epictetus' time. Today, with more materials than ever before, the opportunities for innovation are immense. But advance is possible only if a procedure exists for making a rational choice. This book develops a systematic procedure for selecting materials and processes, leading to the subset which best matches the requirements of a design. It is unique in the way the information it contains has been structured; the structure gives rapid access to data and it gives the user great freedom in exploring the potential of choice. The method is available as software* which allows even greater flexibility.

The approach emphasizes design with materials rather than materials 'science', although the underlying science is used, whenever possible, to help with the structuring of criteria for selection. The first six chapters require little prior knowledge: a first-year engineering knowledge of materials and mechanics is enough. The chapters dealing with shape and multi-objective selection are a little more advanced but can be omitted on a first reading. As far as possible the book integrates materials selection with other aspects of design; the relationship with the stages of design and optimization, and with the mechanics of materials, are developed throughout. At the teaching level, the book is intended as the text for 3rd and 4th year engineering courses on Materials for Design: a 6 to 10 lecture unit can be based on Chapters 1 to 6; a full 20+ lecture course, with associated project work with the associated software, uses the entire book.

Beyond this, the book is intended as a reference text of lasting value. The method, the charts and tables of performance indices have application in real problems of materials and process selection; and the catalogue of 'useful solutions' is particularly helpful in modelling — an essential ingredient of optimal design. The reader can use the book at increasing levels of sophistication as his or her experience grows, starting with the material indices developed in the case studies of the text, and graduating to the modelling of new design problems, leading to new material indices and value functions, and new — and perhaps novel — choices of material. This continuing education aspect is helped by a list of further reading at the end of each chapter, and by a set of problems covering all aspects of the text. Useful reference material is assembled in Appendices at the end of the book.

Like any other book, the contents of this one are protected by copyright. Generally, it is an infringement to copy and distribute material from a copyrighted source. But the best way to use the charts which are a feature of the book is to have a clean copy on which you can draw, try out alternative selection criteria, write comments, and so forth; and presenting the conclusion

---

* The Cambridge Materials Selector (*CMS*), available from Granta Design, Trumpington Mews, 40B High Street, Trumpington, Cambridge CB2 2LS, UK.

of a selection exercise is, often, most easily done in the same way. Although the book itself is copyrighted, the reader is authorized to make copies of the charts, and to reproduce these, with proper reference to their source, as he or she wishes.

M.F. Ashby
Cambridge, August 1998

Acknowledgements

Many colleagues have been generous in discussion, criticism and constructive suggestions. I particularly wish to thank Dr David Cebon, Mr Ken Wallace, Dr Amal Esawi and Dr Ulrike Wegst of the Engineering Design Centre, Engineering Department, Cambridge, Dr Paul Weaver of the Department of Aeronautical Engineering at the University of Bristol and Professor Michael Brown of the Cavendish Laboratory, Cambridge, UK.

CONVERSION OF UNITS — STRESS AND PRESSURE*

	MN/m^2	dyn/cm^2	lb/in^2	kgf/mm^2	bar	$long\ ton/in^2$
MN/m^2	1	10^7	1.45×10^2	0.102	10	6.48×10^{-2}
dyn/cm^2	10^{-7}	1	1.45×10^{-5}	1.02×10^{-8}	10^{-6}	6.48×10^{-9}
lb/in^2	6.89×10^{-3}	6.89×10^4	1	703×10^{-4}	6.89×10^{-2}	4.46×10^{-4}
kgf/mm^2	9.81	9.81×10^7	1.42×10^3	1	98.1	63.5×10^{-2}
bar	0.10	10^6	14.48	1.02×10^{-2}	1	6.48×10^{-3}
$long\ ton/in^2$	15.44	1.54×10^8	2.24×10^3	1.54	1.54×10^2	1

CONVERSION OF UNITS — ENERGY*

	J	erg	cal	eV	Btu	ft lbf
J	1	10^7	0.239	6.24×10^{18}	9.48×10^{-4}	0.738
erg	10^{-7}	1	2.39×10^{-8}	6.24×10^{11}	9.48×10^{-11}	7.38×10^{-8}
cal	4.19	4.19×10^7	1	2.61×10^{19}	3.97×10^{-3}	3.09
eV	1.60×10^{-19}	1.60×10^{-12}	3.38×10^{-20}	1	1.52×10^{-22}	1.18×10^{-19}
Btu	1.06×10^3	1.06×10^{10}	2.52×10^2	6.59×10^{21}	1	7.78×10^2
ft lbf	1.36	1.36×10^7	0.324	8.46×10^{18}	1.29×10^{-3}	1

CONVERSION OF UNITS — POWER*

	kW(kJ/s)	erg/s	hp	ft lbf/s
kW(kJ/s)	1	10^{-10}	1.34	7.38×10^2
erg/s	10^{-10}	1	1.34×10^{-10}	7.38×10^{-8}
hp	7.46×10^{-1}	7.46×10^9	1	5.50×10^2
ft lbf/s	1.36×10^{-3}	1.36×10^7	1.82×10^{-3}	1

*To convert row unit to column unit, multiply by the number at the
column-row intersection, thus 1 MN/m^2 = 10 bar

PHYSICAL CONSTANTS IN SI UNITS

Absolute zero temperature	$-273.2°C$
Acceleration due to gravity, g	9.807 m/s^2
Avogadro's number, N_A	6.022×10^{23}
Base of natural logarithms, e	2.718
Boltzmann's constant, k	$1.381 \times 10^{-23} \text{ J/K}$
Faraday's constant k	$9.648 \times 10^4 \text{ C/mol}$
Gas constant, $\bar{R}$	8.314 J/mol/K
Permeability of vacuum, μ_0	$1.257 \times 10^{-6} \text{ H/m}$
Permittivity of vacuum, ε_0	$8.854 \times 10^{-12} \text{ F/m}$
Planck's constant, h	$6.626 \times 10^{-34} \text{ J/s}$
Velocity of light in vacuum, c	$2.998 \times 10^8 \text{ m/s}$
Volume of perfect gas at STP	$22.41 \times 10^{-3} \text{ m}^3\text{/mol}$

CONVERSION OF UNITS

Angle, θ	1 rad	$57.30°$
Density, ρ	1 lb/ft^3	16.03 kg/m^3
Diffusion coefficient, D	1 cm^3/s	$1.0 \times 10^{-4} \text{ m}^2\text{/s}$
Energy, U	See inside back cover	
Force, F	1 kgf	9.807 N
	1 lbf	4.448 N
	1 dyne	$1.0 \times 10^{-5} \text{ N}$
Length, ℓ	1 ft	304.8 mm
	1 inch	25.40 mm
	1 Å	0.1 nm
Mass, M	1 tonne	1000 kg
	1 short ton	908 kg
	1 long ton	1107 kg
	1 lb mass	0.454 kg
Power, P	See inside back cover	
Stress, σ	See inside back cover	
Specific heat, Cp	1 cal/g$°$C	4.188 kJ/kg$°$C
	Btu/lb$°$F	4.187 kJ/kg$°$C
Stress intensity, K_{1C}	1 ksi$\sqrt{\text{in}}$	$1.10 \text{ MN/m}^{3/2}$
Surface energy γ	1 erg/cm^2	1 mJ/m^2
Temperature, T	1$°$F	0.556$°$K
Thermal conductivity λ	1 cal/s cm$°$C	418.8 W/m$°$C
	1 Btu/h ft$°$F	1.731 W/m$°$C
Volume, V	1 Imperial gall	$4.546 \times 10^{-3} \text{ m}^3$
	1 US gall	$3.785 \times 10^{-3} \text{ m}^3$
Viscosity, η	1 poise	0.1 N s/m^2
	1 lb ft s	0.1517 N s/m^2

Chapter 1 _____

Introduction

1.1 Introduction and synopsis

'Design' is one of those words that means all things to all people. Every manufactured thing, from the most lyrical of ladies' hats to the greasiest of gearboxes, qualifies, in some sense or other, as a design. It can mean yet more. Nature, to some is Divine Design; to others it is design by Natural Selection, the ultimate genetic algorithm. The reader will agree that it is necessary to narrow the field, at least a little.

This book is about mechanical design, and the role of materials in it. Mechanical components have mass; they carry loads; they conduct heat and electricity; they are exposed to wear and to corrosive environments; they are made of one or more materials; they have shape; and they must be manufactured (Figure 1.1). The book describes how these activities are related.

Materials have limited design since man first made clothes, built shelters and waged wars. They still do. But materials and processes to shape them are developing faster now than at any previous time in history; the challenges and opportunities they present are greater than ever before. The book develops a strategy for exploiting materials in design.

1.2 Materials in design

Design is the process of translating a new idea or a market need into the detailed information from which a product can be manufactured. Each of its stages requires decisions about the materials from which the product is to be made and the process for making it. Normally, the choice of material is dictated by the design. But sometimes it is the other way round: the new product, or the evolution of the existing one, was suggested or made possible by the new material. The number of materials available to the engineer is vast: something between 40 000 and 80 000 are at his or her (from here on 'his' means both) disposal. And although standardization strives to reduce the number, the continuing appearance of new materials with novel, exploitable, properties expands the options further.

How, then, does the engineer choose, from this vast menu, the material best suited to his purpose? Must he rely on experience? Or can a *systematic procedure* be formulated for making a rational choice? The question has to be answered at a number of levels, corresponding to the stage the design has reached. At the beginning the design is fluid and the options are wide; all materials must be considered. As the design becomes more focused and takes shape, the selection criteria sharpen and the shortlist of materials which can satisfy them narrows. Then more accurate data are required (although for a lesser number of materials) and a different way of analysing the choice must be used. In the final stages of design, precise data are needed, but for still fewer materials — perhaps only one. The procedure must recognize the initial richness of choice, narrow this to a small subset, and provide the precision and detail on which final design calculations can be based.

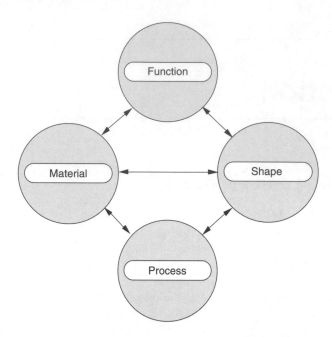

Fig. 1.1 Function, material, process and shape interact. Later chapters deal with each in turn.

The choice of material cannot be made independently of the choice of process by which the material is to be formed, joined, finished, and otherwise treated. Cost enters, both in the choice of material and in the way the material is processed. And — it must be recognized — good engineering design alone is not enough to sell a product. In almost everything from home appliances through automobiles to aircraft, the form, texture, feel, colour, decoration of the product — the satisfaction it gives the person who buys or uses it — are important. This aesthetic aspect (known confusingly as 'industrial design') is not treated in most courses on engineering, but it is one that, if neglected, can lose the manufacturer his market. Good designs work; excellent designs also give pleasure.

Design problems, almost always, are open-ended. They do not have a unique or 'correct' solution, although some solutions will clearly be better than others. They differ from the analytical problems used in teaching mechanics, or structures, or thermodynamics, or even materials, which generally do have single, correct answers. So the first tool a designer needs is an open mind: the willingness to consider all possibilities. But a net cast widely draws in many fish. A procedure is necessary for selecting the excellent from the merely good.

This book deals with the materials aspects of the design process. It develops a methodology which, properly applied, gives guidance through the forest of complex choices the designer faces. The ideas of material and process attributes are introduced. They are mapped on material and process selection charts which show the lay of the land, so to speak, and simplify the initial survey for potential candidate materials. The interaction between material and shape can be built into the method, as can the more complex aspects of optimizing the balance between performance and cost. None of this can be implemented without data for material properties and process attributes: ways to find them are described. The role of aesthetics in engineering design is discussed. The forces driving change in the materials world are surveyed. The Appendices contain useful information.

The methodology has further applications. It suggests a strategy for material development, particularly of composites and structured materials like sandwich panels. It points to a scheme for identifying the most promising applications for new materials. And it lends itself readily to computer implementation, offering the potential for interfaces with computer-aided design, function modelling, optimization routines and so forth.

All this will be found in the following chapters, with case studies illustrating applications. But first, a little history.

1.3 The evolution of engineering materials

Throughout history, materials have limited design. The ages in which man has lived are named for the materials he used: stone, bronze, iron. And when he died, the materials he treasured were buried with him: Tutankhamen with shards of coloured glass in his stone sarcophagus, Agamemnon with his bronze sword and mask of gold, each representing the high technology of his day.

If they had lived and died today, what would they have taken with them? Their titanium watch, perhaps; their carbon-fibre reinforced tennis racquet, their metal-matrix composite mountain bike, their polyether-ethyl-ketone crash helmet. This is not the age of one material; it is the age of an immense range of materials. There has never been an era in which the evolution of materials was faster and the range of their properties more varied. The menu of materials available to the engineer has expanded so rapidly that designers who left college twenty years ago can be forgiven for not knowing that half of them exist. But not-to-know is, for the designer, to risk disaster. Innovative design, often, means the imaginative exploitation of the properties offered by new or improved materials. And for the man in the street, the schoolboy even, not-to-know is to miss one of the great developments of our age: the age of advanced materials.

This evolution and its increasing pace are illustrated in Figure 1.2. The materials of prehistory (>10 000 BC, the Stone Age) were ceramics and glasses, natural polymers and composites. Weapons — always the peak of technology — were made of wood and flint; buildings and bridges of stone and wood. Naturally occurring gold and silver were available locally but played only a minor role in technology. The discovery of copper and bronze and then iron (the Bronze Age, 4000 BC–1000 BC and the Iron Age, 1000 BC–AD 1620) stimulated enormous advances, replacing the older wooden and stone weapons and tools (there is a cartoon on my office door, put there by a student, presenting an aggrieved Celt confronting a swordsmith with the words 'You sold me this bronze sword last week and now I'm supposed to upgrade to iron!'). Cast iron technology (1620s) established the dominance of metals in engineering; and the evolution of steels (1850 onward), light alloys (1940s) and special alloys since then consolidated their position. By the 1960s, 'engineering materials' meant 'metals'. Engineers were given courses in metallurgy; other materials were barely mentioned.

There had, of course, been developments in the other classes of material. Portland cement, refractories, fused silica among ceramics, and rubber, bakelite, and polyethylene among polymers, but their share of the total materials market was small. Since 1960 all that has changed. The rate of development of new metallic alloys is now slow; demand for steel and cast iron has in some countries actually fallen*. The polymer and composite industries, on the other hand, are growing rapidly, and projections of the growth of production of the new high-performance ceramics suggests rapid expansion here also.

---

* Do not, however, imagine that the days of steel are over. Steel production accounts for 90% of all world metal output, and its unique combination of strength, ductility, toughness and low price makes steel irreplaceable.

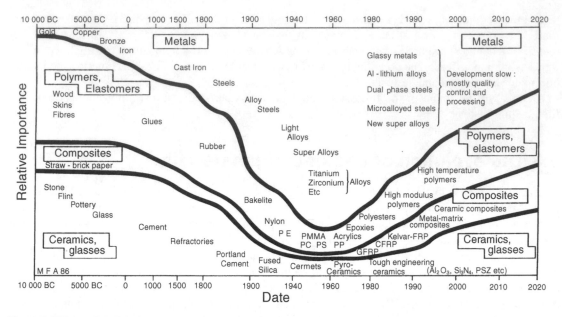

Fig. 1.2 The evolution of engineering materials with time. 'Relative Importance' in the stone and bronze ages is based on assessments of archaeologists; that in 1960 is based on allocated teaching hours in UK and US universities; that in 2020 on predictions of material usage in automobiles by manufacturers. The time scale is non-linear. The rate of change is far faster today than at any previous time in history.

This rapid rate of change offers opportunities which the designer cannot afford to ignore. The following case study is an example. There are more in Chapter 15.

1.4 The evolution of materials in vacuum cleaners

'Sweeping and dusting are homicidal practices: they consist of taking dust from the floor, mixing it in the atmosphere, and causing it to be inhaled by the inhabitants of the house. In reality it would be preferable to leave the dust alone where it was.'

That was a doctor, writing about 100 years ago. More than any previous generation, the Victorians and their contemporaries in other countries worried about dust. They were convinced that it carried disease and that dusting merely dispersed it where, as the doctor said, it became yet more infectious. Little wonder, then, that they invented the vacuum cleaner.

The vacuum cleaners of 1900 and before were human-powered (Figure 1.3(a)). The housemaid, standing firmly on the flat base, pumped the handle of the cleaner, compressing bellows which, with leather flap-valves to give a one-way flow, sucked air through a metal can containing the filter at a flow rate of about 1 litre per second. The butler manipulated the hose. The materials are, by today's standards, primitive: the cleaner is made almost entirely from natural polymers and fibres; wood, canvas, leather and rubber. The only metal is the straps which link the bellows (soft iron) and the can containing the filter (mild steel sheet, rolled to make a cylinder). It reflects the use of materials in 1900. Even a car, in 1900, was mostly made of wood, leather, and rubber; only the engine and drive train had to be metal.

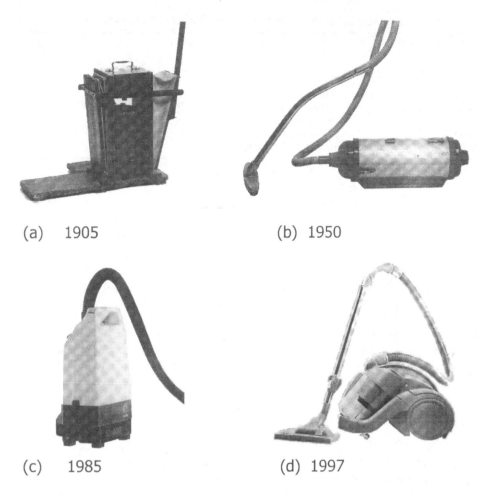

(a) 1905

(b) 1950

(c) 1985

(d) 1997

Fig. 1.3 Vacuum cleaners: (a) The hand-powered bellows cleaner of 1900, largely made of wood and leather. (b) The cylinder cleaner of 1950. (c) The lightweight cleaner of 1985, almost entirely polymer. (d) A centrifugal dust-extraction cleaner of 1997.

The electric vacuum cleaner first appeared around 1908*. By 1950 the design had evolved into the cylinder cleaner shown in Figure 1.3(b) (flow rate about 10 litres per second). Air flow is axial, drawn through the cylinder by an electric fan. The fan occupies about half the length of the cylinder; the rest holds the filter. One advance in design is, of course, the electrically driven air pump. The motor, it is true, is bulky and of low power, but it can function continuously without tea breaks or housemaid's elbow. But there are others: this cleaner is almost entirely made of metal: the case, the endcaps, the runners, even the tube to suck up the dust are mild steel: metals have replaced natural materials entirely.

Developments since then have been rapid, driven by the innovative use of new materials. The 1985 vacuum cleaner of Figure 1.3(c) has the power of roughly 18 housemaids working flat out

---

* Inventors: Murray Spengler and William B. Hoover. The second name has become part of the English language, along with those of such luminaries as John B. Stetson (the hat), S.F.B. Morse (the code), Leo Henrik Baikeland (Bakelite) and Thomas Crapper (the flush toilet).

Table 1.1 Comparison of cost, power and weight of vacuum cleaners

Cleaner and Date	Dominant materials	Power (W)	Weight (kg)	Cost*
Hand powered, 1900	Wood, canvas, leather	50	10	£240/$380
Cylinder, 1950	Mild Steel	300	6	£96/$150
Cylinder, 1985	Moulded ABS and polypropylene	800	4	£60/$95
Dyson, 1995	Polypropylene, polycarbonate, ABS	1200	6.3	£190/$300

*Costs have been adjusted to 1998 values, allowing for inflation.

(800 watts) and a corresponding air flow rate; cleaners with twice that power are now available. Air flow is still axial and dust removal by filtration, but the unit is smaller than the old cylinder cleaners. This is made possible by a higher power-density in the motor, reflecting better magnetic materials and higher operating temperatures (heat-resistant insulation, windings and bearings). The casing is entirely polymeric, and is an example of good design with plastics. The upper part is a single moulding, with all additional bits attached by snap fasteners moulded into the original component. No metal is visible anywhere; even the straight part of the suction tube, metal in all earlier models, is now polypropylene. The number of components is enormously reduced: the casing has just four parts, held together by just one fastener, compared with 11 parts and 28 fasteners for the 1950 cleaner. The saving on weight and cost is enormous, as the comparison in Table 1.1 shows.

It is arguable that this design (and its many variants) is near-optimal for today's needs; that a change of working principle, material or process could increase performance but at a cost penalty unacceptable to the consumer. We will leave the discussion of balancing performance against cost to a later chapter, and merely note here that one manufacturer disagrees. The cleaner shown in Figure 1.3(d) exploits a different concept: that of centrifugal separation, rather than filtration. For this to work, the power and rotation speed have to be high; the product is larger, noisier, heavier and much more expensive than the competition. Yet it sells — a testament to good industrial design and imaginative, aggressive marketing.

All this has happened within one lifetime. Competitive design requires the innovative use of new materials and the clever exploitation of their special properties, both engineering and aesthetic. There have been many manufacturers of vacuum cleaners who failed to innovate and exploit; now they are extinct. That sombre thought prepares us for the chapters which follow, in which we consider what they forgot: the optimum use of materials in design.

1.5 Summary and conclusions

The number of engineering materials is large: estimates range from 40 000 to 80 000. The designer must select from this vast menu the material best suited to his task. This, without guidance, can be a difficult and tedious business, so there is a temptation to choose the material that is 'traditional' for the application: glass for bottles; steel cans. That choice may be safely conservative, but it rejects the opportunity for innovation. Engineering materials are evolving faster, and the choice is wider than ever before. Examples of products in which a novel choice of material has captured a market are as common as — well — as plastic bottles. Or aluminium cans. It is important in the early stage of design, or of re-design, to examine the full materials menu, not rejecting options merely because they are unfamiliar. And that is what this book is about.

1.6 Further reading

The history and evolution of materials

Connoisseurs will tell you that in its 11th edition the *Encyclopaedia Britannica* reached a peak of excellence which has not since been equalled, although subsequent editions are still usable. On matters of general and technical history it, and the seven-volume *History of Technology*, are the logical starting points. More specialized books on the history and evolution of metals, ceramics, glass, and plastics make fascinating browsing. A selection of the most entertaining is given below.

'*Encyclopaedia Britannica*', 11th edition. The Encyclopaedia Britannica Company, New York 1910.
Davey, N. (1960) *A History of Building Materials*. Camelot Press, London, UK.
Delmonte, J. (1985) *Origins of Materials and Processes*. Technomic Publishing Company, Pennsylvania.
Derry, T.K. and Williams, T.I. (1960) *A Short History of Technology*'. Oxford University Press, Oxford.
Dowson, D. (1979) *History of Tribology*'. Longman, London.
Michaelis, R.R. (1992) Gold: art, science and technology, *Interdisciplinary Science Reviews*, **17**(3), 193.
Singer, C., Holmyard, E.J., Hall, A.R. and Williams, T.I. (eds) (1954–1978) *A History of Technology* (7 volumes plus annual supplements). Oxford University Press, Oxford.
Tylecoate, R.F. (1992) *A History of Metallurgy*, 2nd edition. The Institute of Materials, London.

Vacuum cleaners

Forty, A. (1986) *Objects of Desire: Design and Society since 1750*, Thames and Hudson, London, p.174 et seq.

Chapter 2

The design process

2.1 Introduction and synopsis

It is *mechanical design* with which we are primarily concerned here; it deals with the physical principles, the proper functioning and the production of mechanical systems. This does not mean that we ignore *industrial design*, which speaks of pattern, colour, texture, and (above all) consumer appeal — but that comes later. The starting point is good mechanical design, and the role of materials in it.

Our aim is to develop a methodology for selecting materials and processes which is *design-led*; that is, the selection uses, as inputs, the functional requirements of the design. To do so we must first look briefly at design itself. Like most technical fields it is encrusted with its own special jargon; it cannot all be avoided. This chapter introduces some of the words and phrases — the vocabulary — of design, the stages in its implementation, and the ways in which materials selection links with these.

2.2 The design process

Design is an iterative process. The starting point is a market need or a new idea; the end point is the full specifications of a product that fills the need or embodies the idea. It is essential to define the need precisely, that is, to formulate a *need statement*, often in the form: 'a device is required to perform task X'. Writers on design emphasize that the statement should be *solution-neutral* (that is, it should not imply how the task will be done), to avoid narrow thinking limited by pre-conceptions. Between the need statement and the product specification lie the set of stages shown in Figure 2.1: the stages of *conceptual design*, *embodiment design* and *detailed design*.

The product itself is called a *technical system*. A technical system consists of *assemblies*, *sub-assemblies* and *components*, put together in a way that performs the required task, as in the breakdown of Figure 2.2. It is like describing a cat (the system) as made up of one head, one body, one tail, four legs, etc. (the assemblies), each composed of components — femurs, quadriceps, claws, fur. This decomposition is a useful way to analyse an existing design, but it is not of much help in the design process itself, that is, in the synthesis of new designs. Better, for this purpose, is one based on the ideas of systems analysis; it thinks of the inputs, flows and outputs of information, energy and materials, as in Figure 2.3. The design converts the inputs into the outputs. An electric motor converts electrical into mechanical energy; a forging press takes and reshapes material; a burglar alarm collects information and converts it to noise. In this approach, the system is broken down into connected subsystems which perform specific sub-functions, as in Figure 2.3; the resulting arrangement is called the *function structure* or *function decomposition* of the system. It is like describing a cat as an appropriate linkage of a respiratory system, a cardio-vascular system,

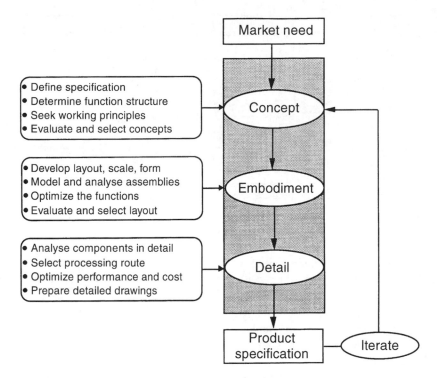

Fig. 2.1 The design flow chart. The design proceeds from an identification and clarification of task through concept, embodiment and detailed analysis to a product specification.

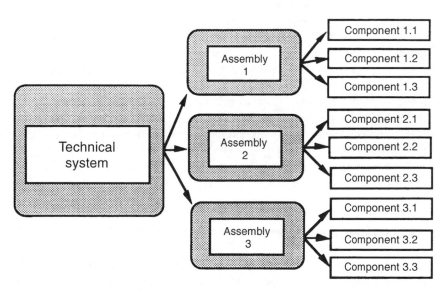

Fig. 2.2 The analysis of a technical system as a breakdown into assemblies and components. Material and process selection is at the component level.

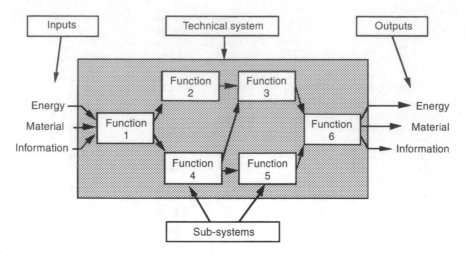

Fig. 2.3 The systems approach to the analysis of a technical system, seen as transformation of energy, materials and information (signals). This approach, when elaborated, helps structure thinking about alternative designs.

a nervous system, a digestive system and so on. Alternative designs link the unit functions in alternative ways, combine functions, or split them. The function-structure gives a systematic way of assessing design options.

The design proceeds by developing concepts to fill each of the sub-functions in the function structure, each based on a *working principle*. At this, the conceptual design stage (Figure 2.1 again), all options are open: the designer considers alternative concepts for the sub-functions and the ways in which these might be separated or combined. The next stage, embodiment, takes each promising concept and seeks to analyse its operation at an approximate level, sizing the components, and selecting materials which will perform properly in the ranges of stress, temperature and environment suggested by the analysis or required by the specification, examining the implications for performance and cost. The embodiment stage ends with a feasible layout which is passed to the detailed design stage. Here specifications for each component are drawn up; critical components may be subjected to precise mechanical or thermal analysis; optimization methods are applied to components and groups of components to maximize performance; a final choice of geometry and material is made, the production is analysed and the design is costed. The stage ends with detailed production specifications.

Described in the abstract, these ideas are not easy to grasp. An example will help — it comes in Section 2.6. First, a look at types of design.

2.3 Types of design

It is not always necessary to start, as it were, from scratch. *Original design* does: it involves a new idea or working principle (the ball-point pen, the compact disc). New materials can offer new, unique combinations of properties which enable original design. High-purity silicon enabled the transistor; high-purity glass, the optical fibre; high coercive-force magnets, the miniature earphone. Sometimes the new material suggests the new product; sometimes instead the new product demands the development of a new material: nuclear technology drove the development of a series of new zirconium-based alloys; space technology stimulated the development of lightweight composites; turbine technology today drives development of high-temperature alloys and ceramics.

Adaptive or *development design* takes an existing concept and seeks an incremental advance in performance through a refinement of the working principle. This, too, is often made possible by developments in materials: polymers replacing metals in household appliances; carbon fibre replacing wood in sports goods. The appliance and the sports-goods market are both large and competitive. Markets here have frequently been won (and lost) by the way in which the manufacturer has exploited new materials.

Variant design involves a change of scale or dimension or detailing without change of function or the method of achieving it: the scaling up of boilers, or of pressure vessels, or of turbines, for instance. Change of scale or range of conditions may require change of material: small boats are made of fibreglass, large ones are made of steel; small boilers are made of copper, large ones of steel; subsonic planes are made of one alloy, supersonic of another; and for good reasons, detailed in later chapters.

2.4 Design tools and materials data

To implement the steps of Figure 2.1, use is made of *design tools*. They are shown as inputs, attached to the left of the main backbone of the design methodology in Figure 2.4. The tools enable the modelling and optimization of a design, easing the routine aspects of each phase. Function modellers suggest viable function structures. Geometric and 3-D solid modelling packages allow visualization and create files which can be downloaded to numerically controlled forming processes. Optimization, DFM, DFA* and cost-estimation software allow details to be refined. Finite element packages allow precise mechanical and thermal analysis even when the geometry is complex. There is a natural progression in the use of the tools as the design evolves: approximate analysis and modelling at the conceptual stage; more sophisticated modelling and optimization at the embodiment stage; and precise ('exact' — but nothing is ever that) analysis at the detailed design stage.

Materials selection enters each stage of the design. The nature of the data needed in the early stages differs greatly in its level of precision and breadth from that needed later on (Figure 2.4, right-hand side). At the concept stage, the designer requires approximate property values, but for the widest possible range of materials. All options are open: a polymer may be the best choice for one concept, a metal for another, even though the function is the same. The problem at this stage is not precision; it is breadth and access: how can the vast range of data be presented to give the designer the greatest freedom in considering alternatives? Selection systems exist which achieve this.

Embodiment design needs data for a subset of materials, but at a higher level of precision and detail. They are found in more specialized handbooks and software which deal with a single class of materials — metals, for instance — and allow choice at a level of detail not possible from the broader compilations which include all materials.

The final stage of detailed design requires a still higher level of precision and detail, but for only one or a very few materials. Such information is best found in the data sheets issued by the material producers themselves. A given material (polyethylene, for instance) has a range of properties which derive from differences in the way different producers make it. At the detailed design stage, a supplier must be identified, and the properties of his product used in the design calculations; that

---

* Design for Manufacture and Design for Assembly

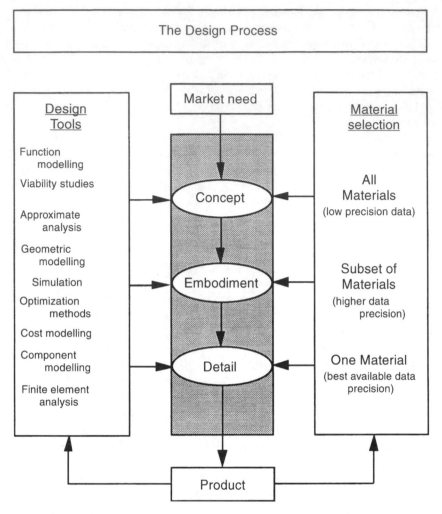

Fig. 2.4 The design flow chart, showing how design tools and materials selection enter the procedure. Information about materials is needed at each stage, but at very different levels of breadth and precision.

from another supplier may have slightly different properties. And sometimes even this is not good enough. If the component is a critical one (meaning that its failure could, in some sense or another, be disastrous) then it may be prudent to conduct in-house tests to measure the critical properties, using a sample of the material that will be used to make the product itself.

It's all a bit like choosing a bicycle. You first decide which concept best suits your requirements (street bike, mountain bike, racing, folding, shopping...), limiting the choice to one subset. Then comes the next level of detail: how many gears you need, what shape of handlebars, which sort of brakes, further limiting the choice. At this point you consider the trade-off between weight and cost, identifying (usually with some compromise) a small subset which meet both your desires and your budget. Finally, if your bicycle is important to you, you seek further information in bike magazines, manufacturers' literature or the views of enthusiasts, and try the candidate bikes out yourself. Only then do you make a final selection.

The materials input into design does not end with the establishment of production. Products fail in service, and failures contain information. It is an imprudent manufacture who does not collect and analyse data on failures. Often this points to the misuse of a material, one which re-design or re-selection can eliminate.

2.5 Function, material, shape and process

The selection of a material and process cannot be separated from the choice of shape. We use the word 'shape' to include the external shape (the macro-shape), and — when necessary — the internal shape, as in a honeycomb or cellular structure (the micro-shape). The achieve the shape, the material is subjected to processes which, collectively, we shall call manufacture: they include primary forming processes (like casting and forging), material removal processes (machining, drilling), finishing processes (such as polishing) and joining processes (welding, for example). Function, material, shape and process interact (Figure 2.5). Function dictates the choice of both material and shape. Process is influenced by the material: by its formability, machinability, weldability, heat-treatability and so on. Process obviously interacts with shape — the process determines the shape, the size, the precision and, of course, the cost. The interactions are two-way: specification of shape restricts the choice of material and process; but equally the specification of process limits the materials you can use and the shapes they can take. The more sophisticated the design, the tighter the specifications

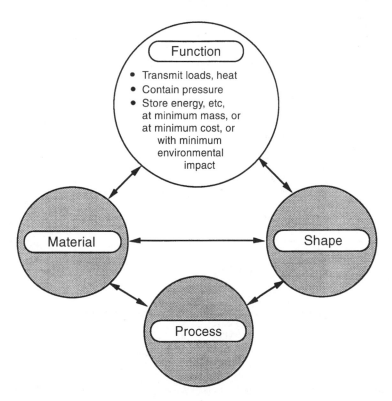

Fig. 2.5 The central problem of materials selection in mechanical design: the interaction between function, material, process and shape.

and the greater the interactions. It is like making wine: to make cooking wine, almost any grape and fermentation process will do; to make champagne, both grape and process must be tightly constrained.

The interaction between function, material, shape and process lies at the heart of the material selection process. But first: a case study to illustrate the design process.

2.6 Devices to open corked bottles

Wine, like cheese, is one of man's improvements on nature. And ever since man has cared about wine, he has cared about cork to keep it safely sealed in flasks and bottles. 'Corticum... demovebit amphorae...' — 'Uncork the amphora...' sang Horace* (27 BC) to celebrate the anniversary of his miraculous escape from death by a falling tree. But how did he do it?

A corked bottle creates a market need: it is the need to gain access to the wine inside. We might state it thus: 'a device is required to pull corks from wine bottles'. But hold on. The need must be expressed in solution-neutral form, and this is not. The aim is to gain access to the wine; our statement implies that this will be done by removing the cork, and that it will be removed by pulling. There could be other ways. So we will try again: 'a device is required to allow access to wine in a corked bottle' (Figure 2.6) and one might add, 'with convenience, at modest cost, and without contaminating the wine'.

Five concepts for doing this are shown in Figure 2.7. In sequence, they are to remove the cork by axial traction (= pulling); to remove it by shear tractions; to push it out from below; to pulverize it; and to by-pass it altogether — by knocking the neck off the bottle, perhaps.

Numerous devices exist to achieve the first three of these. The others are used too, though generally only in moments of desperation. We shall eliminate these on the grounds that they might

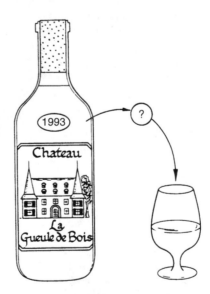

Fig. 2.6 The market need: a device is sought to allow access to wine contained in a corked bottle.

---

* Horace, Q. 27 BC, Odes, BOOK III, Ode 8, line 10.

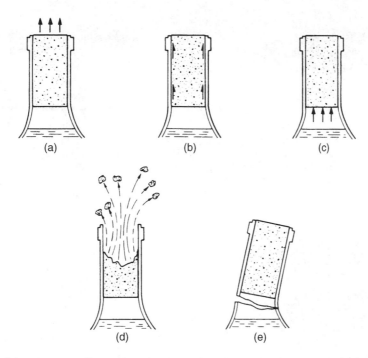

Fig. 2.7 Six possible concepts, illustrating physical principles, to fill the need expressed by Figure 2.6.

contaminate the wine, and examine the others more closely, exploring working principles. Figure 2.8 shows one for each of the first three concepts: in the first, a screw is threaded into the cork to which an axial pull is applied; in the second, slender elastic blades inserted down the sides of the cork apply shear tractions when pulled; and in the third the cork is pierced by a hollow needle through which a gas is pumped to push it out.

Figure 2.9 shows examples of cork removers using these working principles. All are described by the function structure sketched in the upper part of Figure 2.10: create a force, transmit a

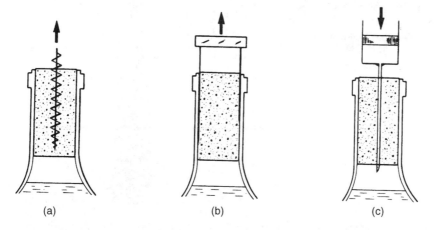

Fig. 2.8 Working principles for implementing the first three schemes of Figure 2.7.

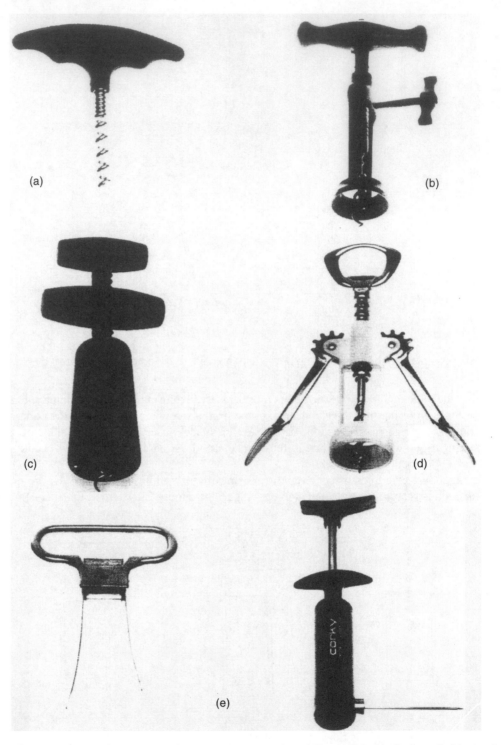

Fig. 2.9 Cork removers which employ the working principles of Figure 2.8: (a) direct pull; (b) gear lever, screw-assisted pull; (c) spring-assisted pull (a spring in the body is compressed as the screw is driven into the cork); (d) shear blade systems; (e) pressure-induced removal systems.

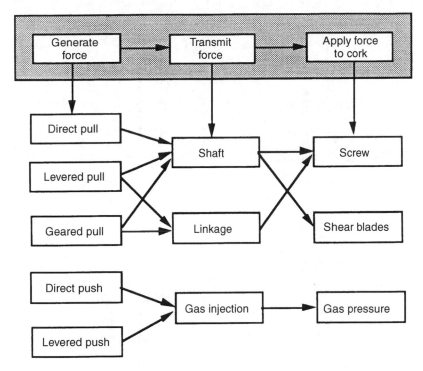

Fig. 2.10 The function structure and working principles of cork removers.

force, apply force to cork. They differ in the working principle by which these functions are achieved, as indicated in the lower part of Figure 2.10. The cork removers in the photos combine working principles in the ways shown by the linking lines. Others could be devised by making other links.

Figure 2.11 shows embodiment sketches for devices based on just one concept — that of axial traction. The first is a direct pull; the other three use some sort of mechanical advantage — levered pull, geared pull and spring-assisted pull; the photos show examples of all of these.

The embodiments of Figure 2.8 identify the *functional requirements* of each component of the device, which might be expressed in statements like:

- a light lever (that is, a beam) to carry a prescribed bending moment;
- a cheap screw to transmit a prescribed load to the cork;
- a slender elastic blade which will not buckle when driven between the cork and bottleneck;
- a thin, hollow needle strong enough to penetrate a cork;

and so on. The functional requirements of each component are the inputs to the materials selection process. They lead directly to the *property limits* and *material indices* of Chapter 5: they are the first step in optimizing the choice of material to fill a given requirement. The procedure developed there takes requirements such as 'light strong beam' or 'slender elastic blade' and uses them to identify a subset of materials which will perform this function particularly well. That is what is meant by *design-led material selection*.

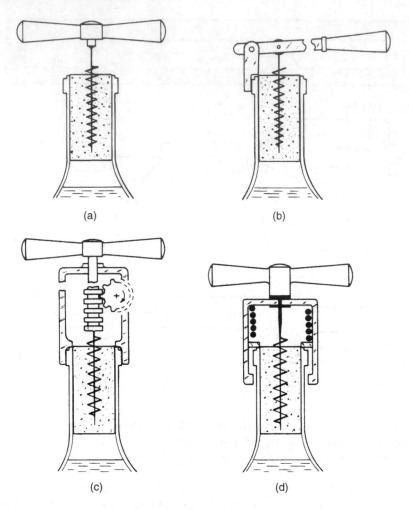

Fig. 2.11 Embodiment sketches for four concepts: direct pull, levered pull, geared pull and spring-assisted pull. Each system is made up of components which perform a sub-function. The requirements of these sub-functions are the inputs to the materials selection method.

2.7 Summary and conclusions

Design is an iterative process. The starting point is a *market need* captured in a *need statement*. A *concept* for a product which meets that need is devised. If initial estimates and exploration of alternatives suggest that the concept is viable, the design proceeds to the *embodiment* stage: working principles are selected, size and layout are decided, and initial estimates of performance and cost are made. If the outcome is successful, the designer proceeds to the *detailed design* stage: optimization of performance, full analysis (using computer methods if necessary) of critical components, preparation of detailed production drawings, specification of tolerance, precision, joining methods, finishing and so forth.

Materials selection enters at each stage, but at different levels of breadth and precision. At the conceptual stage all materials and processes are potential candidates, requiring a procedure which allows rapid access to data for a wide range of each, although without the need for great precision. The preliminary selection passes to the embodiment stage, the calculations and optimizations of which require information at a higher level of precision and detail. They eliminate all but a small shortlist of options which contains the candidate material and processes for the final, detailed stage of the design. For these few, data of the highest quality are necessary.

Data exist at all these levels. Each level requires its own data-management scheme, described in the following chapters. The management is the skill: it must be design-led, yet must recognize the richness of choice and embrace the complex interaction between the material, its shape, the process by which it is given that shape, and the function it is required to perform.

Given this complexity, why not opt for the safe bet: stick to what you (or others) used before? Many have chosen that option. Few are still in business.

2.8 Further reading

A chasm exists between books on Design Methodology and those on Materials Selection: each largely ignores the other. The book by French is remarkable for its insights, but the word 'Material' does not appear in its index. Pahl and Beitz has near-biblical standing in the design camp, but is heavy going. Ullman is a reduced version of Pahl and Beitz, and easier to digest. The book by Charles, Crane and Furness and that by Farag present the materials case well, but are less good on design. Lewis illustrates material selection through case studies, but does not develop a systematic procedure. The best compromise, perhaps, is Dieter.

General texts on design methodology

Ertas, A. and Jones, J.C. (1993) *The Engineering Design Process*. Wiley, New York.
French, M.J., (1985) *Conceptual Design for Engineers*. The Design Council, London, and Springer, Berlin.
Pahl, G. and Beitz, W. (1997) *Engineering Design*, 2nd edition, translated by K. Wallace and L. Blessing. The Design Council, London, and Springer, Berlin.
Ullman, D.G. (1992) *The Mechanical Design Process*. McGraw-Hill, New York.

General texts on materials selection in design

Budinski, K. (1979) *Engineering Materials, Properties and Selection*. Prentice-Hall, Englewood Cliffs, NJ.
Charles, J.A., Crane, F.A.A. and Furness J.A.G. (1987) *Selection and Use of Engineering Materials*, 3rd edition. Butterworth-Heinemann, Oxford.
Dieter, G.E. (1991) *Engineering Design, A Materials and Processing Approach*, 2nd edition. McGraw-Hill, New York.
Farag, M.M. (1989) *Selection of Materials and Manufacturing Processes for Engineering Design*. Prentice-Hall, Englewood Cliffs, NJ.
Lewis, G. (1990) *Selection of Engineering Materials*. Prentice-Hall, Englewood Cliffs, NJ.

Corks and corkscrews

McKearin, H. (1973) On 'stopping', bottling and binning, *International Bottler and Packer*, April, pp. 47–54.
Perry, E. (1980) Corkscrews and Bottle Openers. Shire Publications Ltd, Aylesbury.
The Design Council (1994) *Teaching Aids Program EDTAP DE9*. The Design Council, London.
Watney, B.M. and Babbige, H.D. (1981) *Corkscrews*. Sotheby's Publications, London.

Chapter 3

Engineering materials and their properties

3.1 Introduction and synopsis

Materials, one might say, are the food of design. This chapter presents the menu: the full shopping list of materials. A successful product — one that performs well, is good value for money and gives pleasure to the user — uses the best materials for the job, and fully exploits their potential and characteristics: brings out their flavour, so to speak.

The classes of materials — metals, polymers, ceramics, and so forth — are introduced in Section 3.2. But it is not, in the end, a material that we seek; it is a certain profile of properties. The properties important in thermo-mechanical design are defined briefly in Section 3.3. The reader confident in the definitions of moduli, strengths, damping capacities, thermal conductivities and the like may wish to skip this, using it for reference, when needed, for the precise meaning and units of the data in the selection charts which come later. The chapter ends, in the usual way, with a summary.

3.2 The classes of engineering material

It is conventional to classify the materials of engineering into the six broad classes shown in Figure 3.1: metals, polymers, elastomers, ceramics, glasses and composites. The members of a class have features in common: similar properties, similar processing routes, and, often, similar applications.

Metals have relatively high moduli. They can be made strong by alloying and by mechanical and heat treatment, but they remain ductile, allowing them to be formed by deformation processes. Certain high-strength alloys (spring steel, for instance) have ductilities as low as 2%, but even this is enough to ensure that the material yields before it fractures and that fracture, when it occurs, is of a tough, ductile type. Partly because of their ductility, metals are prey to fatigue and of all the classes of material, they are the least resistant to corrosion.

Ceramics and *glasses*, too, have high moduli, but, unlike metals, they are brittle. Their 'strength' in tension means the brittle fracture strength; in compression it is the brittle crushing strength, which is about 15 times larger. And because ceramics have no ductility, they have a low tolerance for stress concentrations (like holes or cracks) or for high contact stresses (at clamping points, for instance). Ductile materials accommodate stress concentrations by deforming in a way which redistributes the load more evenly; and because of this, they can be used under static loads within a small margin of their yield strength. Ceramics and glasses cannot. Brittle materials always have

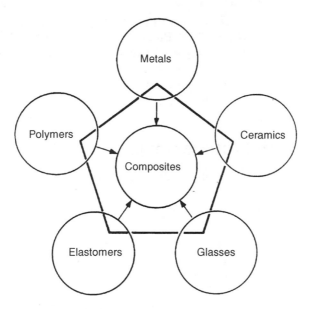

Fig. 3.1 The menu of engineering materials.

a wide scatter in strength and the strength itself depends on the volume of material under load and the time for which it is applied. So ceramics are not as easy to design with as metals. Despite this, they have attractive features. They are stiff, hard and abrasion-resistant (hence their use for bearings and cutting tools); they retain their strength to high temperatures; and they resist corrosion well. They must be considered as an important class of engineering material.

Polymers and *elastomers* are at the other end of the spectrum. They have moduli which are low, roughly 50 times less than those of metals, but they can be strong — nearly as strong as metals. A consequence of this is that elastic deflections can be large. They creep, even at room temperature, meaning that a polymer component under load may, with time, acquire a permanent set. And their properties depend on temperature so that a polymer which is tough and flexible at 20°C may be brittle at the 4°C of a household refrigerator, yet creep rapidly at the 100°C of boiling water. None have useful strength above 200°C. If these aspects are allowed for in the design, the advantages of polymers can be exploited. And there are many. When combinations of properties, such as strength-per-unit-weight, are important, polymers are as good as metals. They are easy to shape: complicated parts performing several functions can be moulded from a polymer in a single operation. The large elastic deflections allow the design of polymer components which snap together, making assembly fast and cheap. And by accurately sizing the mould and pre-colouring the polymer, no finishing operations are needed. Polymers are corrosion resistant, and they have low coefficients of friction. Good design exploits these properties.

Composites combine the attractive properties of the other classes of materials while avoiding some of their drawbacks. They are light, stiff and strong, and they can be tough. Most of the composites at present available to the engineer have a polymer matrix — epoxy or polyester, usually — reinforced by fibres of glass, carbon or Kevlar. They cannot be used above 250°C because the polymer matrix softens, but at room temperature their performance can be outstanding. Composite components are expensive and they are relatively difficult to form and join. So despite their attractive properties the designer will use them only when the added performance justifies the added cost.

The classification of Figure 3.1 has the merit of grouping together materials which have some commonalty in properties, processing and use. But it has its dangers, notably those of specialization (the metallurgist who knows nothing of polymers) and of conservative thinking ('we shall use steel because we have always used steel'). In later chapters we examine the engineering properties of materials from a different perspective, comparing properties across all classes of material. It is the first step in developing the freedom of thinking that the designer needs.

3.3 The definitions of material properties

Each material can be thought of as having a set of attributes: its properties. It is not a material, *per se*, that the designer seeks; it is a specific combination of these attributes: a *property-profile*. The material name is the identifier for a particular property-profile.

The properties themselves are standard: density, modulus, strength, toughness, thermal conductivity, and so on (Table 3.1). For completeness and precision, they are defined, with their limits, in this section. It makes tedious reading. If you think you know how properties are defined, you might jump to Section 3.4, returning to this section only if the need arises.

The *density*, ρ (units: kg/m^3), is the weight per unit volume. We measure it today as Archimedes did: by weighing in air and in a fluid of known density.

The *elastic modulus* (units: GPa or GN/m^2) is defined as 'the slope of the linear-elastic part of the stress−strain curve' (Figure 3.2). Young's modulus, E, describes tension or compression, the shear modulus G describes shear loading and the bulk modulus K describes the effect of hydrostatic pressure. Poisson's ratio, ν, is dimensionless: it is the negative of the ratio of the lateral strain to the

Table 3.1 Design-limiting material properties and their usual SI units*

Class	Property	Symbol and units	
General	Cost	C_m	($/kg)
	Density	ρ	(kg/m^3)
Mechanical	Elastic moduli (Young's, shear, bulk)	E, G, K	(GPa)
	Strength (yield, ultimate, fracture)	σ_f	(MPa)
	Toughness	G_c	(kJ/m^2)
	Fracture toughness	K_{Ic}	(MPa m$^{1/2}$)
	Damping capacity	η	(—)
	Fatigue endurance limit	σ_e	(MPa)
Thermal	Thermal conductivity	λ	(W/mK)
	Thermal diffusivity	a	(m^2/s)
	Specific heat	C_p	(J/kg K)
	Melting point	T_m	(K)
	Glass temperature	T_g	(K)
	Thermal expansion coefficient	α	($^\circ$K^{-1})
	Thermal shock resistance	ΔT	($^\circ$K)
	Creep resistance	-	(—)
Wear	Archard wear constant	k_A	(MPa^{-1})
Corrosion/	Corrosion rate	K	(mm/year)
Oxidation	Parabolic rate constant	k_P	(m^2/s)

*Conversion factors to imperial and cgs units appear inside the back and front covers of this book.

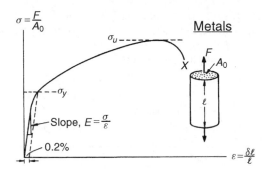

Fig. 3.2 The stress–strain curve for a metal, showing the modulus, E, the 0.2% yield strength, σ_y, and the ultimate strength σ_u.

axial strain, $\varepsilon_2/\varepsilon_1$, in axial loading. In reality, moduli measured as slopes of stress–strain curves are inaccurate (often low by a factor of two or more), because of contributions to the strain from anelasticity, creep and other factors. Accurate moduli are measured dynamically: by exciting the natural vibrations of a beam or wire, or by measuring the velocity of sound waves in the material. In an isotropic material, the moduli are related in the following ways:

$$E = \frac{3G}{1 + G/3K} \qquad G = \frac{E}{2(1 + \nu)} \qquad K = \frac{E}{3(1 - 2\nu)} \qquad (3.1)$$

Commonly

when

$$\left.\begin{array}{c} \nu \approx 1/3 \\ G \approx 3/8E \\ K \approx E \end{array}\right\} \qquad (3.2a)$$

and

Elastomers are exceptional. For these:

when

$$\left.\begin{array}{c} \nu \approx 1/2 \\ G \approx 1/3E \\ K \gg E \end{array}\right\} \qquad (3.2b)$$

and

Data books and databases like those described in Chapter 13 list values for all four moduli. In this book we examine data for E; approximate values for the others can be derived from equations (3.2) when needed.

The *strength*, σ_f, of a solid (units: MPa or MN/m^2) requires careful definition. For metals, we identify σ_f with the 0.2% offset yield strength σ_y (Figure 3.2), that is, the stress at which the stress–strain curve for axial loading deviates by a strain of 0.2% from the linear-elastic line. In metals it is the stress at which dislocations first move large distances, and is the same in tension and compression. For polymers, σ_f is identified as the stress σ_y at which the stress–strain curve becomes markedly non-linear: typically, a strain of 1% (Figure 3.3). This may be caused by 'shear-yielding': the irreversible slipping of molecular chains; or it may be caused by 'crazing': the formation of low density, crack-like volumes which scatter light, making the polymer look white. Polymers are a little stronger ($\approx$20%) in compression than in tension. Strength, for ceramics and glasses, depends strongly on the mode of loading (Figure 3.4). In tension, 'strength' means the fracture strength, σ_f^t.

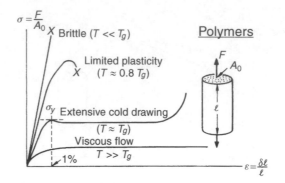

Fig. 3.3 Stress–strain curves for a polymer, below, at and above its glass transition temperature, T_g.

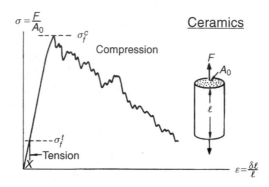

Fig. 3.4 Stress–strain curves for a ceramic in tension and in compression. The compressive strength σ_c is 10 to 15 times greater than the tensile strength σ_t.

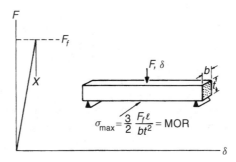

Fig. 3.5 The modulus-of-rupture (MOR) is the surface stress at failure in bending. It is equal to, or slightly larger than the failure stress in tension.

In compression it means the crushing strength σ_f^c which is much larger; typically

$$\sigma_f^c = 10 \text{ to } 15 \times \sigma_f^t \tag{3.3}$$

When the material is difficult to grip (as is a ceramic), its strength can be measured in bending. The *modulus of rupture* or *MOR* (units: MPa or MN/m^2) is the maximum surface stress in a bent beam at the instant of failure (Figure 3.5). One might expect this to be exactly the same as the strength measured in tension, but for ceramics it is larger (by a factor of about 1.3) because the volume subjected to this maximum stress is small and the probability of a large flaw lying in it is small also; in simple tension all flaws see the maximum stress.

The strength of a composite is best defined by a set deviation from linear-elastic behaviour: 0.5% is sometimes taken. Composites which contain fibres (and this includes natural composites like wood) are a little weaker (up to 30%) in compression than tension because fibres buckle. In subsequent chapters, σ_f for composites means the tensile strength.

Strength, then, depends on material class and on mode of loading. Other modes of loading are possible: shear, for instance. Yield under multiaxial loads are related to that in simple tension by a yield function. For metals, the Von Mises yield function is a good description:

$$(\sigma_1 - \sigma_2)^2 + (\sigma_2 - \sigma_3)^2 + (\sigma_3 - \sigma_1)^2 = 2\sigma_f^2 \tag{3.4}$$

where σ_1, σ_2 and σ_3 are the principal stresses, positive when tensile; σ_1, by convention, is the largest or most positive, σ_3 the smallest or least. For polymers the yield function is modified to include the effect of pressure

$$(\sigma_1 - \sigma_2)^2 + (\sigma_2 - \sigma_3)^2 + (\sigma_3 - \sigma_1)^2 = 2\sigma_f^2 \left(1 + \frac{\beta p}{K}\right)^2 \tag{3.5}$$

where K is the bulk modulus of the polymer, β (≈ 2) is a numerical coefficient which characterizes the pressure dependence of the flow strength and the pressure p is defined by

$$p = -\frac{1}{3}(\sigma_1 + \sigma_2 + \sigma_3)$$

For ceramics, a Coulomb flow law is used:

$$\sigma_1 - B\sigma_3 = C \tag{3.6}$$

where B and C are constants.

The *ultimate (tensile) strength* σ_u (units: MPa) is the nominal stress at which a round bar of the material, loaded in tension, separates (Figure 3.2). For brittle solids — ceramics, glasses and brittle polymers — it is the same as the failure strength in tension. For metals, ductile polymers and most composites, it is larger than the strength σ_f, by a factor of between 1.1 and 3 because of work hardening or (in the case of composites) load transfer to the reinforcement.

The *resilience*, R (units: J/m^3), measures the maximum energy stored elastically without any damage to the material, and which is released again on unloading. It is the area under the elastic part of the stress–strain curve:

$$R = \frac{1}{2}\sigma_f \varepsilon_f = \frac{\sigma_f^2}{2E}$$

where σ_f is the failure load, defined as above, ε_f is the corresponding strain and E is Young's modulus. Materials with large values of R make good springs.

The *hardness*, H, of a material (units: MPa) is a crude measure of its strength. It is measured by pressing a pointed diamond or hardened steel ball into the surface of the material. The hardness is defined as the indenter force divided by the projected area of the indent. It is related to the quantity we have defined as σ_f by

$$H \cong 3\sigma_f \tag{3.7}$$

Hardness is often measured in other units, the commonest of which is the Vickers H_v scale with units of kg/mm². It is related to H in the units used here by

$$H = 10H_v$$

The *toughness*, G_c (units: kJ/m²), and the *fracture toughness*, K_c (units: MPa m$^{1/2}$ or MN/m$^{1/2}$) measure the resistance of the material to the propagation of a crack. The fracture toughness is measured by loading a sample containing a deliberately introduced crack of length $2c$ (Figure 3.6), recording the tensile stress σ_c at which the crack propagates. The quantity K_c is then calculated from

$$K_c = Y\frac{\sigma_c}{\sqrt{\pi c}} \tag{3.8}$$

and the toughness from

$$G_c = \frac{K_c^2}{E(1+\nu)} \tag{3.9}$$

where Y is a geometric factor, near unity, which depends on details of the sample geometry, E is Young's modulus and ν is Poisson's ratio. Measured in this way K_c and G_c have well-defined values for brittle materials (ceramics, glasses, and many polymers). In ductile materials a plastic zone develops at the crack tip, introducing new features into the way in which cracks propagate which necessitate more involved characterization. Values for K_c and G_c are, nonetheless, cited, and are useful as a way of ranking materials.

The *loss-coefficient*, η (a dimensionless quantity), measures the degree to which a material dissipates vibrational energy (Figure 3.7). If a material is loaded elastically to a stress σ, it stores an elastic energy

$$U = \int_0^{\sigma_{max}} \sigma\, d\varepsilon = \frac{1}{2}\frac{\sigma^2}{E}$$

per unit volume. If it is loaded and then unloaded, it dissipates an energy

$$\Delta U = \oint \sigma\, d\varepsilon$$

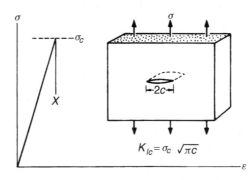

Fig. 3.6 The fracture toughness, K_c, measures the resistance to the propagation of a crack. The failure strength of a brittle solid containing a crack of length $2c$ is $\sigma_f = YK_c\sqrt{\pi c}$ where Y is a constant near unity.

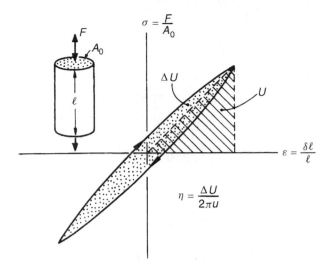

Fig. 3.7 The loss coefficient η measures the fractional energy dissipated in a stress–strain cycle.

The loss coefficient is

$$\eta = \frac{\Delta U}{2\pi U} \tag{3.10}$$

The cycle can be applied in many different ways — some fast, some slow. The value of η usually depends on the timescale or frequency of cycling. Other measures of damping include the *specific damping capacity*, $D = \Delta U/U$, the *log decrement*, Δ (the log of the ratio of successive amplitudes of natural vibrations), the *phase-lag*, δ, between stress and strain, and the *Q-factor* or *resonance factor*, Q. When damping is small ($\eta < 0.01$) these measures are related by

$$\eta = \frac{D}{2\pi} = \frac{\Delta}{\pi} = \tan\delta = \frac{1}{Q} \tag{3.11}$$

but when damping is large, they are no longer equivalent.

Cyclic loading not only dissipates energy; it can also cause a crack to nucleate and grow, culminating in fatigue failure. For many materials there exists a fatigue limit: a stress amplitude below which fracture does not occur, or occurs only after a very large number ($>10^7$) cycles. This information is captured by the *fatigue ratio*, f (a dimensionless quantity). It is the ratio of the fatigue limit to the yield strength, σ_f.

The rate at which heat is conducted through a solid at steady state (meaning that the temperature profile does not change with time) is measured by the *thermal conductivity*, λ (units: W/mK). Figure 3.8 shows how it is measured: by recording the heat flux q(W/m^2) flowing from a surface at temperature T_1 to one at T_2 in the material, separated by a distance X. The conductivity is calculated from Fourier's law:

$$q = -\lambda\frac{dT}{dX} = \frac{(T_1 - T_2)}{X} \tag{3.12}$$

The measurement is not, in practice, easy (particularly for materials with low conductivities), but reliable data are now generally available.

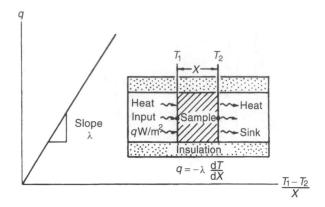

Fig. 3.8 The thermal conductivity λ measures the flux of heat driven by a temperature gradient dT/dX.

When heat flow is transient, the flux depends instead on the *thermal diffusivity*, a (units: m^2/s), defined by

$$a = \frac{\lambda}{\rho C_p} \tag{3.13}$$

where ρ is the density and C_p is the *specific heat at constant pressure* (units: J/kg.K). The thermal diffusivity can be measured directly by measuring the decay of a temperature pulse when a heat source, applied to the material, is switched off; or it can be calculated from λ, via the last equation. This requires values for C_p (virtually identical, for solids, with C_v, the specific heat at constant volume). They are measured by the technique of calorimetry, which is also the standard way of measuring the *melting temperature*, T_m, and the *glass temperature*, T_g (units for both: K). This second temperature is a property of non-crystalline solids, which do not have a sharp melting point; it characterizes the transition from true solid to very viscous liquid. It is helpful, in engineering design, to define two further temperatures: the *maximum service temperature* T_{max} and the *softening temperature*, T_s (both: K). The first tells us the highest temperature at which the material can reasonably be used without oxidation, chemical change or excessive creep becoming a problem; and the second gives the temperature needed to make the material flow easily for forming and shaping.

Most materials expand when they are heated (Figure 3.9). The thermal strain per degree of temperature change is measured by the *linear thermal expansion coefficient*, α (units: K^{-1}). If the material is thermally isotropic, the volume expansion, per degree, is 3α. If it is anisotropic, two or more coefficients are required, and the volume expansion becomes the sum of the principal thermal strains.

The *thermal shock resistance* (units: K) is the maximum temperature difference through which a material can be quenched suddenly without damage. It, and the *creep resistance*, are important in high-temperature design. Creep is the slow, time-dependent deformation which occurs when materials are loaded above about $\frac{1}{3}T_m$ or $\frac{2}{3}T_g$ (Figure 3.10). It is characterized by a set of *creep constants*: a creep exponent n (dimensionless), an activation energy Q (units: kJ/mole), a kinetic factor $\dot{\varepsilon}_0$ (units: s^{-1}), and a reference stress σ_0 (units: MPa or MN/m^2). The creep strain-rate $\dot{\varepsilon}$ at a temperature T caused by a stress σ is described by the equation

$$\dot{\varepsilon} = \dot{\varepsilon}_0 \left(\frac{\sigma}{\sigma_0}\right)^n \exp - \left(\frac{Q}{RT}\right) \tag{3.14}$$

where R is the gas constant (8.314 J/mol K).

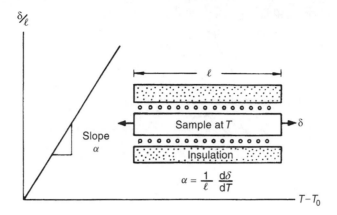

Fig. 3.9 The linear-thermal expansion coefficient α measures the change in length, per unit length, when the sample is heated.

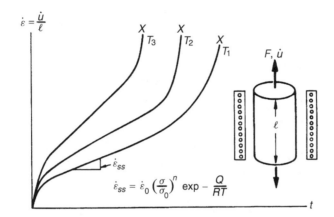

Fig. 3.10 Creep is the slow deformation with time under load. It is characterized by the creep constants, $\dot{\varepsilon}_0$, σ_0 and Q.

Wear, oxidation and corrosion are harder to quantify, partly because they are surface, not bulk, phenomena, and partly because they involve interactions between two materials, not just the properties of one. When solids slide (Figure 3.11) the volume of material lost from one surface, per unit distance slid, is called the wear rate, W. The wear resistance of the surface is characterized by the *Archard wear constant*, k_A (units: m/MN or MPa), defined by the equation

$$\frac{W}{A} = k_A P \tag{3.15}$$

where A is the area of the surface and P the pressure (i.e. force per unit area) pressing them together. Data for k_A are available, but must be interpreted as the property of the sliding couple, not of just one member of it.

Dry corrosion is the chemical reaction of a solid surface with dry gases (Figure 3.12). Typically, a metal, M, reacts with oxygen, O_2, to give a surface layer of the oxide MO_2:

$$M + O_2 = MO_2$$

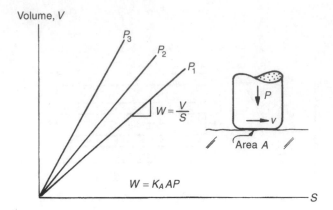

Fig. 3.11 Wear is the loss of material from surfaces when they slide. The wear resistance is measured by the Archard wear constant K_A.

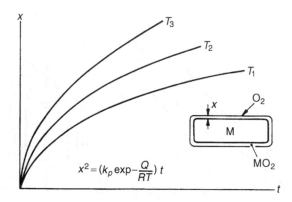

Fig. 3.12 Corrosion is the surface reaction of the material with gases or liquids — usually aqueous solutions. Sometimes it can be described by a simple rate equation, but usually the process is too complicated to allow this.

If the oxide is protective, forming a continuous, uncracked film (thickness x) over the surface, the reaction slows down with time t:

$$\frac{dx}{dt} = \frac{k_p}{x} \left\{ \exp - \left(\frac{Q}{RT} \right) \right\} \tag{3.16}$$

or, on integrating,

$$x^2 = k_p \left\{ \exp - \left(\frac{Q}{RT} \right) \right\} t$$

Here R is the gas constant, T the absolute temperature, and the oxidation behaviour is characterized by the *parabolic rate constant for oxidation* k_p (units: m^2/s) and an activation energy Q (units: kJ/mole).

    Wet corrosion — corrosion in water, brine, acids or alkalis — is much more complicated and cannot be captured by rate equations with simple constants. It is more usual to catalogue corrosion resistance by a simple scale such as A (very good) to E (very bad).

3.4 Summary and conclusions

There are six important classes of materials for mechanical design: metals, polymers elastomers, ceramics, glasses, and composites which combine the properties of two or more of the others. Within a class there is certain common ground: ceramics as a class are hard, brittle and corrosion resistant; metals as a class are ductile, tough and electrical conductors; polymers as a class are light, easily shaped and electrical insulators, and so on — that is what makes the classification useful. But, in design, we wish to escape from the constraints of class, and think, instead, of the material name as an identifier for a certain property-profile — one which will, in later chapters, be compared with an 'ideal' profile suggested by the design, guiding our choice. To that end, the properties important in thermo-mechanical design were defined in this chapter. In the next we develop a way of displaying properties so as to maximize the freedom of choice.

3.5 Further reading

Definitions of material properties can be found in numerous general texts on engineering materials, among them those listed here.

Ashby, M.F. and Jones, D.R.H. (1997; 1998) *Engineering Materials Parts 1 and 2*, 2nd editions. Pergamon Press, Oxford.

Charles, J.A., Crane, F.A.A. and Furness J.A.G. (1987) *Selection and Use of Engineering Materials*, 3rd edition. Butterworth-Heinemann, Oxford.

Farag, M.M. (1989) *Selection of Materials and Manufacturing Processes for Engineering Design* Prentice-Hall, Englewood Cliffs, NJ.

Fontana, M.G. and Greene, N.D. (1967) *Corrosion Engineering*. McGraw-Hill, New York.

Hertzberg, R.W. (1989) *Deformation and Fracture of Engineering Materials*, 3rd edition. Wiley, New York.

Van Vlack, L.H. (1982) *Materials for Engineering*. Addison-Wesley, Reading, MA.

Chapter 4

Materials selection charts

4.1 Introduction and synopsis

Material properties limit performance. We need a way of surveying properties, to get a feel for the values design-limiting properties can have. One property can be displayed as a ranked list or bar-chart. But it is seldom that the performance of a component depends on just one property. Almost always it is a combination of properties that matter: one thinks, for instance, of the strength-to-weight ratio, σ_f/ρ, or the stiffness-to-weight ratio, E/ρ, which enter lightweight design. This suggests the idea of plotting one property against another, mapping out the fields in property-space occupied by each material class, and the sub-fields occupied by individual materials.

The resulting charts are helpful in many ways. They condense a large body of information into a compact but accessible form; they reveal correlations between material properties which aid in checking and estimating data; and they lend themselves to a performance-optimizing technique, developed in Chapter 5, which becomes the basic step of the selection procedure.

The idea of a materials selection chart is described briefly in the following section. The section after that is not so brief: it introduces the charts themselves. There is no need to read it all, but it is helpful to persist far enough to be able to read and interpret the charts fluently, and to understand the meaning of the design guide lines that appear on them. If, later, you use one chart a lot, you should read the background to it, given here, to be sure of interpreting it correctly.

A compilation of all the charts, with a brief explanation of each, is contained in Appendix C of this text. It is intended for reference — that is, as a tool for tackling real design problems. As explained in the Preface, you may copy and distribute these charts without infringing copyright.

4.2 Displaying material properties

The properties of engineering materials have a characteristic span of values. The span can be large: many properties have values which range over five or more decades. One way of displaying this is as a bar-chart like that of Figure 4.1 for thermal conductivity. Each bar represents a single material. The length of the bar shows the range of conductivity exhibited by that material in its various forms. The materials are segregated by class. Each class shows a characteristic range: metals, have high conductivities; polymers have low; ceramics have a wide range, from low to high.

Much more information is displayed by an alternative way of plotting properties, illustrated in the schematic of Figure 4.2. Here, one property (the modulus, E, in this case) is plotted against another (the density, ρ) on logarithmic scales. The range of the axes is chosen to include all materials, from the lightest, flimsiest foams to the stiffest, heaviest metals. It is then found that data for a given class of materials (polymers for example) cluster together on the chart; the *sub-range* associated with one material class is, in all cases, much smaller than the *full* range of that property. Data for

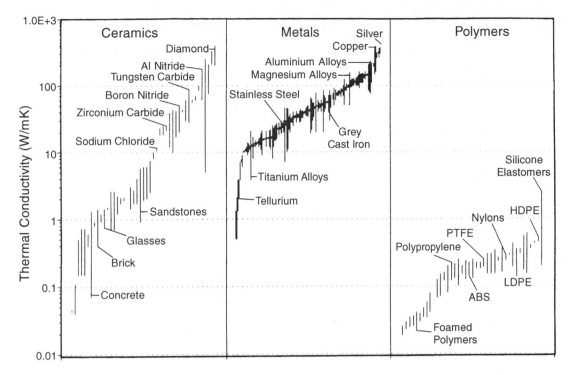

Fig. 4.1 A bar-chart showing thermal conductivity for three classes of solid. Each bar shows the range of conductivity offered by a material, some of which are labelled.

one class can be enclosed in a property envelope, as the figure shows. The envelope encloses all members of the class.

All this is simple enough — just a helpful way of plotting data. But by choosing the axes and scales appropriately, more can be added. The speed of sound in a solid depends on the modulus, E, and the density, ρ; the longitudinal wave speed v, for instance, is

$$v = \left(\frac{E}{\rho}\right)^{1/2}$$

or (taking logs)

$$\log E = \log \rho + 2 \log v$$

For a fixed value of v, this equation plots as a straight line of slope 1 on Figure 4.2. This allows us to add *contours of constant wave velocity* to the chart: they are the family of parallel diagonal lines, linking materials in which longitudinal waves travel with the same speed. All the charts allow additional fundamental relationships of this sort to be displayed. And there is more: design-optimizing parameters called *material indices* also plot as contours on to the charts. But that comes in Chapter 5.

Among the mechanical and thermal properties, there are 18 which are of primary importance, both in characterizing the material, and in engineering design. They were listed in Table 3.1: they include density, modulus, strength, toughness, thermal conductivity, diffusivity and expansion. The charts display data for these properties, for the nine classes of materials listed in Table 4.1. The

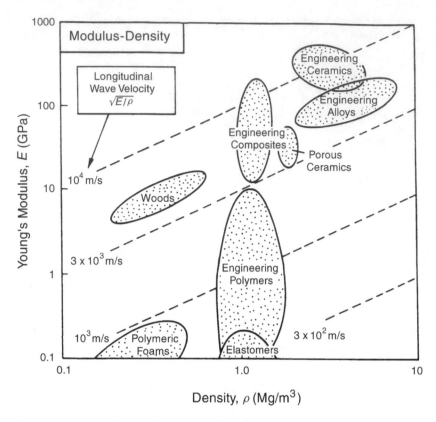

Fig. 4.2 The idea of a Materials Property Chart: Young's modulus, E, is plotted against the density, ρ, on log scales. Each class of material occupies a characteristic part of the chart. The log scales allow the longitudinal elastic wave velocity $v = (E/\rho)^{1/2}$ to be plotted as a set of parallel contours.

class-list is expanded from the original six of Figure 3.1 by distinguishing *engineering composites* from *foams* and from *woods* though all, in the most general sense, are composites; by distinguishing the high-strength *engineering ceramics* (like silicon carbide) from the low-strength *porous ceramics* (like brick); and by distinguishing elastomers (like rubber) from rigid polymers (like nylon). Within each class, data are plotted for a representative set of materials, chosen both to span the full range of behaviour for the class, and to include the most common and most widely used members of it. In this way the envelope for a class encloses data not only for the materials listed in Table 4.1, but for virtually all other members of the class as well.

The charts which follow show a *range* of values for each property of each material. Sometimes the range is narrow: the modulus of copper, for instance, varies by only a few per cent about its mean value, influenced by purity, texture and such like. Sometimes it is wide: the strength of alumina-ceramic can vary by a factor of 100 or more, influenced by porosity, grain size and so on. Heat treatment and mechanical working have a profound effect on yield strength and toughness of metals. Crystallinity and degree of cross-linking greatly influence the modulus of polymers, and so on. These *structure-sensitive* properties appear as elongated bubbles within the envelopes on the charts. A bubble encloses a typical range for the value of the property for a single material. Envelopes (heavier lines) enclose the bubbles for a class.

Table 4.1 Material classes and members of each class

Class	Members	Short name
Engineering Alloys (The metals and alloys of engineering)	Aluminium alloys	Al alloys
	Copper alloys	Cu alloys
	Lead alloys	Lead alloys
	Magnesium alloys	Mg alloys
	Molybdenum alloys	Mo alloys
	Nickel alloys	Ni alloys
	Steels	Steels
	Tin alloys	Tin alloys
	Titanium alloys	Ti alloys
	Tungsten alloys	W alloys
	Zinc alloys	Zn alloys
Engineering Polymers (The thermoplastics and thermosets of engineering)	Epoxies	EP
	Melamines	MEL
	Polycarbonate	PC
	Polyesters	PEST
	Polyethylene, high density	HDPE
	Polyethylene, low density	LDPE
	Polyformaldehyde	PF
	Polymethylmethacrylate	PMMA
	Polypropylene	PP
	Polytetrafluorethylene	PTFE
	Polyvinylchloride	PVC
Engineering Ceramics (Fine ceramics capable of load-bearing application)	Alumina	Al_2O_3
	Diamond	C
	Sialons	Sialons
	Silicon Carbide	SiC
	Silicon Nitride	Si_3N_4
	Zirconia	ZrO_2
Engineering Composites (The composites of engineering practice.) A distinction is drawn between the properties of a ply — 'UNIPLY' — and of a laminate — 'LAMINATES'	Carbon fibre reinforced polymer	CFRP
	Glass fibre reinforced polymer	GFRP
	Kevlar fibre reinforced polymer	KFRP
Porous Ceramics (Traditional ceramics, cements, rocks and minerals)	Brick	Brick
	Cement	Cement
	Common rocks	Rocks
	Concrete	Concrete
	Porcelain	Pcln
	Pottery	Pot
Glasses (Ordinary silicate glass)	Borosilicate glass	B-glass
	Soda glass	Na-glass
	Silica	SiO_2
Woods (Separate envelopes describe properties parallel to the grain and normal to it, and wood products)	Ash	Ash
	Balsa	Balsa
	Fir	Fir
	Oak	Oak
	Pine	Pine
	Wood products (ply, etc)	Woods

(*continued overleaf*)

Table 4.1 (*continued*)

Class	Members	Short name
Elastomers	Natural rubber	Rubber
(Natural and artificial rubbers)	Hard Butyl rubber	Hard Butyl
	Polyurethanes	PU
	Silicone rubber	Silicone
	Soft Butyl rubber	Soft Butyl
Polymer Foams	These include:	
(Foamed polymers of	Cork	Cork
engineering)	Polyester	PEST
	Polystyrene	PS
	Polyurethane	PU

The data plotted on the charts have been assembled from a variety of sources, documented in Chapter 13.

4.3 The material property charts

The modulus–density chart (Chart 1, Figure 4.3)

Modulus and density are familiar properties. Steel is stiff, rubber is compliant: these are effects of modulus. Lead is heavy; cork is buoyant: these are effects of density. Figure 4.3 shows the full range of Young's modulus, E, and density, ρ, for engineering materials.

Data for members of a particular class of material cluster together and can be enclosed by an envelope (heavy line). The same class envelopes appear on all the diagrams: they correspond to the main headings in Table 4.1.

The *density* of a solid depends on three factors: the atomic weight of its atoms or ions, their size, and the way they are packed. The size of atoms does not vary much: most have a volume within a factor of two of $2 \times 10^{-29}\,\text{m}^3$. Packing fractions do not vary much either — a factor of two, more or less: close-packing gives a packing fraction of 0.74; open networks (like that of the diamond-cubic structure) give about 0.34. The spread of density comes mainly from that of atomic weight, from 1 for hydrogen to 238 for uranium. Metals are dense because they are made of heavy atoms, packed densely; polymers have low densities because they are largely made of carbon (atomic weight: 12) and hydrogen in a linear 2 or 3-dimensional network. Ceramics, for the most part, have lower densities than metals because they contain light 0, N or C atoms. Even the lightest atoms, packed in the most open way, give solids with a density of around $1\,\text{Mg/m}^3$. Materials with lower densities than this are foams — materials made up of cells containing a large fraction of pore space.

The *moduli* of most materials depend on two factors: bond stiffness, and the density of bonds per unit area. A bond is like a spring: it has a spring constant, S (units: N/m). Young's modulus, E, is roughly

$$E = \frac{S}{r_0} \qquad (4.1)$$

where r_0 is the 'atom size' (r_0^3 is the mean atomic or ionic volume). The wide range of moduli is largely caused by the range of values of S. The covalent bond is stiff ($S = 20\text{--}200\,\text{N/m}$); the metallic and the ionic a little less so ($S = 15\text{--}100\,\text{N/m}$). Diamond has a very high modulus because the carbon atom is small (giving a high bond density) and its atoms are linked by very strong

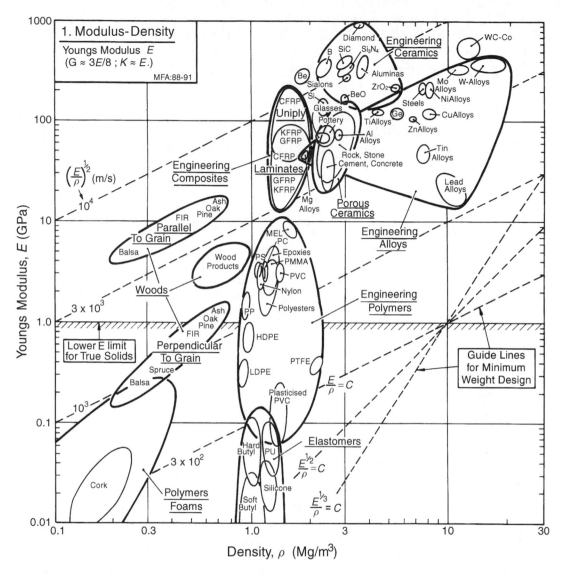

Fig. 4.3 Chart 1: Young's modulus, E, plotted against density, ρ. The heavy envelopes enclose data for a given class of material. The diagonal contours show the longitudinal wave velocity. The guide lines of constant E/ρ, $E^{1/2}/\rho$ and $E^{1/3}/\rho$ allow selection of materials for minimum weight, deflection-limited, design.

springs ($S = 200\,\text{N/m}$). Metals have high moduli because close-packing gives a high bond density and the bonds are strong, though not as strong as those of diamond. Polymers contain both strong diamond-like covalent bonds and weak hydrogen or Van der Waals bonds ($S = 0.5-2\,\text{N/m}$); it is the weak bonds which stretch when the polymer is deformed, giving low moduli.

But even large atoms ($r_0 = 3 \times 10^{-10}\,\text{m}$) bonded with weak bonds ($S = 0.5\,\text{N/m}$) have a modulus of roughly

$$E = \frac{0.5}{3 \times 10^{-10}} \approx 1\,\text{GPa} \tag{4.2}$$

This is the *lower limit* for true solids. The chart shows that many materials have moduli that are lower than this: they are either elastomers or foams. Elastomers have a low E because the weak secondary bonds have melted (their glass temperature T_g is below room temperature) leaving only the very weak 'entropic' restoring force associated with tangled, long-chain molecules; and foams have low moduli because the cell walls bend (allowing large displacements) when the material is loaded.

The chart shows that the modulus of engineering materials spans five decades[*], from 0.01 GPa (low-density foams) to 1000 GPa (diamond); the density spans a factor of 2000, from less than 0.1 to $20\,\mathrm{Mg/m^3}$. At the level of approximation of interest here (that required to reveal the relationship between the properties of materials classes) we may approximate the shear modulus G by $3E/8$ and the bulk modulus K by E, for all materials except elastomers (for which $G = E/3$ and $K \gg E$) allowing the chart to be used for these also.

The log-scales allow more information to be displayed. The velocity of elastic waves in a material, and the natural vibration frequencies of a component made of it, are proportional to $(E/\rho)^{1/2}$; the quantity $(E/\rho)^{1/2}$ itself is the velocity of longitudinal waves in a thin rod of the material. Contours of constant $(E/\rho)^{1/2}$ are plotted on the chart, labelled with the longitudinal wave speed. It varies from less than 50 m/s (soft elastomers) to a little more than 10^4 m/s (fine ceramics). We note that aluminium and glass, because of their low densities, transmit waves quickly despite their low moduli. One might have expected the sound velocity in foams to be low because of the low modulus, but the low density almost compensates. That in wood, across the grain, is low; but along the grain, it is high — roughly the same as steel — a fact made use of in the design of musical instruments.

The chart helps in the common problem of material selection for applications in which weight must be minimized. Guide lines corresponding to three common geometries of loading are drawn on the diagram. They are used in the way described in Chapters 5 and 6 to select materials for elastic design at minimum weight.

The strength–density chart (Chart 2, Figure 4.4)

The modulus of a solid is a well-defined quantity with a sharp value. The strength is not. It is shown, plotted against density, ρ, in Figure 4.4.

The word 'strength' needs definition (see also Chapter 3, Section 3.3). For metals and polymers, it is the *yield strength*, but since the range of materials includes those which have been worked, the range spans initial yield to ultimate strength; for most practical purposes it is the same in tension and compression. For brittle ceramics, the strength plotted here is the *crushing strength in compression*, not that in tension which is 10 to 15 times smaller; the envelopes for brittle materials are shown as broken lines as a reminder of this. For elastomers, strength means the *tear strength*. For composites, it is the *tensile failure strength* (the compressive strength can be less by up to 30% because of fibre buckling). We will use the symbol σ_f for all of these, despite the different failure mechanisms involved.

The considerable vertical extension of the strength bubble for an individual material reflects its wide range, caused by degree of alloying, work hardening, grain size, porosity and so forth. As before, members of a class cluster together and can be enclosed in an envelope (heavy line), and each occupies a characteristic area of the chart.

---

[*] Very low density foams and gels (which can be thought of as molecular-scale, fluid-filled, foams) can have moduli far lower than this. As an example, gelatin (as in Jello) has a modulus of about 5×10^{-5} GPa. Their strengths and fracture toughness, too, can be below the lower limit of the charts.

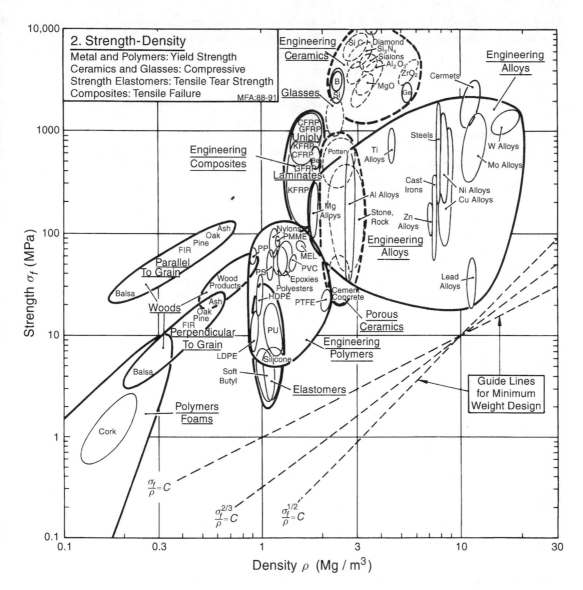

Fig. 4.4 Chart 2: Strength, σ_f, plotted against density, ρ (yield strength for metals and polymers, compressive strength for ceramics, tear strength for elastomers and tensile strength for composites). The guide lines of constant σ_f/ρ, $\sigma_f^{2/3}/\rho$ and $\sigma_f^{1/2}/\rho$ are used in minimum weight, yield-limited, design.

The range of strength for engineering materials, like that of the modulus, spans about five decades: from less than 0.1 MPa (foams, used in packaging and energy-absorbing systems) to 10^4 MPa (the strength of diamond, exploited in the diamond-anvil press). The single most important concept in understanding this wide range is that of the *lattice resistance* or *Peierls stress*: the intrinsic resistance of the structure to plastic shear. Plastic shear in a crystal involves the motion of dislocations. Metals are soft because the non-localized metallic bond does little to prevent dislocation motion, whereas ceramics are hard because their more localized covalent and ionic bonds (which must be broken and reformed when the structure is sheared), lock the dislocations in place. In non-crystalline solids we think instead of the energy associated with the unit step of the flow process: the relative slippage of two segments of a polymer chain, or the shear of a small molecular cluster in a glass network. Their strength has the same origin as that underlying the lattice resistance: if the unit step involves breaking strong bonds (as in an inorganic glass), the materials will be strong; if it involves only the rupture of weak bonds (the Van der Waals bonds in polymers for example), it will be weak. Materials which fail by fracture do so because the lattice resistance or its amorphous equivalent is so large that atomic separation (fracture) happens first.

When the lattice resistance is low, the material can be strengthened by introducing obstacles to slip: in metals, by adding alloying elements, particles, grain boundaries and even other dislocations ('work hardening'); and in polymers by cross-linking or by orienting the chains so that strong covalent as well as weak Van der Waals bonds are broken. When, on the other hand, the lattice resistance is high, further hardening is superfluous — the problem becomes that of suppressing fracture (next section).

An important use of the chart is in materials selection in lightweight plastic design. Guide lines are shown for materials selection in the minimum weight design of ties, columns, beams and plates, and for yield-limited design of moving components in which inertial forces are important. Their use is described in Chapters 5 and 6.

The fracture toughness–density chart (Chart 3, Figure 4.5)

Increasing the plastic strength of a material is useful only as long as it remains plastic and does not fail by fast fracture. The resistance to the propagation of a crack is measured by the *fracture toughness*, K_{Ic}. It is plotted against density in Figure 4.5. The range is large: from 0.01 to over $100 \, \text{MPa m}^{1/2}$. At the lower end of this range are brittle materials which, when loaded, remain elastic until they fracture. For these, linear-elastic fracture mechanics works well, and the fracture toughness itself is a well-defined property. At the upper end lie the super-tough materials, all of which show substantial plasticity before they break. For these the values of K_{Ic} are approximate, derived from critical J-integral (J_c) and critical crack-opening displacement (δ_c) measurements (by writing $K_{Ic} = (EJ_c)^{1/2}$, for instance). They are helpful in providing a ranking of materials. The guidelines for minimum weight design are explained in Chapter 5. The figure shows one reason for the dominance of metals in engineering; they almost all have values of K_{Ic} above $20 \, \text{MPa m}^{1/2}$, a value often quoted as a minimum for conventional design.

The modulus–strength chart (Chart 4, Figure 4.6)

High tensile steel makes good springs. But so does rubber. How is it that two such different materials are both suited for the same task? This and other questions are answered by Figure 4.6, the most useful of all the charts.

It shows Young's modulus E plotted against strength σ_f. The qualifications on 'strength' are the same as before: it means yield strength for metals and polymers, compressive crushing strength for ceramics, tear strength for elastomers, and tensile strength for composite and woods; the symbol σ_f is used for them all. The ranges of the variables, too, are the same. Contours of *failure strain*, σ_f/E (meaning the strain at which the material ceases to be linearly elastic), appear as a family of straight parallel lines.

Examine these first. Engineering polymers have large failure strains of between 0.01 and 0.1; the values for metals are at least a factor of 10 smaller. Even ceramics, in compression, are not as

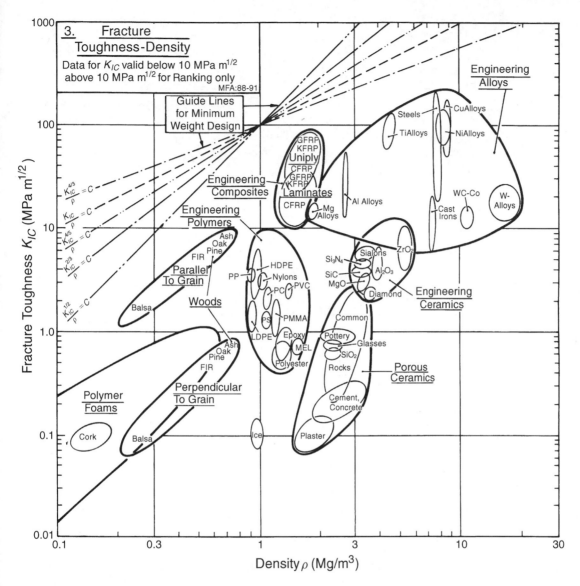

Fig. 4.5 Chart 3: Fracture toughness, K_{Ic}, plotted against density, ρ. The guide lines of constant K_{Ic}, $K_{Ic}^{2/3}/\rho$ and $K_{Ic}^{1/2}/\rho$, etc., help in minimum weight, fracture-limited design.

strong, and in tension they are far weaker (by a further factor of 10 to 15). Composites and woods lie on the 0.01 contour, as good as the best metals. Elastomers, because of their exceptionally low moduli, have values of σ_f/E larger than any other class of material: 0.1 to 10.

The distance over which inter-atomic forces act is small — a bond is broken if it is stretched to more than about 10% of its original length. So the force needed to break a bond is roughly

$$F \approx \frac{Sr_0}{10} \tag{4.3}$$

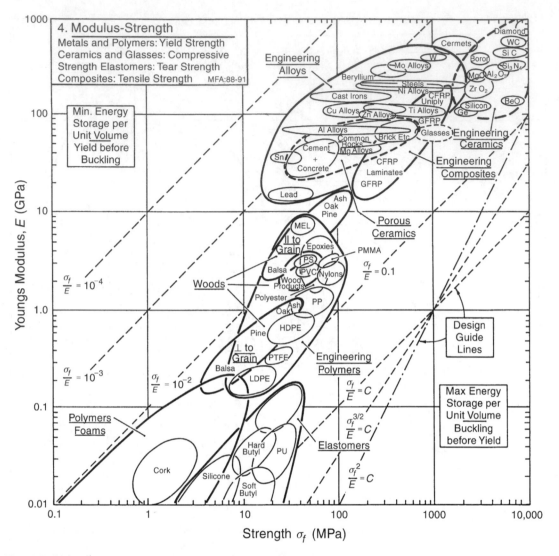

Fig. 4.6 Chart 4: Young's modulus, E, plotted against strength σ_f. The design guide lines help with the selection of materials for springs, pivots, knife-edges, diaphragms and hinges; their use is described in Chapters 5 and 6.

where S, as before, is the bond stiffness. If shear breaks bonds, the strength of a solid should be roughly

$$\sigma_f \approx \frac{F}{r_0^2} = \frac{S}{10r_0} = \frac{E}{10}$$

or

$$\frac{\sigma_f}{E} \approx \frac{1}{10} \tag{4.4}$$

The chart shows that, for some polymers, the failure strain is as large as this. For most solids it is less, for two reasons.

First, non-localized bonds (those in which the cohesive energy derives from the interaction of one atom with large number of others, not just with its nearest neighbours) are not broken when the structure is sheared. The metallic bond, and the ionic bond for certain directions of shear, are like this; very pure metals, for example, yield at stresses as low as $E/10\,000$, and strengthening mechanisms are needed to make them useful in engineering. The covalent bond *is* localized; and covalent solids do, for this reason, have yield strength which, at low temperatures, are as high as $E/10$. It is hard to measure them (although it can sometimes be done by indentation) because of the second reason for weakness: they generally contain defects — concentrators of stress — from which shear or fracture can propagate, often at stresses well below the 'ideal' $E/10$. Elastomers are anomalous (they have strengths of about E) because the modulus does not derive from bond-stretching, but from the change in entropy of the tangled molecular chains when the material is deformed.

This has not yet explained how to choose good materials to make springs. The way in which the chart helps with this is described in Section 6.9.

The specific stiffness–specific strength chart (Chart 5, Figure 4.7)

Many designs — particularly those for things which move — call for stiffness and strength at minimum weight. To help with this, the data of Chart 4 are replotted in Chart 5 (Figure 4.7) after dividing, for each material, by the density; it shows E/ρ plotted against σ_f/ρ.

Ceramics lie at the top right: they have exceptionally high stiffnesses and compressive strengths per unit weight, but their tensile strengths are much smaller. Composites then emerge as the material class with the most attractive specific properties, one of the reasons for their increasing use in aerospace. Metals are penalized because of their relatively high densities. Polymers, because their densities are low, are favoured.

The chart has application in selecting materials for light springs and energy-storage devices. But that too has to wait until Section 6.9.

The fracture toughness–modulus chart (Chart 6, Figure 4.8)

As a general rule, the fracture toughness of polymers is less than that of ceramics. Yet polymers are widely used in engineering structures; ceramics, because they are 'brittle', are treated with much more caution. Figure 4.8 helps resolve this apparent contradiction. It shows the *fracture toughness*, K_{Ic}, plotted against Young's modulus, E. The restrictions described earlier apply to the values of K_{Ic}: when small, they are well defined; when large, they are useful only as a ranking for material selection.

Consider first the question of the *necessary condition for fracture*. It is that sufficient external work be done, or elastic energy released, to supply the surface energy, γ per unit area, of the two new surfaces which are created. We write this as

$$G \geq 2\gamma \tag{4.5}$$

where G is the energy release rate. Using the standard relation $K \approx (EG)^{1/2}$ between G and stress intensity K, we find

$$K \geq (2E\gamma)^{1/2} \tag{4.6}$$

Now the surface energies, γ, of solid materials scale as their moduli; to an adequate approximation $\gamma = Er_0/20$, where r_0 is the atom size, giving

$$K \geq E \left(\frac{r_0}{20}\right)^{1/2} \tag{4.7}$$

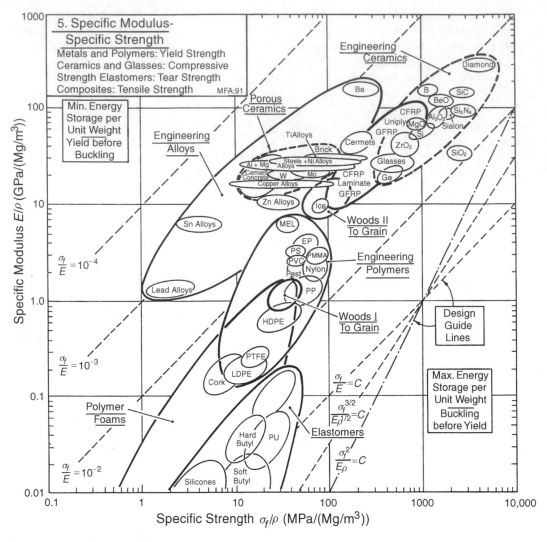

Fig. 4.7 Chart 5: Specific modulus, E/ρ, plotted against specific strength σ_f/ρ. The design guide lines help with the selection of materials for lightweight springs and energy-storage systems.

We identify the right-hand side of this equation with a lower-limiting value of K_{Ic}, when, taking r_0 as 2×10^{-10} m,

$$\frac{(K_{Ic})_{min}}{E} = \left(\frac{r_0}{20}\right)^{1/2} \approx 3 \times 10^{-6} \, \text{m}^{1/2} \tag{4.8}$$

This criterion is plotted on the chart as a shaded, diagonal band near the lower right corner. It defines a *lower limit* on values of K_{Ic}: it cannot be less than this unless some other source of energy such as a chemical reaction, or the release of elastic energy stored in the special dislocation structures caused by fatigue loading, is available, when it is given a new symbol such as $(K_{Ic})_{scc}$. meaning 'K_{Ic} for stress-corrosion cracking'. We note that the most brittle ceramics lie close to the threshold: when they fracture, the energy absorbed is only slightly more than the surface energy. When metals

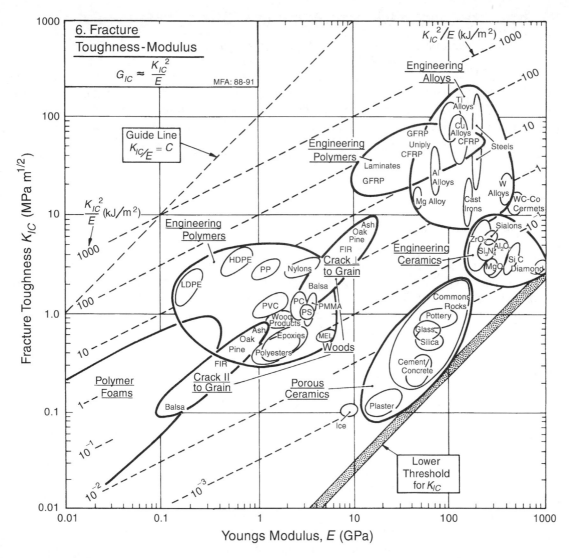

Fig. 4.8 Chart 6: Fracture toughness, K_{Ic}, plotted against Young's modulus, E. The family of lines are of constant K_{Ic}^2/E (approximately G_{Ic}, the fracture energy). These, and the guide line of constant K_{Ic}/E, help in design against fracture. The shaded band shows the 'necessary condition' for fracture. Fracture can, in fact, occur below this limit under conditions of corrosion, or cyclic loading.

and polymers and composites fracture, the energy absorbed is vastly greater, usually because of plasticity associated with crack propagation. We come to this in a moment, with the next chart.

Plotted on Figure 4.8 are contours of *toughness*, G_{Ic}, a measure of the apparent fracture surface energy ($G_{Ic} \approx K_{Ic}/E$). The true surface energies, γ, of solids lie in the range 10^{-4} to 10^{-3} kJ/m². The diagram shows that the values of the toughness start at 10^{-3} kJ/m² and range through almost six decades to 10^3 kJ/m². On this scale, ceramics ($10^{-3}–10^{-1}$ kJ/m²) are much lower than polymers ($10^{-1}–10$ kJ/m²); and this is part of the reason polymers are more widely used in engineering than ceramics. This point is developed further in Section 6.14.

The fracture toughness–strength chart (Chart 7, Figure 4.9)

The stress concentration at the tip of a crack generates a *process zone*: a plastic zone in ductile solids, a zone of micro-cracking in ceramics, a zone of delamination, debonding and fibre pull-out in composites. Within the process zone, work is done against plastic and frictional forces; it is this which accounts for the difference between the measured fracture energy G_{Ic} and the true surface energy 2γ. The amount of energy dissipated must scale roughly with the strength of the material, within the process zone, and with its size, d_y. This size is found by equating the stress field of the crack ($\sigma = K/\sqrt{2\pi r}$) at $r = d_y/2$ to the strength of the material, σ_f, giving

$$d_y = \frac{K_{Ic}^2}{\pi \sigma_f^2} \tag{4.9}$$

Figure 4.9 — fracture toughness against strength — shows that the size of the zone, d_y (broken lines), varies enormously, from atomic dimensions for very brittle ceramics and glasses to almost 1 m for the most ductile of metals. At a constant zone size, fracture toughness tends to increase with strength (as expected): it is this that causes the data plotted in Figure 4.9 to be clustered around the diagonal of the chart.

   The diagram has application in selecting materials for the safe design of load bearing structures. They are described in Sections 6.14 and 6.15.

The loss coefficient–modulus chart (Chart 8, Figure 4.10)

Bells, traditionally, are made of bronze. They can be (and sometimes are) made of glass; and they could (if you could afford it) be made of silicon carbide. Metals, glasses and ceramics all, under the right circumstances, have low intrinsic damping or 'internal friction', an important material property when structures vibrate. Intrinsic damping is measured by the *loss coefficient*, η, which is plotted in Figure 4.10.

   There are many mechanisms of intrinsic damping and hysteresis. Some (the 'damping' mechanisms) are associated with a process that has a specific time constant; then the energy loss is centred about a characteristic frequency. Others (the 'hysteresis' mechanisms) are associated with time-independent mechanisms; they absorb energy at all frequencies. In metals a large part of the loss is hysteretic, caused by dislocation movement: it is high in soft metals like lead and pure aluminium. Heavily alloyed metals like bronze and high-carbon steels have low loss because the solute pins the dislocations; these are the materials for bells. Exceptionally high loss is found in the Mn–Cu alloys because of a strain-induced martensite transformation, and in magnesium, perhaps because of reversible twinning. The elongated bubbles for metals span the large range accessible by alloying and working. Engineering ceramics have low damping because the enormous lattice resistance pins dislocations in place at room temperature. Porous ceramics, on the other hand, are filled with cracks, the surfaces of which rub, dissipating energy, when the material is loaded; the high damping of some cast irons has a similar origin. In polymers, chain segments slide against each other when loaded; the relative motion dissipates energy. The ease with which they slide depends on the ratio of the temperature (in this case, room temperature) to the glass temperature, T_g, of the polymer. When $T/T_g < 1$, the secondary bonds are 'frozen', the modulus is high and the damping is relatively low. When $T/T_g > 1$, the secondary bonds have melted, allowing easy chain slippage; the modulus is low and the damping is high. This accounts for the obvious inverse dependence of

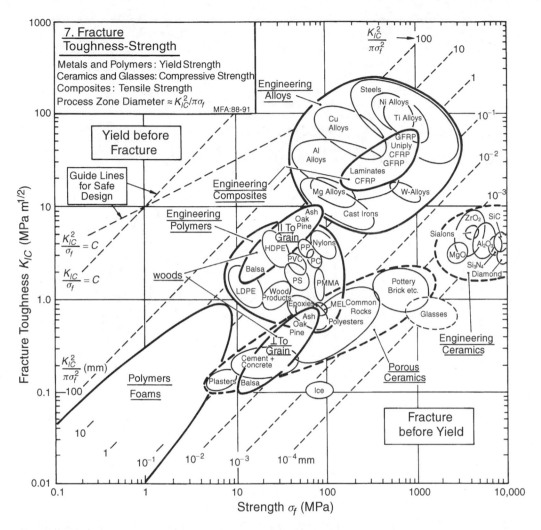

Fig. 4.9 Chart 7: Fracture toughness, K_{Ic}, plotted against strength, σ_f. The contours show the value of $K_{Ic}^2/\pi\sigma_f$ — roughly, the diameter of the process zone at a crack tip. The design guide lines are used in selecting materials for damage-tolerant design.

η on E for polymers in Figure 4.10; indeed, to a first approximation,

$$\eta = \frac{4 \times 10^{-2}}{E} \qquad (4.10)$$

with E in GPa.

The thermal conductivity–thermal diffusivity chart (Chart 9, Figure 4.11)

The material property governing the flow of heat through a material at steady-state is the *thermal conductivity*, λ (units: J/mK); that governing transient heat flow is the *thermal diffusivity*, a

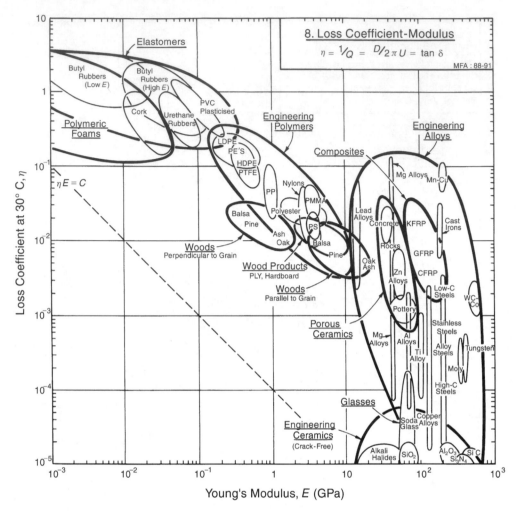

Fig. 4.10 Chart 8: The loss coefficient, η, plotted against Young's modulus, E. The guide line corresponds to the condition $\eta = C/E$.

(units: m²/s). They are related by

$$a = \frac{\lambda}{\rho C_p} \tag{4.11}$$

where ρ in kg/m³ is the density and C_p the specific heat in J/kg K; the quantity ρC_p is the *volumetric specific heat*. Figure 4.11 relates thermal conductivity, diffusivity and volumetric specific heat, at room temperature.

The data span almost five decades in λ and a. Solid materials are strung out along the line*

$$\rho C_p \approx 3 \times 10^6 \text{ J/m}^3\text{K} \tag{4.12}$$

---

* This can be understood by noting that a solid containing N atoms has $3N$ vibrational modes. Each (in the classical approximation) absorbs thermal energy kT at the absolute temperature T, and the vibrational specific heat is $C_p \approx C_v = 3Nk$ (J/K) where k is Boltzmann's constant (1.34×10^{-23} J/K). The volume per atom, Ω, for almost all solids lies within a factor

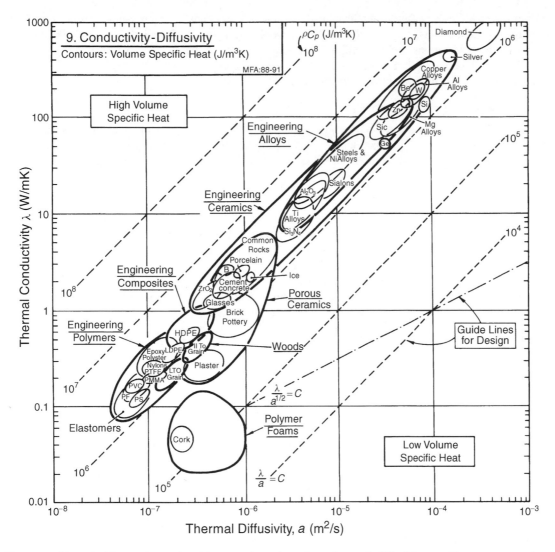

Fig. 4.11 Chart 9: Thermal conductivity, λ, plotted against thermal diffusivity, a. The contours show the volume specific heat, ρC_p. All three properties vary with temperature; the data here are for room temperature.

For solids, C_p and C_v differ very little; at the level of approximation of interest here we can assume them to be equal. As a general rule, then,

$$\lambda = 3 \times 10^6 \, a \qquad (4.13)$$

(λ in J/mK and a in m²/s). Some materials deviate from this rule: they have lower-than-average volumetric specific heat. For a few, like diamond, it is low because their Debye temperatures lie

of two of 1.4×10^{-29} m³; thus the volume of N atoms is $(N\Omega)$ m³. The volume specific heat is then (as the Chart shows):

$$\rho C_v \cong 3Nk/N\Omega = \frac{3k}{\Omega} = 3 \times 10^6 \text{ J/m}^3\text{K}$$

well above room temperature when heat absorption is not classical. The largest deviations are shown by porous solids: foams, low density firebrick, woods and the like. Their low density means that they contain fewer atoms per unit volume and, averaged over the volume of the structure, ρC_v is low. The result is that, although foams have low *conductivities* (and are widely used for insulation because of this), their thermal *diffusivities* are not necessarily low: they may not transmit much heat, but they reach a steady-state quickly. This is important in design — a point brought out by the Case Study of Section 6.17.

The range of both λ and a reflect the mechanisms of heat transfer in each class of solid. Electrons conduct the heat in pure metals such as copper, silver and aluminium (top right of chart). The conductivity is described by

$$\lambda = \frac{1}{3} C_e \bar{c} \ell \tag{4.14}$$

where C_e is the electron specific heat per unit volume, $\bar{c}$ is the electron velocity (2×10^5 m/s) and ℓ the electron mean free path, typically 10^{-7} m in pure metals. In solid solution (steels, nickel-based and titanium alloys) the foreign atoms scatter electrons, reducing the mean free path to atomic dimensions ($\approx 10^{-10}$ m), much reducing λ and a.

Electrons do not contribute to conduction in ceramics and polymers. Heat is carried by phonons — lattice vibrations of short wavelength. They are scattered by each other (through an anharmonic interaction) and by impurities, lattice defects and surfaces; it is these which determine the phonon mean free path, ℓ. The conductivity is still given by equation (4.14) which we write as

$$\lambda = \frac{1}{3} \rho C_p \bar{c} \ell \tag{4.15}$$

but now $\bar{c}$ is the elastic wave speed (around 10^3 m/s — see Chart 1) and ρC_p is the volumetric specific heat again. If the crystal is particularly perfect, and the temperature is well below the Debye temperature, as in diamond at room temperature, the phonon conductivity is high: it is for this reason that single crystal diamond, silicon carbide, and even alumina have conductivities almost as high as copper. The low conductivity of glass is caused by its irregular amorphous structure; the characteristic length of the molecular linkages (about 10^{-9} m) determines the mean free path. Polymers have low conductivities because the elastic wave speed $\bar{c}$ is low (Chart 1), and the mean free path in the disordered structure is small.

The lowest thermal conductivities are shown by highly porous materials like firebrick, cork and foams. Their conductivity is limited by that of the gas in their cells.

The thermal expansion–thermal conductivity chart (Chart 10, Figure 4.12)

Almost all solids expand on heating. The bond between a pair of atoms behaves like a linear elastic spring when the relative displacement of the atoms is small; but when it is large, the spring is non-linear. Most bonds become stiffer when the atoms are pushed together, and less stiff when they are pulled apart, and for that reason they are anharmonic. The thermal vibrations of atoms, even at room temperature, involves large displacements; as the temperature is raised, the anharmonicity of the bond pushes the atoms apart, increasing their mean spacing. The effect is measured by the linear *expansion coefficient*

$$\alpha = \frac{1}{\ell} \frac{d\ell}{dT} \tag{4.16}$$

where ℓ is a linear dimension of the body.

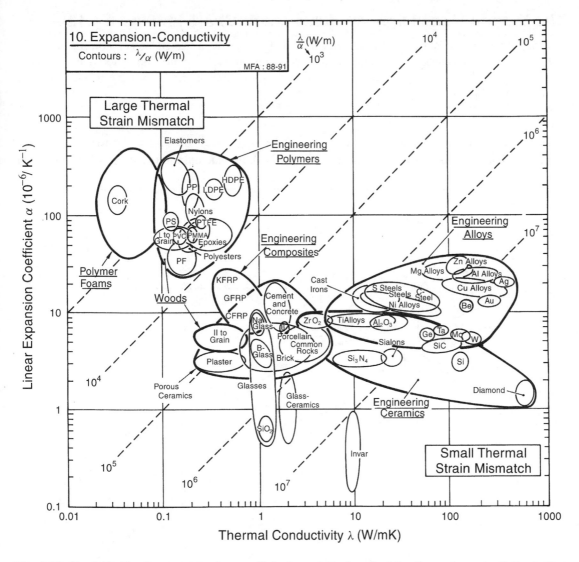

Fig. 4.12 Chart 10: The linear expansion coefficient, α, plotted against the thermal conductivity, λ. The contours show the thermal distortion parameter λ/α.

The expansion coefficient is plotted against the conductivity in Chart 10 (Figure 4.12). It shows that polymers have large values of α, roughly 10 times greater than those of metals and almost 100 times greater than ceramics. This is because the Van-der-Waals bonds of the polymer are very anharmonic. Diamond, silicon, and silica (SiO_2) have covalent bonds which have low anharmonicity (that is, they are almost linear-elastic even at large strains), giving them low expansion coefficients. Composites, even though they have polymer matrices, can have low values of α because the reinforcing fibres — particularly carbon — expand very little.

The charts shows contours of λ/α, a quantity important in designing against thermal distortion. A design application which uses this is developed in Section 6.20.

The thermal expansion–modulus chart (Chart 11, Figure 4.13)

Thermal stress is the stress which appears in a body when it is heated or cooled, but prevented from expanding or contracting. It depends on the expansion coefficient of the material, α, and on its modulus, E. A development of the theory of thermal expansion (see, for example, Cottrell (1964)) leads to the relation

$$\alpha = \frac{\gamma_G \rho C_v}{3E} \tag{4.17}$$

where γ_G is Gruneisen's constant; its value ranges between about 0.4 and 4, but for most solids it is near 1. Since ρC_v is almost constant (equation (4.12)), the equation tells us that α is proportional

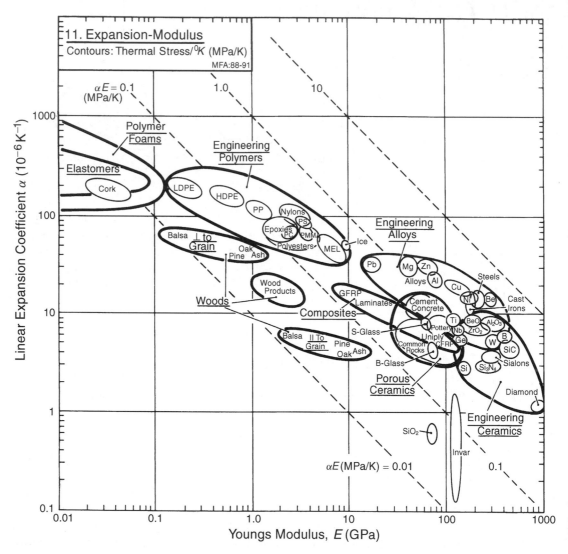

Fig. 4.13 Chart 11: The linear expansion coefficient, α, plotted against Young's modulus, E. The contours show the thermal stress created by a temperature change of 1°C if the sample is axially constrained. A correction factor C is applied for biaxial or triaxial constraint (see text).

to $1/E$. Figure 4.13 shows that this is so. Diamond, with the highest modulus, has one of the lowest coefficients of expansion; elastomers with the lowest moduli expand the most. Some materials with a low coordination number (silica, and some diamond-cubic or zinc-blende structured materials) can absorb energy preferentially in transverse modes, leading to very small (even a negative) value of γ_G and a low expansion coefficient — silica, SiO_2, is an example. Others, like Invar, contract as they lose their ferromagnetism when heated through the Curie temperature and, over a narrow range of temperature, they too show near-zero expansion, useful in precision equipment and in glass–metal seals.

One more useful fact: the moduli of materials scale approximately with their melting point, T_m:

$$E \approx \frac{100\, kT_m}{\Omega} \tag{4.18}$$

where k is Boltzmann's constant and Ω the volume-per-atom in the structure. Substituting this and equation (4.13) for ρC_v into equation (4.17) for α gives

$$\alpha = \frac{\gamma_G}{100\, T_m} \tag{4.19}$$

The expansion coefficient varies inversely with the melting point, or (equivalently stated) for all solids the thermal strain, just before they melt, depends only on γ_G, and this is roughly a constant. Equations (4.18) and (4.19) are examples of property correlations, useful for estimating and checking material properties (Chapter 13).

Whenever the thermal expansion or contraction of a body is prevented, thermal stresses appear; if large enough, they cause yielding, fracture, or elastic collapse (buckling). It is common to distinguish between thermal stress caused by external constraint (a rod, rigidly clamped at both ends, for example) and that which appears without external constraint because of temperature gradients in the body. All scale as the quantity αE, shown as a set of diagonal contours in Figure 4.13. More precisely: the stress $\Delta\sigma$ produced by a temperature change of 1°C in a constrained system, or the stress per °C caused by a sudden change of surface temperature in one which is not constrained, is given by

$$C\Delta\sigma = \alpha E \tag{4.20}$$

where $C = 1$ for axial constraint, $(1 - \nu)$ for biaxial constraint or normal quenching, and $(1 - 2\nu)$ for triaxial constraint, where ν is Poisson's ratio. These stresses are large: typically 1 MPa/K; they can cause a material to yield, or crack, or spall, or buckle, when it is suddenly heated or cooled. The resistance of materials to such damage is the subject of the next section.

The normalized strength–thermal expansion chart (Chart 12, Figure 4.14)

When a cold ice-cube is dropped into a glass of gin, it cracks audibly. The ice is failing by thermal shock. The ability of a material to withstand this is measured by its *thermal shock resistance*. It depends on its thermal expansion coefficient, α, and its normalized tensile strength, σ_t/E. They are the axes of Figure 4.14, on which contours of constant $\sigma_t/\alpha E$ are plotted. The tensile strength, σ_t, requires definition, just as σ_f did. For brittle solids, it is the tensile fracture strength (roughly equal to the modulus of rupture, or MOR). For ductile metals and polymers, it is the tensile yield strength; and for composites it is the stress which first causes permanent damage in the form of delamination, matrix cracking or fibre debonding.

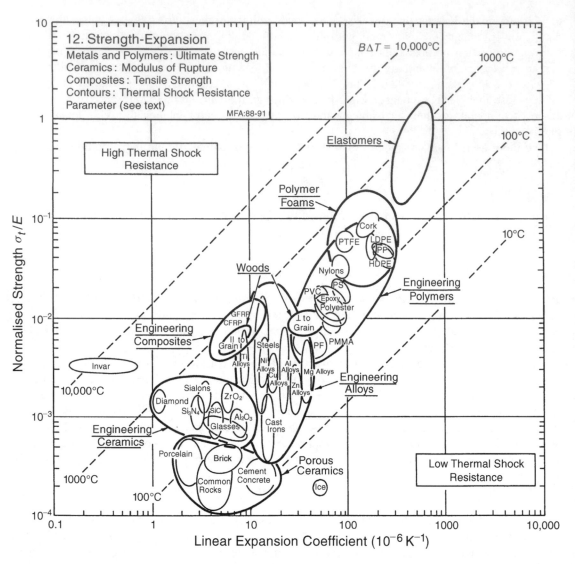

Fig. 4.14 Chart 12: The normalized tensile strength, σ_t/E, plotted against linear coefficient of expansion, α. The contours show a measure of the thermal shock resistance, ΔT. Corrections must be applied for constraint, and to allow for the effect of thermal conduction during quenching.

To use the chart, we note that a temperature change of ΔT, applied to a constrained body — or a sudden change ΔT of the surface temperature of a body which is unconstrained — induces a stress

$$\sigma = \frac{E\alpha\Delta T}{C} \tag{4.21}$$

where C was defined in the last section. If this stress exceeds the local tensile strength σ_t of the material, yielding or cracking results. Even if it does not cause the component to fail, it weakens it.

Table 4.2 Values for the factor A (section $T = 10$ mm)

Conditions	Foams	Polymers	Ceramics	Metals
Slow air flow ($h = 10$ W/m^{2}K)	0.75	0.5	3×10^{-2}	3×10^{-3}
Black body radiation 500 to 0C ($h = 40$ W/m^{2}K)	0.93	0.6	0.12	1.3×10^{-2}
Fast air flow ($h = 10^2$ W/m^{2}K)	1	0.75	0.25	3×10^{-2}
Slow water quench ($h = 10^3$ W/m^{2}K)	1	1	0.75	0.23
Fast water quench ($h = 10^4$ W/m^{2}K)	1	1	1	0.1–0.9

Then a measure of the thermal shock resistance is given by

$$\frac{\Delta T}{C} = \frac{\sigma_t}{\alpha E} \tag{4.22}$$

This is not quite the whole story. When the constraint is internal, the thermal conductivity of the material becomes important. 'Instant' cooling when a body is quenched requires an infinite rate of heat transfer at its surface. Heat transfer rates are measured by the heat transfer coefficient, h, and are never infinite. Water quenching gives a high h, and then the values of ΔT calculated from equation (4.22) give an approximate ranking of thermal shock resistance. But when heat transfer at the surface is poor and the thermal conductivity of the solid is high (thereby reducing thermal gradients) the thermal stress is less than that given by equation (4.21) by a factor A which, to an adequate approximation, is given by

$$A = \frac{th/\lambda}{1 + th/\lambda} \tag{4.23}$$

where t is a typical dimension of the sample in the direction of heat flow; the quantity th/λ is usually called the Biot modulus. Table 4.2 gives typical values of A, for each class, using a section size of 10 mm. The equation defining the thermal shock resistance, ΔT, now becomes

$$B\Delta T = \frac{\sigma_t}{\alpha E} \tag{4.24}$$

where $B = C/A$. The contours on the diagram are of $B\Delta T$. The table shows that, for rapid quenching, A is unity for all materials except the high-conductivity metals: then the thermal shock resistance is simply read from the contours, with appropriate correction for the constraint (the factor C). For slower quenches, ΔT is larger by the factor $1/A$, read from the table.

The strength–temperature chart (Chart 13, Figure 4.15)

As the temperature of a solid is raised, the amplitude of thermal vibration of its atoms increases and solid expands. Both the expansion and the vibration makes plastic flow easier. The strengths of solids fall, slowly at first and then more rapidly, as the temperature increases. Chart 13 (Figure 4.15) captures some of this information. It shows the range of yield strengths of families of materials plotted against temperature. The near-horizontal part of each lozenge shows the strength in the regime in which temperature has little effect; the downward-sloping part shows the more precipitate drop as the maximum service temperature is reached.

There are better ways of describing *high-temperature strength* than this, but they are much more complicated. The chart gives a birds-eye view of the regimes of stress and temperature in which each material class, and material, is usable. Note that even the best polymers have little strength

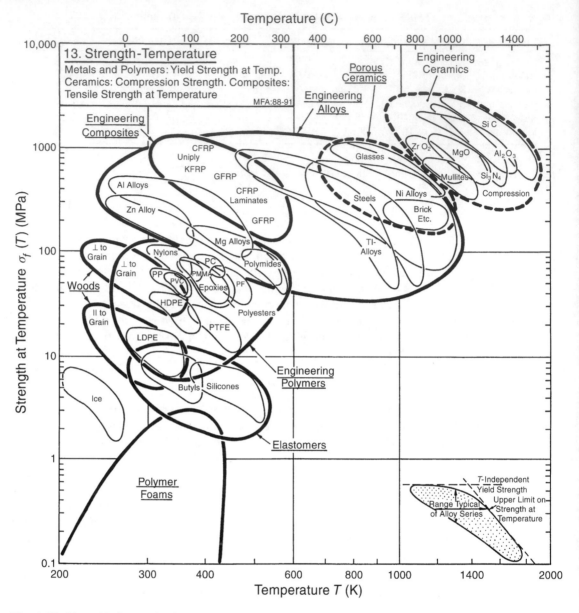

Fig. 4.15 Chart 13: Strength plotted against temperature. The inset explains the shape of the lozenges.

above 200°C; most metals become very soft by 800°C; and only ceramics offer strength above 1500°C.

The modulus–relative cost chart (Chart 14, Figure 4.16)

Properties like modulus, strength or conductivity do not change with time. Cost is bothersome because it does. Supply, scarcity, speculation and inflation contribute to the considerable fluctuations

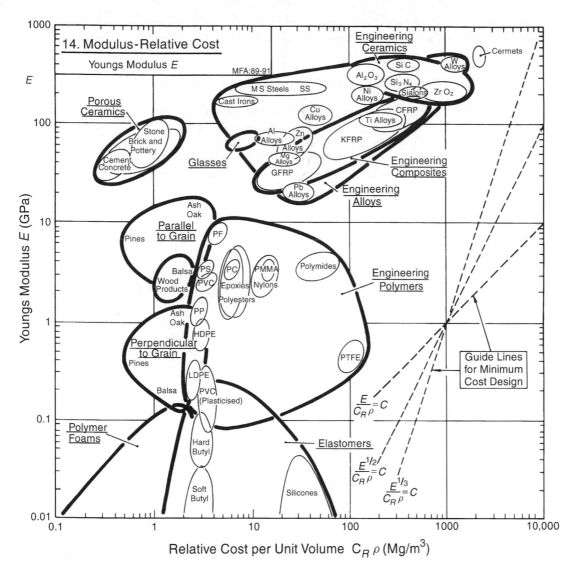

Fig. 4.16 Chart 14: Young's modulus, E, plotted against relative cost per unit volume, $C_p\rho$. The design guide lines help selection to maximize stiffness per unit cost.

in the cost-per-kilogram of a commodity like copper or silver. Data for cost-per-kg are tabulated for some materials in daily papers and trade journals; those for others are harder to come by. To make some correction for the influence of inflation and the units of currency in which cost is measured, we define a *relative cost* C_R:

$$C_R = \frac{\text{cost-per-kg of the material}}{\text{cost-per-kg of mild steel rod}}$$

At the time of writing, steel reinforcing rod costs about £0.2/kg (US$ 0.3/kg).

Chart 14 (Figure 4.16) shows the modulus E plotted against relative cost per unit volume $C_R\rho$, where ρ is the density. Cheap stiff materials lie towards the bottom right.

The strength–relative cost chart (Chart 15, Figure 4.17)

Cheap strong materials are selected using Chart 15 (Figure 4.17). It shows strength, defined as before, plotted against relative cost, defined above. The qualifications on the definition of strength, given earlier, apply here also.

It must be emphasized that the data plotted here and on Chart 14 are less reliable than those of previous charts, and subject to unpredictable change. Despite this dire warning, the two charts are

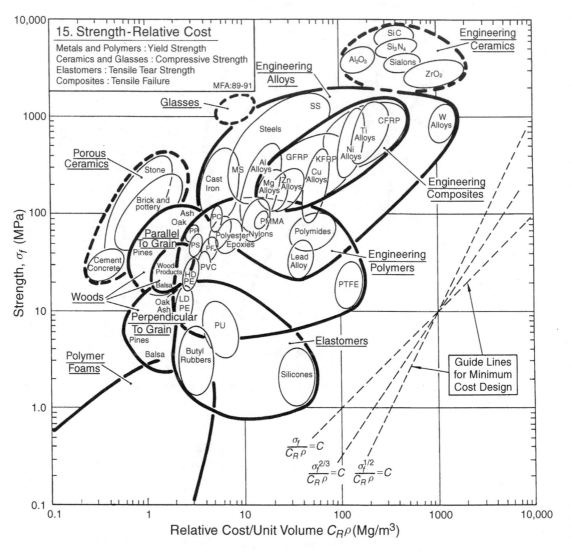

Fig. 4.17 Chart 15: Strength, σ_f, plotted against relative cost per unit volume, $C_p\rho$. The design guide lines help selection to maximize strength per unit cost.

genuinely useful. They allow selection of materials, using the criterion of 'function per unit cost'. An example is given in Section 6.5.

The wear rate/bearing pressure chart (Charts 16, Figures 4.18)

God, it is said, created solids; it was the devil who made surfaces. When surfaces touch and slide, there is friction; and where there is friction, there is wear. Tribologists — the collective noun for those who study friction and wear — are fond of citing the enormous cost, through lost energy and worn equipment, for which these two phenomena are responsible. It is certainly true that if friction could be eliminated, the efficiency of engines, gear boxes, drive trains and the like would increase; and if wear could be eradicated, they would also last longer. But before accepting this totally black image, one should remember that, without wear, pencils would not write on paper or chalk on blackboards; and without friction, one would slither off the slightest incline.

Tribological properties are not attributes of one material alone, but of one material sliding on another with — almost always — a third in between. The number of combinations is far too great to allow choice in a simple, systematic way. The selection of materials for bearings, drives, and sliding seals relies heavily on experience. This experience is captured in reference sources (for which see Chapter 13); in the end it is these which must be consulted. But it does help to have a feel for the magnitude of friction coefficients and wear rates, an idea of how these relate to material class.

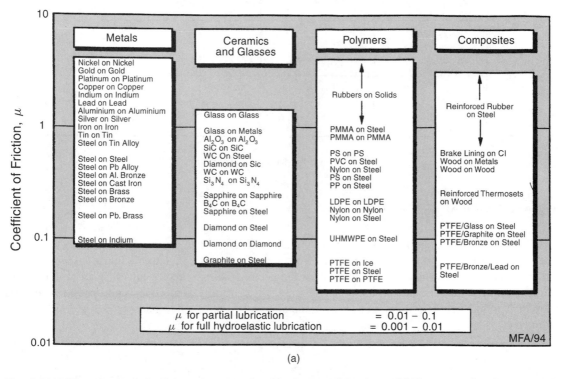

(a)

Fig. 4.18 (a) The friction coefficient for common bearing combinations. (b) The normalized wear rate, k_A, plotted against hardness, H. The chart gives an overview of the way in which common engineering materials behave. Selection to resist wear is discussed further in Chapter 13.

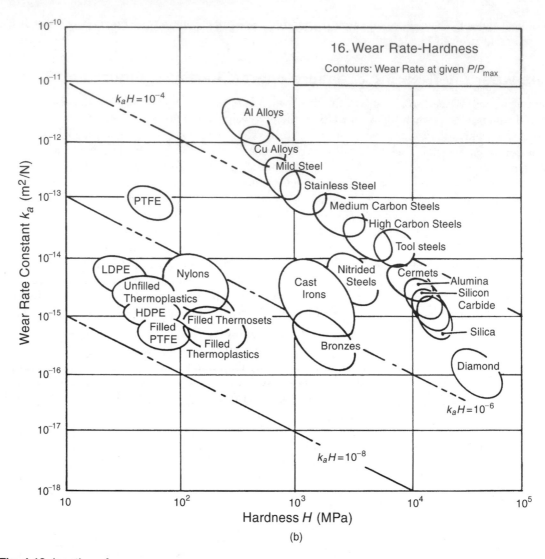

Fig. 4.18 (continued)

When two surfaces are placed in contact under a normal load F_n and one is made to slide over the other, a force F_s opposes the motion. This force is proportional to F_n but does not depend on the area of the surface — and this is the single most significant result of studies of friction, since it implies that surfaces do not contact completely, but only touch over small patches, the area of which is independent of the apparent, nominal area of contact A_n. The *coefficient friction* μ is defined by

$$\mu = \frac{F_s}{F_n} \tag{4.25}$$

Values for μ for dry sliding between surfaces are shown in Figure 4.18(a) Typically, $\mu \approx 0.5$. Certain materials show much higher values, either because they seize when rubbed together (a soft metal rubbed on itself with no lubrication, for instance) or because one surface has a sufficiently low modulus that it conforms to the other (rubber on rough concrete). At the other extreme are sliding combinations with exceptionally low coefficients of friction, such as PTFE, or bronze bearings loaded graphite, sliding on polished steel. Here the coefficient of friction falls as low as 0.04, though this is still high compared with friction for lubricated surfaces, as indicated at the bottom of the diagram.

When surfaces slide, they wear. Material is lost from both surfaces, even when one is much harder than the other. The *wear-rate, W*, is conventionally defined as

$$W = \frac{\text{Volume of material removed from contact surface}}{\text{Distance slid}} \qquad (4.26)$$

and thus has units of m^2. A more useful quantity, for our purposes, is the specific wear-rate

$$\Omega = \frac{W}{A_n} \qquad (4.27)$$

which is dimensionless. It increases with bearing pressure P (the normal force F_n divided by the nominal area A_n), such that the ratio

$$k_a = \frac{W}{F_n} = \frac{\Omega}{P} \qquad (4.28)$$

with units of (MPa)$^{-1}$, is roughly constant. The quantity k_a is a measure of the propensity of a sliding couple for wear: high k_a means rapid wear at a given bearing pressure.

The bearing pressure P is the quantity specified by the design. The ability of a surface to resist a static pressure is measured by its hardness, so we anticipate that the maximum bearing pressure P_{max} should scale with the hardness H of the softer surface:

$$P_{\text{max}} = CH$$

where C is a constant. Thus the wear-rate of a bearing surface can be written:

$$\Omega = k_a P = C \left(\frac{P}{P_{\text{max}}} \right) k_a H \qquad (4.29)$$

Two material properties appear in this equation: the wear constant k_a and the hardness H. They are plotted in Chart 16, Figure 4.18(b), which allows selection procedure for materials to resist wear at low sliding rates. Note, first, that materials of a given class (metals, for instance) tend to lie along a downward sloping diagonal across the figure, reflecting the fact that low wear rate is associated with high hardness. The best materials for bearings for a given bearing pressure P are those with the lowest value of k_a, that is, those nearest the bottom of the diagram. On the other hand, an efficient bearing, in terms of size or weight, will be loaded to a safe fraction of its maximum bearing pressure, that is, to a constant value of P/P_{max}, and for these, materials with the lowest values of the product $k_a H$ are best. The diagonal contours on the figure show constant values of this quantity.

The environmental attack chart (Chart 17, Figure 4.19)

All engineering materials are reactive chemicals. Their long-term properties — particularly strength properties — depend on the rate and nature of their reaction with their environment. The reaction can take many forms, of which the commonest are *corrosion* and *oxidation*. Some of these produce

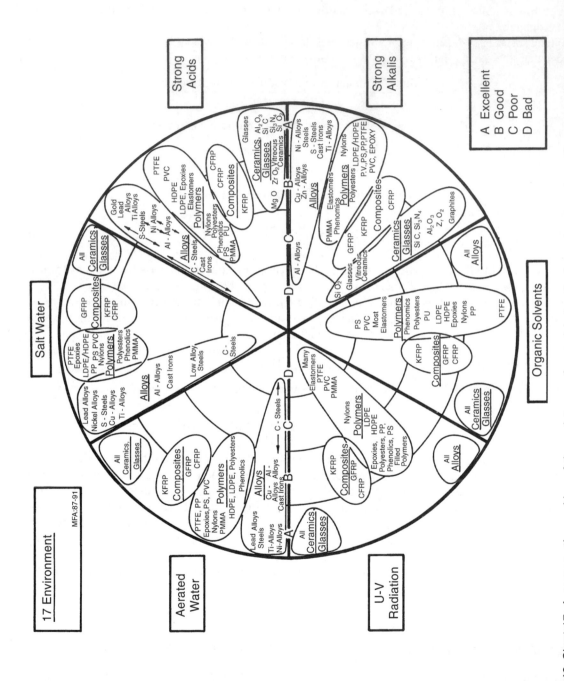

Fig. 4.19 Chart 17: A comparative ranking of the resistance of materials to attack by six common environments. It is an introduction to a problem which requires detailed and complex expertise, and should be used for the broadest guidance only. Selection to resist corrosion is discussed further in Chapter 13.

a thin, stable, adherent film with negligible loss of base material; they are, in general, protective. Others are more damaging, either because they reduce the section by steady dissolution or spalling-off of solid corrosion products, or because, by penetrating grain boundaries (in metals) or inducing chemical change by inter-diffusion (in polymers) they reduce the effective load-bearing capacity without apparent loss of section. And among these, the most damaging are those for which the loss of load-bearing capacity increases *linearly*, rather than parabolically, with time — that is, the damage rate (at a fixed temperature) is constant.

The considerable experience of environmental attack and its prevention is captured in reference sources listed in Chapter 13. Once a candidate material has been chosen, information about its reaction to a given environment can be found in these. Commonly, they rank the resistance of a material to attack in a given environment according to a scale such as 'A' (excellent) to 'D' (awful). This information is shown, for six environments, in Chart 17 (Figure 4.19). Its usefulness is very limited; at best it gives warning of a potential environmental hazard associated with the use of a given material. The proper way to select material to resist corrosion requires the methods of Chapter 13.

4.4 Summary and conclusions

The engineering properties of materials are usefully displayed as material selection charts. The charts summarize the information in a compact, easily accessible way; and they show the range of any given property accessible to the designer and identify the material class associated with segments of that range. By choosing the axes in a sensible way, more information can be displayed: a chart of modulus E against density ρ reveals the longitudinal wave velocity $(E/\rho)^{1/2}$; a plot of fracture toughness K_{Ic} against modulus E shows the fracture surface energy G_{Ic}; a diagram of thermal conductivity λ against diffusivity, a, also gives the volume specific heat ρC_v; expansion, α, against normalized strength, σ_t/E, gives thermal shock resistance ΔT.

The most striking feature of the charts is the way in which members of a material class cluster together. Despite the wide range of modulus and density associated with metals (as an example), they occupy a field which is distinct from that of polymers, or that of ceramics, or that of composites. The same is true of strength, toughness, thermal conductivity and the rest: the fields sometimes overlap, but they always have a characteristic place within the whole picture.

The position of the fields and their relationship can be understood in simple physical terms: the nature of the bonding, the packing density, the lattice resistance and the vibrational modes of the structure (themselves a function of bonding and packing), and so forth. It may seem odd that so little mention has been made of micro-structure in determining properties. But the charts clearly show that the first-order difference between the properties of materials has its origins in the mass of the atoms, the nature of the inter-atomic forces and the geometry of packing. Alloying, heat treatment and mechanical working all influence micro-structure, and through this, properties, giving the elongated bubbles shown on many of the charts; but the magnitude of their effect is less, by factors of 10, than that of bonding and structure.

The charts have numerous applications. One is the checking and validation of data (Chapter 13); here use is made both of the range covered by the envelope of material properties, and of the numerous relations between material properties (like $E\Omega = 100\,kT_m$), described in Section 4.3. Another concerns the development of, and identification of uses for, new materials; materials which fill gaps in one or more of the charts generally offer some improved design potential. But most important of all, the charts form the basis for a procedure for materials selection. That is developed in the following chapters.

4.5 Further reading

The best book on the physical origins of the mechanical properties of materials remains that by Cottrell. Values for the material properties which appear on the charts derive from sources documented in Chapter 13.

Material properties: general

Cottrell, A.H. (1964) *Mechanical Properties of Matter*. Wiley, New York.
Tabor, D. (1978) *Properties of Matter*, Penguin Books, London.

Chapter 5

Materials selection — the basics

5.1 Introduction and synopsis

This chapter sets out the basic procedure for selection, establishing the link between material and function (Figure 5.1). A material has *attributes*: its density, strength, cost, resistance to corrosion, and so forth. A design demands a certain profile of these: a low density, a high strength, a modest cost and resistance to sea water, perhaps. The problem is that of identifying the desired attribute profile and then comparing it with those of real engineering materials to find the best match. This we do by, first, *screening and ranking* the candidates to give a shortlist, and then seeking detailed *supporting information* for each shortlisted candidate, allowing a final choice. It is important to start with the full menu of materials in mind; failure to do so may mean a missed opportunity. If an innovative choice is to be made, it must be identified early in the design process. Later, too many decisions have been taken and commitments made to allow radical change: it is now or never.

The immensely wide choice is narrowed, first, by applying *property limits* which screen out the materials which cannot meet the design requirements. Further narrowing is achieved by ranking the candidates by their ability to maximize performance. Performance is generally limited not by a single property, but by a combination of them. The best materials for a light stiff tie-rod are those with the greatest value of the 'specific stiffness', E/ρ, where E is Young's modulus and ρ the density. The best materials for a spring, regardless of its shape or the way it is loaded, are those with the greatest value of σ_f^2/E, where σ_f is the failure stress. The materials which best resist thermal shock are those with the largest value of $\sigma_f/E\alpha$, where α is the thermal coefficient of expansion; and so forth. Combinations such as these are called *material indices*: they are groupings of material properties which, when maximized, maximize some aspect of performance. There are many such indices. They are derived from the design requirements for a component by an analysis of *function*, *objectives* and *constraints*. This chapter explains how to do this.

The materials property charts introduced in Chapter 4 are designed for use with these criteria. Property limits and material indices are plotted onto them, isolating the subset of materials which are the best choice for the design. The procedure is fast, and makes for lateral thinking. Examples of the method are given in Chapter 6.

5.2 The selection strategy

Material attributes

Figure 5.2 illustrates how the Kingdom of Materials can be subdivided into families, classes, subclasses and members. Each member is characterized by a set of attributes: its properties. As an example, the Materials Kingdom contains the family 'Metals' which in turn contains the class 'Aluminium alloys', the subclass '5000 series' and finally the particular member 'Alloy 5083 in the

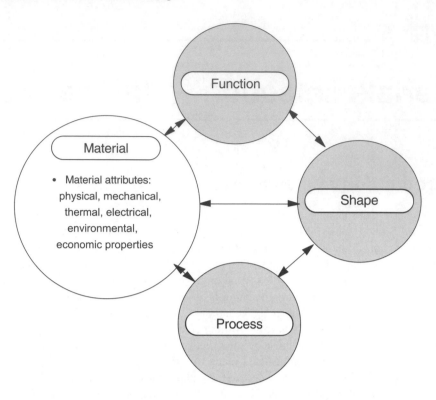

Fig. 5.1 Material selection is determined by function. Shape sometimes influences the selection. This chapter and the next deal with materials selection when this is independent of shape.

Kingdom	Family	Class	Sub-class	Member	Attributes
			1000	5005-0	Density
			2000	5005-H4	Modulus
	Ceramics	Steels	3000	5005-H6	Strength
	Glasses	Cu alloys	4000	5083-0	Toughness
Material	Metals	Al alloys	5000	5083-H2	T-conductivity
	Polymers	Ti-alloys	6000	5083-H4	T-expansion
	Elastomers	Ni-alloys	7000	5154-0	Resistivity
	Composites	Zn-alloys	8000	5154-H2...	Cost
					Corrosion
					Oxidation ....

Fig. 5.2 The taxonomy of the kingdom of materials and their attributes.

H2 heat treatment condition'. It, and every other member of the materials kingdom, is characterized by a set of attributes which include its mechanical, thermal, electrical and chemical properties, its processing characteristics, its cost and availability, and the environmental consequences of its use. We call this its *property-profile*. Selection involves seeking the best match between the property-profile of materials in the kingdom and that required by the design.

There are two main steps which we here call *screening and ranking*, and *supporting information* (Figure 5.3). The two steps can be likened to those in selecting a candidate for a job. The job is first advertised, defining essential skills and experience ('essential attributes'), screening-out potential

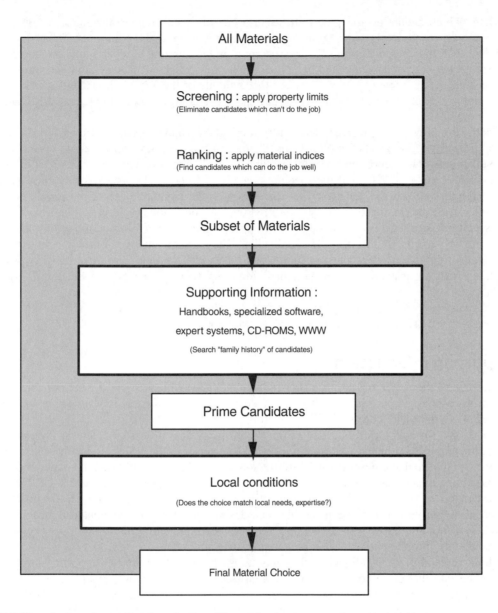

Fig. 5.3 The strategy for materials selection. The main steps are enclosed in bold boxes.

applicants whose attribute-profile does not match the job requirements and allowing a shortlist to be drawn up. References and interviews are then sought for the shortlisted candidates, building a file of supporting information.

Screening and ranking

Unbiased selection requires that all materials are considered to be candidates until shown to be otherwise, using the steps detailed in the boxes of Figure 5.3. The first of these, *screening*, eliminates candidates which cannot do the job at all because one or more of their attributes lies outside the limits imposed by the design. As examples, the requirement that 'the component must function at 250°C', or that 'the component must be transparent to light' imposes obvious limits on the attributes of *maximum service temperature* and *optical transparency* which successful candidates must meet. We refer to these as property limits. They are the analogue of the job advertisement which requires that the applicant 'must have a valid driving licence', or 'a degree in computer science', eliminating anyone who does not.

Property limits do not, however, help with ordering the candidates that remain. To do this we need optimization criteria. They are found in the material indices, developed below, which measure how well a candidate which has passed the limits can do the job. Familiar examples of indices are the specific stiffness E/ρ and the specific strength σ_f/ρ (E is the Young's modulus, σ_f is the failure strength and ρ is the density). The materials with the largest values of these indices are the best choice for a light, stiff tie-rod, or a light, strong tie-rod respectively. There are many others, each associated with maximizing some aspect of performance*. They allow ranking of materials by their ability to perform well in the given application. They are the analogue of the job advertisement which states that 'typing speed and accuracy are a priority', or that 'preference will be given to candidates with a substantial publication list', implying that applicants will be ranked by these criteria.

To summarize: property limits isolate candidates which are capable of doing the job; material indices identify those among them which can do the job well.

Supporting information

The outcome of the screening step is a shortlist of candidates which satisfy the quantifiable requirements of the design. To proceed further we seek a detailed profile of each: its *supporting information* (Figure 5.3, second heavy box).

Supporting information differs greatly from the property data used for screening. Typically, it is descriptive, graphical or pictorial: case studies of previous uses of the material, details of its corrosion behaviour in particular environments, information of availability and pricing, experience of its environmental impact. Such information is found in handbooks, suppliers data sheets, CD-based data sources and the World-Wide Web. Supporting information helps narrow the shortlist to a final choice, allowing a definitive match to be made between design requirements and material attributes. The parallel, in filling a job, is that of taking up references and conducting interviews — an opportunity to probe deeply into the character and potential of the candidate.

Without screening, the candidate-pool is enormous; there is an ocean of supporting information, and dipping into this gives no help with selection. But once viable candidates have been identified by screening, supporting information is sought for these few alone. The *Encyclopaedia Britannica* is an example of a source of supporting information; it is useful if you know what you are looking for, but overwhelming in its detail if you do not.

Local conditions

The final choice between competing candidates will often depend on local conditions: on the existing in-house expertise or equipment, on the availability of local suppliers, and so forth. A systematic procedure cannot help here — the decision must instead be based on local knowledge. This does

---

* Maximizing performance often means *minimizing* something: cost is the obvious example; mass, in transport systems, is another. A low-cost or light component, here, improves performance. Chapter 6 contains examples of both.

not mean that the result of the systematic procedure is irrelevant. It is always important to know which material is best, even if, for local reasons, you decide not to use it.

We will explore supporting information more fully in Chapter 13. Here we focus on the derivation of property limits and indices.

5.3 Deriving property limits and material indices

How are the design requirements for a component (which define what it must do) translated into a prescription for a material? To answer this we must look at the *function* of the component, the *constraints* it must meet, and the *objectives* the designer has selected to optimize its performance.

Function, objectives and constraints

Any engineering component has one or more *functions*: to support a load, to contain a pressure, to transmit heat, and so forth. In designing the component, the designer has an *objective*: to make it as cheap as possible, perhaps, or as light, or as safe, or perhaps some combination of these. This must be achieved subject to *constraints*: that certain dimensions are fixed, that the component must carry the given load or pressure without failure, that it can function in a certain range of temperature, and in a given environment, and many more. Function, objective and constraints (Table 5.1) define the boundary conditions for selecting a material and — in the case of load-bearing components — a shape for its cross-section.

Let us elaborate a little using the simplest of mechanical components as examples, helped by Figure 5.4. The loading on a component can generally be decomposed into some combination of axial tension or compression, bending, and torsion. Almost always, one mode dominates. So common is this that the functional name given to the component describes the way it is loaded: *ties* carry tensile loads; *beams* carry bending moments; *shafts* carry torques; and *columns* carry compressive axial loads. The words 'tie', 'beam', 'shaft' and 'column' each imply a function. Many simple engineering functions can be described by single words or short phrases, saving the need to explain the function in detail. In designing any one of these the designer has an objective: to make it as light as possible, perhaps (aerospace), or as safe (nuclear-reactor components), or as cheap — if there is no other objective, there is always that of minimizing cost. This must be achieved while meeting constraints: that the component carries the design loads without failing; that it survives in the chemical and thermal environment in which it must operate; and that certain limits on its dimensions must be met. The first step in relating design requirements to material properties is a clear statement of function, objectives and constraints.

Table 5.1 Function, objectives and constraints

Function	What does component do?
Objective	What is to be maximized or minimized?
Constraints*	What non-negotiable conditions must be met?
	What negotiable but desirable conditions ...?

---

* It is sometimes useful to distinguish between 'hard' and 'soft' constraints. Stiffness and strength might be absolute requirements (hard constraints); cost might be negotiable (a soft constraint).

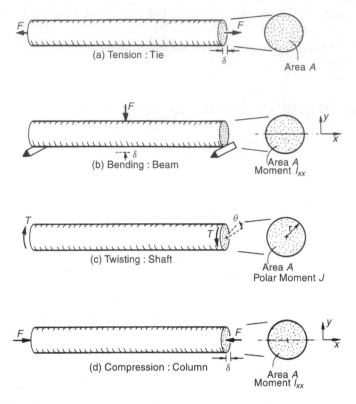

Fig. 5.4 A cylindrical tie-rod loaded (a) in tension, (b) in bending, (c) in torsion and (d) axially, as a column. The best choice of materials depends on the mode of loading and on the design goal; it is found by deriving the appropriate material index.

Property limits

Some constraints translate directly into simple *limits on material properties*. If the component must operate at 250°C, then all materials with a maximum service temperature less than this are eliminated. If it must be electrically insulating, then all material with a resistivity below $10^{20}\,\mu\Omega\,\mathrm{cm}$ are rejected. The screening step of the procedure of Figure 5.3 uses property limits derived in this way to reduce the kingdom of materials to an initial shortlist.

Constraints on stiffness, strength and many other component characteristics are used in a different way. This is because stiffness (to take an example) can be achieved in more than one way: by choosing a material with a high modulus, certainly; but also by simply increasing the cross-section; or, in the case of bending-stiffness or stiffness in torsion, by giving the section an efficient shape (a box or I-section, or tube). Achieving a specified stiffness (the constraint) involves a trade-off between these, and to resolve it we need to invoke an objective. The outcome of doing so is a material index. They are keys to optimized material selection. So how do you find them?

Material indices

A *material index* is a combination of material properties which characterizes the performance of a material in a given application.

First, a general statement of the scheme; then examples. *Structural elements* are components which perform a physical function: they carry loads, transmit heat, store energy and so on; in short, they satisfy *functional requirements*. The functional requirements are specified by the design: a tie must carry a specified tensile load; a spring must provide a given restoring force or store a given energy, a heat exchanger must transmit heat with a given heat flux, and so on.

The design of a structural element is specified by three things: the functional requirements, the geometry and the properties of the material of which it is made. The performance of the element is described by an equation of the form

$$p = f \left[\left(\begin{array}{c} \text{Functional} \\ \text{requirements,} \quad F \end{array} \right), \left(\begin{array}{c} \text{Geometric} \\ \text{parameters,} \quad G \end{array} \right), \left(\begin{array}{c} \text{Material} \\ \text{properties,} \quad M \end{array} \right) \right] \qquad (5.1)$$

or $\qquad p = f(F, G, M)$

where p describes some aspect of the performance of the component: its mass, or volume, or cost, or life for example; and 'f' means 'a function of'. *Optimum design* is the selection of the material and geometry which maximize or minimize p, according to its desirability or otherwise.

The three groups of parameters in equation (5.1) are said to be *separable* when the equation can be written

$$p = f_1(F)f_2(G)f_3(M) \qquad (5.2)$$

where f_1, f_2 and f_3 are separate functions which are simply multiplied together. When the groups are separable, as they generally are, the optimum choice of material becomes independent of the details of the design; it is the same for all geometries, G, and for all the values of the functional requirement, F. Then the optimum subset of materials can be identified without solving the complete design problem, or even knowing all the details of F and G. This enables enormous simplification: the performance for all F and G is maximized by maximizing f_3 (M), which is called the material efficiency coefficient, or *material index* for short*. The remaining bit, $f_1(F)f_2(G)$, is related to the *structural efficiency coefficient*, or *structural index*. We don't need it now, but will examine it briefly in Section 5.5.

Each combination of function, objective and constraint leads to a material index (Figure 5.5); the index is characteristic of the combination. The following examples show how some of the indices are derived. The method is general, and, in later chapters, is applied to a wide range of problems. A catalogue of indices is given in Appendix C.

Example 1: The material index for a light, strong, tie

A design calls for a cylindrical tie-rod of specified length ℓ, to carry a tensile force F without failure; it is to be of minimum mass. Here, 'maximizing performance' means 'minimizing the mass while still carrying the load F safely'. Function, objective and constraints are listed in Table 5.2.

We first seek an equation describing the quantity to be maximized or minimized. Here it is the mass m of the tie, and it is a minimum that we seek. This equation, called the *objective function*, is

$$m = A\ell\rho \qquad (5.3)$$

where A is the area of the cross-section and ρ is the density of the material of which it is made. The length ℓ and force F are specified and are therefore fixed; the cross-section A, is free. We can

---

* Also known as the 'merit index', 'performance index', or 'material factor'. In this book it is called the 'material index' throughout.

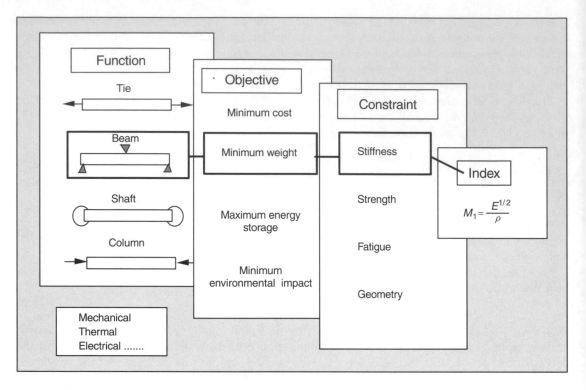

Fig. 5.5 The specification of function, objective and constraint leads to a materials index. The combination in the highlighted boxes leads to the index $E^{1/2}/\rho$.

Table 5.2 Design requirements for the light tie

Function	Tie-rod
Objective	Minimize the mass
Constraints	(a) Length ℓ specified
	(b) Support tensile load F without failing

reduce the mass by reducing the cross-section, but there is a constraint: the section-area A must be sufficient to carry the tensile load F, requiring that

$$\frac{F}{A} \le \sigma_f \qquad (5.4)$$

where σ_f is the failure strength. Eliminating A between these two equations gives

$$m \ge (F)(\ell)\left(\frac{\rho}{\sigma_f}\right) \qquad (5.5)$$

Note the form of this result. The first bracket contains the specified load F. The second bracket contains the specified geometry (the length ℓ of the tie). The last bracket contains the material properties. The lightest tie which will carry F safely* is that made of the material with the smallest value of ρ/σ_f. It is more natural to ask what must be *maximized* in order to maximize performance; we therefore invert the material properties in equation (5.5) and define the material index M as:

$$M = \frac{\sigma_f}{\rho}$$

(5.6)

The lightest tie-rod which will safely carry the load F without failing is that with the largest value of this index, the 'specific strength', mentioned earlier. A similar calculation for a light *stiff* tie leads to the index

$$M = \frac{E}{\rho}$$

(5.7)

where E is Young's modulus. This time the index is the 'specific stiffness'. But things are not always so simple. The next example shows how this comes about.

Example 2: The material index for a light, stiff beam

The mode of loading which most commonly dominates in engineering is not tension, but bending — think of floor joists, of wing spars, of golf-club shafts. Consider, then, a light beam of square section $b \times b$ and length ℓ loaded in bending which must meet a constraint on its stiffness S, meaning that it must not deflect more than δ under a load F (Figure 5.6). Table 5.3 itemizes the function, the objective and the constraints.

Appendix A of this book catalogues useful solutions to a range of standard problems. The stiffness of beams is one of these. Turning to Section A3 we find an equation for the stiffness of an elastic

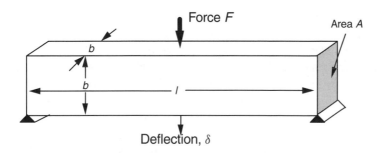

Fig. 5.6 A beam of square section, loaded in bending. Its stiffness is $S = F/\delta$, where F is the load and δ is the deflection. In Example 2, the active constraint is that of stiffness, S; it is this which determines the section area A. In Example 3, the active constraint is that of strength; it now determines the section area A.

---

* In reality a safety factor, S_f, is always included in such a calculation, such that equation (5.4) becomes $F/A \le \sigma_f/S_f$. If the same safety factor is applied to each material, its value does not influence the choice. We omit it here for simplicity.

Table 5.3 Design requirements for the light stiff beam

Function	Beam
Objective	Minimize the mass
Constraints	(a) Length ℓ specified
	(b) Support bending load F without deflecting too much

beam. The constraint requires that $S = F/\delta$ be greater than this:

$$S = \frac{F}{\delta} \geq \frac{C_1 EI}{\ell^3} \tag{5.8}$$

where E is Young's modulus, C_1 is a constant which depends on the distribution of load and I is the second moment of the area of the section, which, for a beam of square section ('Useful Solutions', Appendix A, Section A2), is

$$I = \frac{b^4}{12} = \frac{A^2}{12} \tag{5.9}$$

The stiffness S and the length ℓ are specified; the section A is free. We can reduce the mass of the beam by reducing A, but only so far that the stiffness constraint is still met. Using these two equations to eliminate A in equation (5.3) gives

$$m \geq \left(\frac{12S}{C_1 \ell}\right)^{1/2} \ell^3 \left(\frac{\rho}{E^{1/2}}\right) \tag{5.10}$$

The brackets are ordered as before: functional requirement, geometry and material. The best materials for a light, stiff beam are those with large values of the material index

$$\boxed{M = \frac{E^{1/2}}{\rho}} \tag{5.11}$$

Here, as before, the properties have been inverted; to minimize the mass, we must maximize M. Note the procedure. The length of the rod or beam is specified but we are free to choose the section area A. The *objective* is to minimize its mass, m. We write an equation for m; it is called the *objective function*. But there is a *constraint*: the rod must carry the load F without yielding in tension (in the first example) or bending too much (in the second). Use this to eliminate the free variable A. Arrange the result in the format

$$p = f_1(F)f_2(G)f_3(M)$$

and read off the combination of properties, M, to be maximized. It sounds easy, and it is so long as you are clear from the start what you are trying to maximize or minimize, what the constraints are, which parameters are specified, and which are free. In deriving the index, we have assumed that the section of the beam remained square so that both edges changed in length when A changed. If one of the two dimensions is held fixed, the index changes. If only the height is free, it becomes

(via an identical derivation)

$$M = \frac{E^{1/3}}{\rho}$$

(5.12)

and if only the width is free, it becomes

$$M = \frac{E}{\rho}$$

(5.13)

Example 3: The material index for a light, strong beam

In stiffness-limited applications, it is elastic deflection which is the active constraint: it limits performance. In strength-limited applications, deflection is acceptable provided the component does not fail; strength is the active constraint. Consider the selection of a beam for a strength-limited application. The dimensions are the same as before. Table 5.4 itemizes the design requirements.

The objective function is still equation (5.3), but the constraint is now that of strength: the beam must support F without failing. The failure load of a beam (Appendix A, Section A4) is:

$$F_f = C_2 \frac{I\sigma_f}{y_m \ell}$$

(5.14)

where C_2 is a constant and y_m is the distance between the neutral axis of the beam and its outer filament ($C_2 = 4$ and $y_m = t/2$ for the configuration shown in the figure). Using this and equation (5.9) to eliminate A in equation (5.3) gives the mass of the beam which will just support the load F_f:

$$m_2 = \left(\frac{6}{C_2} \frac{F_f}{\ell^2}\right)^{2/3} \ell^3 \left[\frac{\rho}{\sigma_y^{2/3}}\right]$$

(5.15)

The mass is minimized by selecting materials with the largest values of the index

$$M = \frac{\sigma_f^{3/2}}{\rho}$$

(5.16)

This is the moment to distinguish more clearly between a *constraint* and an *objective*. A constraint is a feature of the design which must be met at a specified level (stiffness in the last example). An

Table 5.4 Design requirements for the light strong beam

Function	Beam
Objective	Minimize the mass
Constraints	(a) Length ℓ specified
	(b) Support bending load F without failing by yield or fracture

objective is a feature for which an extremum is sought (mass, just now). An important judgement is that of deciding which is to be which. It is not always obvious: for a racing bicycle, as an example, mass might be minimized with a constraint on cost; for a shopping bicycle, cost might be minimized with a constraint on the mass. It is the objective which gives the objective function; the constraints set the free variables it contains.

So far the objective has been that of minimizing weight. There are many others. In the selection of a material for a spring, the objective is that of maximizing the elastic energy it can store. In seeking materials for thermal-efficient insulation for a furnace, the best are those with the lowest thermal conductivity and heat capacity. And most common of all is the wish to minimize cost. So here is an example involving cost.

Example 4: The material index for a cheap, stiff column

Columns support compressive loads: the legs of a table; the pillars of the Parthenon. We seek materials for the cheapest cylindrical column of specified height, ℓ, which will safely support a load F (Figure 5.7). Table 5.5 lists the requirements.

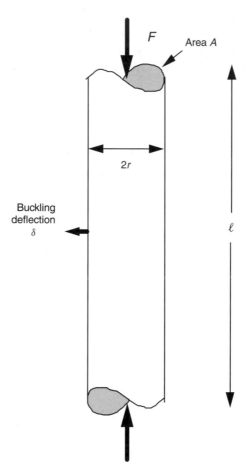

Fig. 5.7 A column carrying a compressive load F. The constraint that it must not buckle determines the section area A.

Table 5.5 Design requirements for the cheap column

Function	Column
Objective	Minimize the cost
Constraints	(a) Length ℓ specified
	(b) Support compressive load F without buckling

A slender column uses less material than a fat one, and thus is cheaper; but it must not be so slender that it will buckle under the design load, F. The objective function is the cost

$$C = A\ell C_m \rho \tag{5.17}$$

where C_m is the cost/kg of the material* of the column. It will buckle elastically if F exceeds the Euler load, F_{crit}, found in Appendix A, 'Useful Solutions', Section A5. The design is safe if

$$F \leq F_{\text{crit}} = \frac{n\pi^2 EI}{\ell^2} \tag{5.18}$$

where n is a constant that depends on the end constraints and $I = \pi r^2/4 = A^2/4\pi$ is the second moment of area of the column (see Appendix A for both). The load F and the length ℓ are specified; the free variable is the section-area A. Eliminating A between the last two equations, using the definition of I, gives:

$$C \geq \left(\frac{4}{n\pi}\right)^{1/2} \left(\frac{F}{\ell^2}\right)^{1/2} \ell^3 \left(\frac{C_m \rho}{E^{1/2}}\right) \tag{5.19}$$

The pattern is the usual one: functional requirement, geometry, material. The cost of the column is minimized by choosing materials with the largest value of the index

$$\boxed{M = \frac{E^{1/2}}{C_m \rho}} \tag{5.20}$$

From all this we distil the procedure for deriving a material index. It is shown in Table 5.6.

Table 5.7 summarizes a few of the indices obtained in this way. Appendix D contains a more complete catalogue. We now examine how to use them to select materials.

5.4 The selection procedure

Property limits: go/no-go conditions and geometric restrictions

Any design imposes certain non-negotiable demands on the material of which it is made. Temperature is one: a component which is to carry load at 500°C cannot be made of a polymer since all polymers lose their strength and decompose at lower temperatures than this. Electrical conductivity is another: components which must insulate cannot be made of metals because all metals conduct well. Corrosion resistance can be a third. Cost is a fourth: 'precious' metals are not used in structural applications simply because they cost too much.

---

* C_m is the cost/kg of the *processed* material, here, the material in the form of a circular rod or column.

Table 5.6 Procedure for deriving material indices

Step	Action
1	*Define the design requirements:* (a) Function: what does the component do? (b) Objective: what is to be maximized or minimized? (c) Constraints: essential requirements which must be met: stiffness, strength, corrosion resistance, forming characteristics...
2	Develop an *equation* for the objective in terms of the functional requirements, the geometry and the material properties (the *objective function*).
3	Identify the *free* (unspecified) *variables*.
4	Develop *equations* for the constraints (no yield; no fracture; no buckling, etc.).
5	*Substitute* for the free variables from the constraint equations into the objective function.
6	*Group the variables* into three groups: functional requirements, F, geometry, G, and material properties, M, thus $$\text{Performance characteristic} \leq f_1(F)f_2(G)f_3(M)$$ or $\quad\quad\quad$ $$\text{Performance characteristic} \geq f_1(F)f_2(G)f_3(M)$$
7	*Read off* the material index, expressed as a quantity M, which optimizes the performance characteristic.

Table 5.7 Examples of material indices

Function, Objective and Constraint	Index
Tie, minimum weight, stiffness prescribed	$\dfrac{E}{\rho}$
Beam, minimum weight, stiffness prescribed	$\dfrac{E^{1/2}}{\rho}$
Beam, minimum weight, strength prescribed	$\dfrac{\sigma_y^{2/3}}{\rho}$
Beam, minimum cost, stiffness prescribed	$\dfrac{E^{1/2}}{C_m\rho}$
Beam, minimum cost, strength prescribed	$\dfrac{\sigma_y^{2/3}}{C_m\rho}$
Column, minimum cost, buckling load prescribed	$\dfrac{E^{1/2}}{C_m\rho}$
Spring, minimum weight for given energy storage	$\dfrac{\sigma_y^2}{E\rho}$
Thermal insulation, minimum cost, heat flux prescribed	$\dfrac{1}{\lambda C_m\rho}$
Electromagnet, maximum field, temperature rise prescribed	$\kappa C_p\rho$

(ρ = density; E = Young's modulus; σ_y = elastic limit; C_m = cost/kg; λ = thermal conductivity; κ = electrical conductivity; C_p = specific heat)

Geometric constraints also generate property limits. In the examples of the last section the length ℓ was constrained. There can be others. Here are two examples. The tie of Example 1, designed to carry a tensile force F without yielding (equation 5.4), requires a section

$$A \geq \frac{F}{\sigma_f}$$

If, to fit into a confined space, the section is limited to $A \leq A^*$, then the only possible candidate materials are those with strengths greater than

$$\sigma_f^* = \frac{F}{A^*} \tag{5.21}$$

Similarly, if the column of Example 4, designed to carry a load F without buckling, is constrained to have a diameter less than $2r^*$, it will require a material with modulus (found by inverting equation (5.18)) greater than

$$E^* = \frac{4F\ell^2}{n\pi^3 r^{*4}} \tag{5.22}$$

Property limits plot as horizontal or vertical lines on material selection charts. The restriction on r leads to a lower bound for E, given by equation (5.22). An upper limit on density (if one were desired) requires that

$$\rho < \rho^* \tag{5.23}$$

One way of applying the limits is illustrated in Figure 5.8. It shows a schematic $E-\rho$ chart, in the manner of Chapter 4, with a pair of limits for E and ρ plotted on it. The optimizing search is restricted to the window between the limits within which the next steps of the procedure operate. Less quantifiable properties such as corrosion resistance, wear resistance or formability can all appear as primary limits, which take the form

$$P > P^*$$

or
$$P < P^* \tag{5.24}$$

where P is a property (service temperature, for instance) and P^* is a critical value of that property, set by the design, which must be exceeded, or (in the case of cost or corrosion rate) must *not* be exceeded.

One should not be too hasty in applying property limits; it may be possible to engineer a route around them. A component which gets too hot can be cooled; one that corrodes can be coated with a protective film. Many designers apply property limits for fracture toughness, K_{Ic}, and ductility ε_f, insisting on materials with, as rules of thumb, $K_{Ic} > 15\,\mathrm{MPa\,m}^{1/2}$ and $\varepsilon_f > 2\%$ in order to guarantee adequate tolerance to stress concentrations. By doing this they eliminate materials which the more innovative designer is able to use to good purpose (the limits just cited for K_{Ic} and ε_f eliminate most polymers and all ceramics, a rash step too early in the design). At this stage, keep as many options open as possible.

Performance maximizing criteria

The next step is to seek, from the subset of materials which meet the property limits, those which maximize the performance of the component. We will use the design of light, stiff components as an example; the other material indices are used in a similar way.

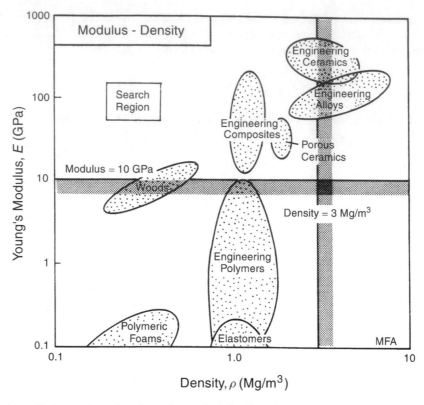

Fig. 5.8 A schematic $E-\rho$ chart showing a lower limit for E and an upper one for ρ.

Figure 5.9 shows, as before, the modulus E, plotted against density ρ, on log scales. The material indices E/ρ, $E^{1/2}/\rho$ and $E^{1/3}/\rho$ can be plotted onto the figure. The condition

$$E/\rho = C$$

or taking logs

$$\log E = \log \rho + \log C \tag{5.25}$$

is a family of straight parallel lines of slope 1 on a plot of $\log E$ against $\log \rho$; each line corresponds to a value of the constant C. The condition

$$E^{1/2}/\rho = C \tag{5.24}$$

gives another set, this time with a slope of 2; and

$$E^{1/3}/\rho = C \tag{5.25}$$

gives yet another set, with slope 3. We shall refer to these lines as selection *guide lines*. They give the slope of the family of parallel lines belonging to that index.

It is now easy to read off the subset materials which optimally maximize performance for each loading geometry. All the materials which lie on a line of constant $E^{1/2}/\rho$ perform equally well as a light, stiff beam (Example 2); those above the line are better, those below, worse. Figure 5.10 shows

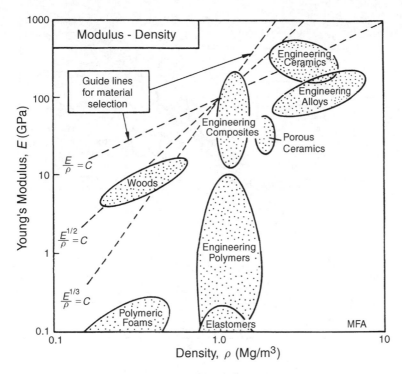

Fig. 5.9 A schematic $E-\rho$ chart showing guide lines for the three material indices for stiff, lightweight design.

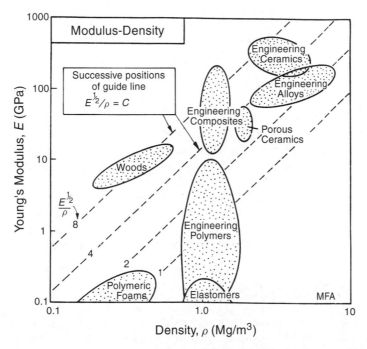

Fig. 5.10 A schematic $E-\rho$ chart showing a grid of lines for the material index $M = E^{1/2}/\rho$. The units are $(GPa)^{1/2}/(Mg/m^3)$.

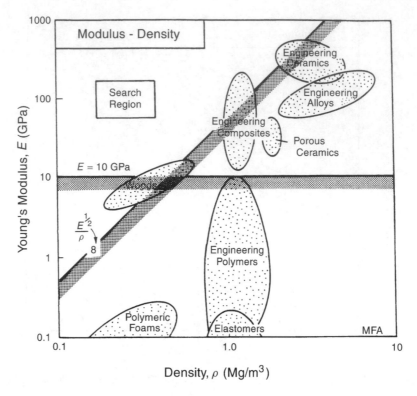

Fig. 5.11 A selection based on the index $M = E^{1/2}/\rho$, together with the property limit $E > 10\,\text{GPa}$. The shaded band with slope 2 has been positioned to isolate a subset of materials with high $E^{1/2}/\rho$; the horizontal ones lie at $E = 10\,\text{GPa}$. The materials contained in the Search Region become the candidates for the next stage of the selection process.

a grid of lines corresponding to values of $M = E^{1/2}/\rho$ from 1 to 8 in units of $\text{GPa}^{1/2}/(\text{Mg}\,\text{m}^{-3})$. A material with $M = 4$ in these units gives a beam which has half the weight of one with $M = 2$. One with $M = 8$ weighs one quarter as much. The subset of materials with particularly good values of the index is identified by picking a line which isolates a *search area* containing a reasonably small number of candidates, as shown schematically in Figure 5.11. Properly limits can be added, narrowing the search window: that corresponding to $E > 10\,\text{GPa}$ is shown. The shortlist of candidate materials is expanded or contracted by moving the index line.

   The procedure is extended in Chapters 7 and 9 to include section shape and to deal with multiple constraints and objectives. Before moving on to these, it is a good idea to consolidate the ideas so far by applying them to a number of Case Studies. They follow in Chapter 6. But first a word about the structural index.

5.5 The structural index

Books on optimal design of structures (e.g. Shanley, 1960) make the point that the efficiency of material usage in mechanically loaded components depends on the product of three factors: the material index, as defined here; a factor describing section shape, the subject of our Chapter 7; and a *structural index**, which contains elements of the F and G of equation (5.1). The subjects of this book — material and process selection — focus attention on the material index and on shape; but we should examine the structural index briefly, partly to make the connection with the classical theory of optimal design, and partly because it becomes useful (even to us) when structures are scaled in size.

Consider, as an example, the development of the index for a cheap, stiff column, given as Example 4 in Section 5.2. The objective was that of minimizing cost. The *mechanical efficiency* is a measure of the load carried divided by the 'objective' — in this case, cost per unit length. Using equation (5.19) the efficiency of the column is given by

$$\frac{F}{(C/\ell)} = \left(\frac{n\pi}{4}\right)^{1/2} \left[\frac{F}{\ell^2}\right]^{1/2} \left[\frac{E^{1/2}}{C_m\rho}\right] \tag{5.26}$$

The first bracketed term on the right is merely a constant. The last is the material index. The structural index is the middle one: F/ℓ^2. It has the dimensions of stress; it is a measure of the intensity of loading. Design proportions which are optimal, minimizing material usage, are optimal for structures of any size provided they all have the same structural index. The performance equations (5.5), (5.10), (5.15) and (5.19) were all written in a way which isolated the structural index

The structural index for a column of minimum weight is the same as that for one which minimizes material cost; it is F/ℓ^2 again. For beams of minimum weight, or cost, or energy content, it is the same: F/ℓ^2. For ties it is simply 1 (try it: use equation (5.5) to calculate the load F divided by the mass per unit length, m/ℓ). For panels loaded in bending or such that they buckle it is $F/\ell b$ where ℓ and b are the (fixed) dimensions of the panel.

5.6 Summary and conclusions

The design requirements of a component which performs mechanical, thermal or electrical functions can be formulated in terms of one or more objective functions, limited by constraints. The objective function describes the quantity to be maximized or minimized in the design. One or more of the variables describing the geometry is 'free', that is, it (or they) can be varied to optimize the design. If the number of constraints is equal to the number of free variables, the problem is fully constrained; the constraints are substituted into the objective function identifying the group of material properties (the 'material index') to be maximized or minimized in selecting a material. The charts allow this using the method outlined in this chapter. Often, the index characterizes an entire class of designs, so that the details of shape or loading become unimportant in deriving it. The commonest of these indices are assembled in Appendix C of this book, but there are more. New problems throw up new indices, as the Case Studies of the next chapter will show.

5.7 Further reading

The books listed below discuss optimization methods and their application in materials engineering. None contains the approach developed here.

---

* Also called the 'structural loading coefficient', the 'strain number' or the 'strain index'.

Dieter, G.E. (1991) *Engineering Design, A Materials and Processing Approach*, 2nd edition, Chapter 5, McGraw-Hill, New York.

Gordon, J.E. (1978) *Structures, or Why Things don't Fall through the Floor*, Penguin Books, Harmondsworth.

Johnson, R.C. (1980) *Optimum Design of Mechanical Elements*, 2nd edition, Wiley, New York.

Shanley, F.R. (1960) *Weight–Strength Analysis of Aircraft Structures*, 2nd edition, Dover Publications, New York.

Siddall, J.N. (1982) *Optimal Engineering Design*, Marcel Dekker, New York.

Chapter 6

Materials selection — case studies

6.1 Introduction and synopsis

Here we have a collection of case studies* illustrating the screening methods† of Chapter 5. Each is laid out in the same way:

(a) *the problem statement*, setting the scene;
(b) *the model*, identifying function, objectives and constraints from which emerge the property limits and material indices;
(c) *the selection* in which the full menu of materials is reduced by screening and ranking to a short-list of viable candidates; and
(d) *the postscript*, allowing a commentary on results and philosophy.

Techniques for seeking further information are left to later chapters.

The first few examples are simple but illustrate the method well. Later examples are less obvious and require clear identification of the objectives, the constraints, and the free variables. Confusion here can lead to bizarre and misleading conclusions. Always apply common sense: does the selection include the traditional materials used for that application? Are some members of the subset obviously unsuitable? If they are, it is usually because a constraint has been overlooked: it must be formulated and applied.

The case studies are deliberately simplified to avoid obscuring the method under layers of detail. In most cases nothing is lost by this: the best choice of material for the simple example is the same as that for the more complex, for the reasons given in Chapter 5.

6.2 Materials for oars

Credit for inventing the rowed boat seems to belong to the Egyptians. Boats with oars appear in carved relief on monuments built in Egypt between 3300 and 3000 BC. Boats, before steam power, could be propelled by poling, by sail and by oar. Oars gave more control than the other two, the military potential of which was well understood by the Romans, the Vikings and the Venetians.

---

* A computer-based exploration of these and other case studies can be found in *Case Studies in Materials Selection* by M.F. Ashby and D. Cebon, published by Granta Design, Trumpington Mews, 40B High Street, Trumpington CB2 2LS, UK (1996).

† The material properties used here are taken from the *CMS* compilation published by Granta Design, Trumpington Mews, 40B High Street, Trumpington CB2 2LS, UK.

Records of rowing races on the Thames in London extend back to 1716. Originally the competitors were watermen, rowing the ferries used to carry people and goods across the river. Gradually gentlemen became involved (notably the young gentlemen of Oxford and Cambridge), sophisticating both the rules and the equipment. The real stimulus for development of boat and oar came in 1900 with the establishment of rowing as an Olympic sport. Since then both have exploited to the full the craftsmanship and materials of their day. Consider, as an example, the oar.

The model

Mechanically speaking, an oar is a beam, loaded in bending. It must be strong enough to carry the bending moment exerted by the oarsman without breaking, it must have just the right stiffness to match the rower's own characteristics and give the right 'feel', and — very important — it must be as light as possible. Meeting the strength constraint is easy. Oars are designed on stiffness, that is, to give a specified elastic deflection under a given load. The upper part of Figure 6.1 shows an oar: a blade or 'spoon' is bonded to a shaft or 'loom' which carries a sleeve and collar to give positive location in the rowlock. The lower part of the figure shows how the oar stiffness is measured: a 10 kg weight is hung on the oar 2.05 m from the collar and the deflection at this point is measured. A soft oar will deflect nearly ·50 mm; a hard one only 30. A rower, ordering an oar, will specify how hard it should be.

The oar must also be light; extra weight increases the wetted area of the hull and the drag that goes with it. So there we have it: an oar is a beam of specified stiffness and minimum weight. The material index we want was derived in Chapter 5 as equation (5.11). It is that for a light, stiff beam:

$$M = \frac{E^{1/2}}{\rho} \tag{6.1}$$

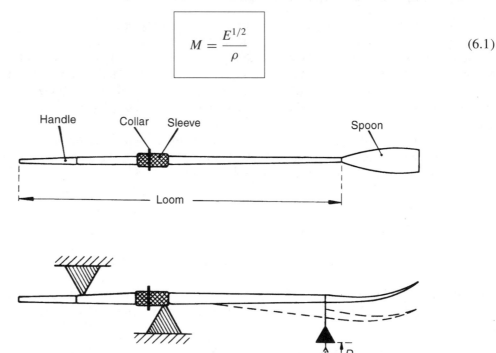

Fig. 6.1 An oar. Oars are designed on stiffness, measured in the way shown in the lower figure, and they must be light.

Table 6.1 Design requirements for the oar

Function	Oar, meaning light, stiff beam
Objective	Minimize the mass
Constraints	(a) Length L specified
	(b) Bending stiffness S specified
	(c) Toughness $G_c > 1 \, \text{kJ/m}^2$
	(d) Cost $C_m < \$100/\text{kg}$

There are other obvious constraints. Oars are dropped, and blades sometimes clash. The material must be tough enough to survive this, so brittle materials (those with a toughness less than $1 \, \text{kJ/m}^2$) are unacceptable. And, while sportsmen will pay a great deal for the ultimate in equipment, there *are* limits on cost. Given these requirements, summarized in Table 6.1, what materials should make good oars?

The selection

Figure 6.2 shows the appropriate chart: that in which Young's modulus, E, is plotted against density, ρ. The selection line for the index M has a slope of 2, as explained in Section 5.3; it is positioned so that a small group of materials is left above it. They are the materials with the largest values of M, and it is these which are the best choice, provided they satisfy the other constraints (simple property limits on toughness and cost). They contain three classes of material: woods, carbon and glass-fibre reinforced polymers, and certain ceramics (Table 6.2). Ceramics are brittle; their toughnesses fail to meet that required by the design. The recommendation is clear. Make your oars out of wood or, better, out of CFRP.

Postscript

Now we know what oars *should* be made of. What, in reality, is used? Racing oars and sculls are made either of wood or of a high performance composite: carbon-fibre reinforced epoxy.

Wooden oars are made today, as they were 100 years ago, by craftsmen working largely by hand. The shaft and blade are of Sitka spruce from the northern US or Canada, the further north the better because the short growing season gives a finer grain. The wood is cut into strips, four of which are laminated together (leaving a hollow core) to average the stiffness. A strip of hardwood is bonded to the compression side of the shaft to add stiffness and the blade is glued to the shaft. The rough oar is then shelved for some weeks to settle down, and finished by hand cutting and polishing. The final spruce oar weighs between 4 and 4.3 kg, and costs (in 1998) about £150 or $250.

Composite blades are a little lighter than wood for the same stiffness. The component parts are fabricated from a mixture of carbon and glass fibres in an epoxy matrix, assembled and glued. The advantage of composites lies partly in the saving of weight (typical weight: 3.9 kg) and partly in the greater control of performance: the shaft is moulded to give the stiffness specified by the purchaser. Until recently a CFRP oar cost more than a wooden one, but the price of carbon fibres has fallen sufficiently that the two cost about the same.

Could we do better? The chart shows that wood and CFRP offer the lightest oars, at least when normal construction methods are used. Novel composites, not at present shown on the chart, might permit further weight saving; and functional-grading (a thin, very stiff outer shell with a low density core) might do it. But both appear, at present, unlikely.

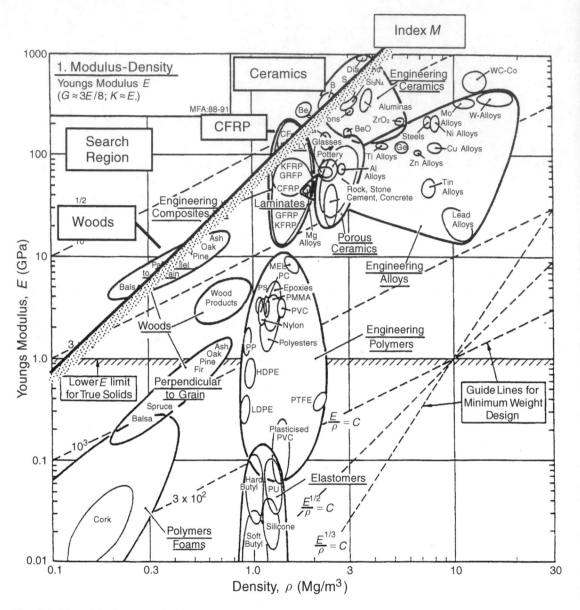

Fig. 6.2 Materials for oars. CFRP is better than wood because the structure can be controlled.

Table 6.2 Materials for oars

Material	M $(GPa)^{1/2}/(Mg/m^3)$	Comment
Woods	5–8	Cheap, traditional, but with natural variability
CFRP	4–8	As good as wood, more control of properties
GFRP	2–3.5	Cheaper than CFRP but lower M, thus heavier
Ceramics	4–8	Good M but toughness low and cost high

Further reading

Redgrave, S. (1992) *Complete Book of Rowing*, Partridge Press, London.

Related case studies

Case Study 6.3: Mirrors for large telescopes
Case Study 6.4: Table legs

6.3 Mirrors for large telescopes

There are some very large optical telescopes in the world. The newer ones employ complex and cunning tricks to maintain their precision as they track across the sky — more on that in the Postscript. But if you want a simple telescope, you make the reflector as a single rigid mirror. The largest such telescope is sited on Mount Semivodrike, near Zelenchukskaya in the Caucasus Mountains of Russia. The mirror is 6 m (236 inches) in diameter. To be sufficiently rigid, the mirror, which is made of glass, is about 1 m thick and weighs 70 tonnes.

The total cost of a large (236-inch) telescope is, like the telescope itself, astronomical — about UK £150 m or US $240 m. The mirror itself accounts for only about 5% of this cost; the rest is that of the mechanism which holds, positions and moves it as it tracks across the sky. This mechanism must be stiff enough to position the mirror relative to the collecting system with a precision about equal to that of the wavelength of light. It might seem, at first sight, that doubling the mass m of the mirror would require that the sections of the support structure be doubled too, so as to keep the stresses (and hence the strains and displacements) the same; but the heavier structure then deflects under its own weight. In practice, the sections have to increase as m^2, and so does the cost.

Before the turn of the century, mirrors were made of speculum metal (density: about 8 Mg/m³). Since then, they have been made of glass (density: 2.3 Mg/m³), silvered on the front surface, so none of the optical properties of the glass are used. Glass is chosen for its mechanical properties only; the 70 tonnes of glass is just a very elaborate support for 100 nm (about 30 g) of silver. Could one, by taking a radically new look at materials for mirrors, suggest possible routes to the construction of lighter, cheaper telescopes?

The model

At its simplest, the mirror is a circular disc, of diameter $2a$ and mean thickness t, simply supported at its periphery (Figure 6.3). When horizontal, it will deflect under it own weight m; when vertical it will not deflect significantly. This distortion (which changes the focal length and introduces aberrations into the mirror) must be small enough that it does not interfere with performance; in practice, this means that the deflection δ of the midpoint of the mirror must be less than the wavelength of light. Additional requirements are: high dimensional stability (no creep), and low thermal expansion (Table 6.3).

The mass of the mirror (the property we wish to minimize) is

$$m = \pi a^2 t \rho \tag{6.2}$$

where ρ is the density of the material of the disc. The elastic deflection, δ, of the centre of a horizontal disc due to its own weight is given, for a material with Poisson's ratio of 0.3 (Appendix A: 'Useful

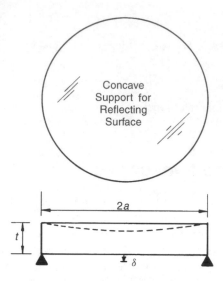

Fig. 6.3 The mirror of a large optical telescope is modelled as a disc, simply supported at its periphery. It must not sag by more than a wavelength of light at its centre.

Table 6.3 Design requirements for the telescope mirror

Function	Precision mirror
Objective	Minimize the mass
Constraints	(a) Radius a specified
	(b) Must not distort more than δ under its own weight
	(c) High dimensional stability: no creep, no moisture take-up, low thermal expansion

Solutions'), by

$$\delta = \frac{3}{4\pi} \frac{mga^2}{Et^3} \qquad (6.3)$$

The quantity g in this equation is the acceleration due to gravity: 9.81 m/s^2; E, as before, is Young's modulus. We require that this deflection be less than (say) 10 µm. The diameter of the disc is specified by the telescope design, but the thickness is a free variable. Solving for t and substituting this into the first equation gives

$$m = \left(\frac{3g}{4\delta}\right)^{1/2} \pi a^4 \left[\frac{\rho}{E^{1/3}}\right]^{3/2} \qquad (6.4)$$

The lightest mirror is the one with the greatest value of the material index

$$\boxed{M = \frac{E^{1/3}}{\rho}} \qquad (6.5)$$

We treat the remaining constraints as property limits, requiring a melting point greater than 1000 K to avoid creep, zero moisture take up, and a low thermal expansion coefficient ($\alpha < 20 \times 10^{-6}$/K).

The selection

Here we have another example of elastic design for minimum weight. The appropriate chart is again that relating Young's modulus E and density ρ — but the line we now construct on it has a slope of 3, corresponding to the condition $M = E^{1/3}/\rho = \text{constant}$ (Figure 6.4). Glass lies on the line $M = 2\,(\text{GPa})^{1/3}\text{m}^3/\text{Mg}$. Materials which lie above it are better, those below, worse. Glass is much better than steel or speculum metal (that is why most mirrors are made of glass); but it is less

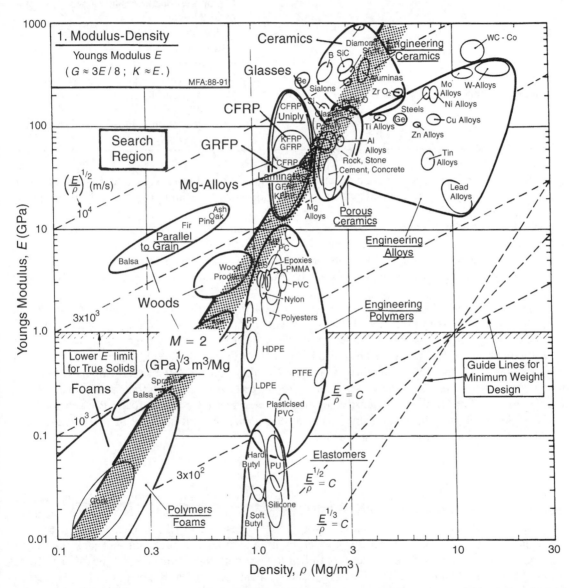

Fig. 6.4 Materials for telescope mirrors. Glass is better than most metals, among which magnesium is a good choice. Carbon-fibre reinforced polymers give, potentially, the lowest weight of all, but may lack adequate dimensional stability. Foamed glass is a possible candidate.

Table 6.4 Mirror backing for 200-inch telescope

Material	$M = E^{1/3}/\rho$ $(GPa)^{1/3}m^3/Mg$	m (tonne) $a = 6$ m	Comment
Steel (or Speculum)	0.7	158	Very heavy. The original choice.
Concrete	1.4	56	Heavy. Creep, thermal distortion a problem.
Al-alloys	1.5	53	Heavy, high thermal expansion.
Glass	1.6	48	The present choice.
GFRP	1.7	44	Not dimensionally stable enough — use for radio telescope.
Mg-alloys	2.1	38	Lighter than glass but high thermal expansion.
Wood	3.6	14	Dimensionally unstable.
Beryllium	3.65	14	Very expensive — good for small mirrors.
Foamed polystyrene	3.9	13	Very light, but dimensionally unstable. Foamed glass?
CFRP	4.3	11	Very light, but not dimensionally stable; use for radio telescopes.

good than magnesium, several ceramics, carbon-fibre and glass-fibre reinforced polymers, or — an unexpected finding — stiff foamed polymers. The shortlist before applying the property limits is given in Table 6.4.

One must, of course, examine other aspects of this choice. The mass of the mirror can be calculated from equation (6.5) for the materials listed in the table. Note that the polystyrene foam and the CFRP mirrors are roughly one-fifth the weight of the glass one, and that the support structure could thus be as much as 25 times less expensive than that for an orthodox glass mirror. But could they be made?

Some of the choices — the polystyrene foam or the CFRP — may at first seem impractical. But the potential cost saving (the factor of 25) is so vast that they are worth examining. There are ways of casting a thin film of silicone rubber or of epoxy onto the surface of the mirror-backing (the polystyrene or the CFRP) to give an optically smooth surface which could be silvered. The most obvious obstacle is the lack of stability of polymers — they change dimensions with age, humidity, temperature and so on. But glass itself can be reinforced with carbon fibres; and it can also be foamed to give a material with a density not much greater than polystyrene foam. Both foamed and carbon-reinforced glass have the same chemical and environmental stability as solid glass. They could provide a route to large cheap mirrors.

Postscript

There are, of course, other things you can do. The stringent design criterion ($\delta > 10\,\mu$m) can be partially overcome by engineering design without reference to the material used. The 8.2 m Japanese telescope on Mauna Kea, Hawaii and the Very Large Telescope (VLT) at Cerro Paranal Silla in Chile each have a thin glass reflector supported by little hydraulic or piezo-electric jacks that exert distributed forces over its back surface, controlled to vary with the attitude of the mirror. The Keck telescope, also on Mauna Kea, is segmented, each segment independently positioned to give optical focus. But the limitations of this sort of mechanical system still require that the mirror meet a stiffness target. While stiffness at minimum weight is the design requirement, the material-selection criteria remain unchanged.

Radio telescopes do not have to be quite as precisely dimensioned as optical ones because they detect radiation with a longer wavelength. But they are much bigger (60 metres rather than 6) and they suffer from similar distortional problems. Microwaves have wavelengths in the mm band, requiring precision over the mirror face of 0.25 mm. A recent 45 m radio telescope built for the University of Tokyo achieves this, using CFRP. Its parabolic surface is made of 6000 CFRP panels, each servo controlled to compensate for macro-distortion. Recent telescopes have been made from CFRP, for exactly the reasons we deduced. Beryllium appears on our list, but is impractical for large mirrors because of its cost. Small mirrors for space applications must be light for a different reason (to reduce take-off weight) and must, in addition, be as immune as possible to temperature change. Here beryllium comes into its own.

Related case studies

Case Study 6.5:   Materials for table legs
Case Study 6.20: Materials to minimize thermal distortion

6.4 Materials for table legs

Luigi Tavolino, furniture designer, conceives of a lightweight table of daring simplicity: a flat sheet of toughened glass supported on slender, unbraced, cylindrical legs (Figure 6.5). The legs must be solid (to make them thin) and as light as possible (to make the table easier to move). They must support the table top and whatever is placed upon it without buckling. What materials could one recommend?

Fig. 6.5 A lightweight table with slender cylindrical legs. Lightness and slenderness are independent design goals, both constrained by the requirement that the legs must not buckle when the table is loaded. The best choice is a material with high values of both $E^{1/2}/\rho$ and E.

Table 6.5 Design requirements for table legs

Function	Column (supporting compressive loads)
Objective	(a) Minimize the mass
	(b) Maximize slenderness
Constraints	(a) Length ℓ specified
	(b) Must not buckle under design loads
	(c) Must not fracture if accidentally struck

The model

This is a problem with two objectives*: weight is to be minimized, and slenderness maximized. There is one constraint: resistance to buckling. Consider minimizing weight first.

The leg is a slender column of material of density ρ and modulus E. Its length, ℓ, and the maximum load, P, it must carry are determined by the design: they are fixed. The radius r of a leg is a free variable. We wish to minimize the mass m of the leg, given by the objective function

$$m = \pi r^2 \ell \rho \tag{6.6}$$

subject to the constraint that it supports a load P without buckling. The elastic load P_{crit} of a column of length ℓ and radius r (see Appendix A, 'Useful Solutions') is

$$P_{\text{crit}} = \frac{\pi^2 EI}{\ell^2} = \frac{\pi^3 E r^4}{4\ell^2} \tag{6.7}$$

using $I = \pi r^4/4$ where I is the second moment of area of the column. The load P must not exceed P_{crit}. Solving for the free variable, r, and substituting it into the equation for m gives

$$m \geq \left(\frac{4P}{\pi}\right)^{1/2} (\ell)^2 \left[\frac{\rho}{E^{1/2}}\right] \tag{6.8}$$

The material properties are grouped together in the last pair of brackets. The weight is minimized by selecting the subset of materials with the greatest value of the material index

$$\boxed{M_1 = \frac{E^{1/2}}{\rho}}$$

(a result we could have taken directly from Appendix B).

Now slenderness. Inverting equation (6.7) with $P = P_{\text{crit}}$ gives an equation for the thinnest leg which will not buckle:

$$r = \left(\frac{4P}{\pi^3}\right)^{1/4} (\ell)^{1/2} \left[\frac{1}{E}\right]^{1/4} \tag{6.9}$$

The thinnest leg is that made of the material with the largest value of the material index

$$\boxed{M_2 = E}$$

---

* Formal methods for dealing with multiple objectives are developed in Chapter 9.

The selection

We seek the subset of materials which have high values of $E^{1/2}/\rho$ and E. Figure 6.6 shows the appropriate chart: Young's modulus, E, plotted against density, ρ. A guideline of slope 2 is drawn on the diagram; it defines the slope of the grid of lines for values of $E^{1/2}/\rho$. The guideline is displaced upwards (retaining the slope) until a reasonably small subset of materials is isolated above it; it is shown at the position $M_1 = 6\,\mathrm{GPa}^{1/2}/(\mathrm{Mg/m}^3)$. Materials above this line have higher values of

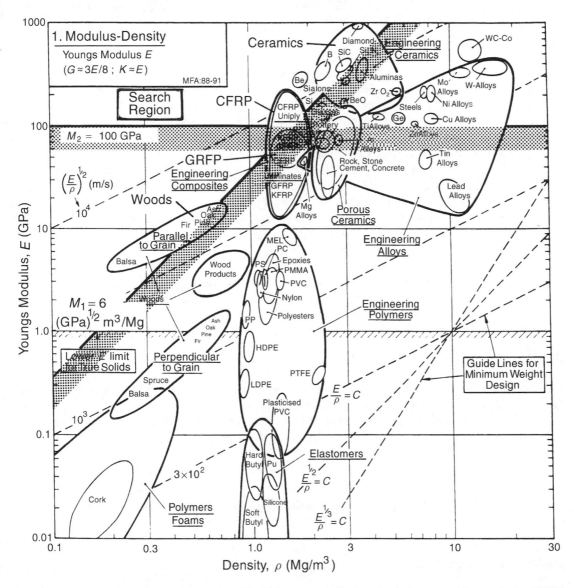

Fig. 6.6 Materials for light, slender legs. Wood is a good choice; so is a composite such as CFRP, which, having a higher modulus than wood, gives a column which is both light and slender. Ceramics meet the stated design goals, but are brittle.

Table 6.6 Materials for table legs

Material	M_1 $(GPa^{1/2}m^3/Mg)$	M_2 (GPa)	Comment
Woods	5–8	4–20	Outstanding M_1; poor M_2. Cheap, traditional, reliable.
CFRP	4–8	30–200	Outstanding M_1 and M_2, but expensive.
GFRP	3.5–5.5	20–90	Cheaper than CFRP, but lower M_1 and M_2.
Ceramics	4–8	150–1000	Outstanding M_1 and M_2. Eliminated by brittleness.

M_1. They are identified on the figure: *woods* (the traditional material for table legs), *composites* (particularly CFRP) and certain special *engineering ceramics*. Polymers are out: they are not stiff enough; metals too: they are too heavy (even magnesium alloys, which are the lightest). The choice is further narrowed by the requirement that, for slenderness, E must be large. A horizontal line on the diagram links materials with equal values of E; those above are stiffer. Figure 6.6 shows that placing this line at $M_1 = 100\,GPa$ eliminates woods and GFRP. If the legs must be really thin, then the shortlist is reduced to CFRP and ceramics: they give legs which weigh the same as the wooden ones but are much thinner. Ceramics, we know, are brittle: they have low values of fracture toughness. Table legs are exposed to abuse — they get knocked and kicked; common sense suggests that an additional constraint is needed, that of adequate toughness. This can be done using Chart 6 (Figure 4.7); it eliminates ceramics, leaving CFRP. The cost of CFRP (Chart 14, Figure 4.15) may cause Snr. Tavolino to reconsider his design, but that is another matter: he did not mention cost in his original specification.

It is a good idea to lay out the results as a table, showing not only the materials which are best, but those which are second-best — they may, when other considerations are involved become the best choice. Table 6.6 shows one way of doing it.

Postscript

Tubular legs, the reader will say, must be lighter than solid ones. True; but they will also be fatter. So it depends on the relative importance Mr Tavolino attaches to his two objectives — lightness and slenderness — and only he can decide that. If he can be persuaded to live with fat legs, tubing can be considered — and the material choice may be different. Materials selection when section-shape is a variable comes in Chapter 7.

Ceramic legs were eliminated because of low toughness. If (improbably) the goal was to design a light, slender-legged table for use at high temperatures, ceramics should be reconsidered. The brittleness problem can be by-passed by protecting the legs from abuse, or by pre-stressing them in compression.

Related case studies

Case Study 6.3: Mirrors for large telescopes
Case Study 8.2: Spars for man-powered planes
Case Study 8.3: Forks for a racing bicycle

6.5 Cost — structural materials for buildings

The most expensive thing that most people buy is the house they live in. Roughly half the cost of a house is the cost of the materials of which it is made, and they are used in large quantities (family house: around 200 tonnes; large apartment block: around 20 000 tonnes). The materials are used in three ways (Figure 6.7): structurally to hold the building up; as cladding, to keep the weather out; and as 'internals', to insulate against heat, sound, and so forth).

Consider the selection of materials for the structure. They must be stiff, strong, and cheap. Stiff, so that the building does not flex too much under wind loads or internal loading. Strong, so that there is no risk of it collapsing. And cheap, because such a lot of material is used. The structural frame of a building is rarely exposed to the environment, and is not, in general, visible. So criteria of corrosion resistance, or appearance , are not important here. The design goal is simple: strength and stiffness at minimum cost. To be more specific: consider the selection of material for floor beams. Table 6.7 summarizes the requirements.

The model

The way of deriving material indices for cheap, stiff and strong beams was developed in Chapter 5. The results we want are listed in Table 5.7. The critical components in building are loaded either

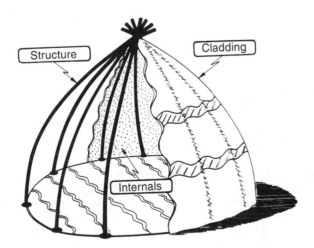

Fig. 6.7 The materials of a building perform three broad roles. The frame gives mechanical support; the cladding excludes the environment; and the internal surfacing controls heat, light and sound.

Table 6.7 Design requirements for floor beams

Function	Floor beams
Objective	Minimize the cost
Constraints	(a) Length L specified
	(b) Stiffness: must not deflect too much under design loads
	(c) Strength: must not fail under design loads

in bending (floor joists, for example) or as columns (the vertical members). The two indices that we want to maximize are:

$$M_1 = \frac{E^{1/2}}{\rho C_m}$$

and

$$M_2 = \frac{\sigma_f^{2/3}}{\rho C_m}$$

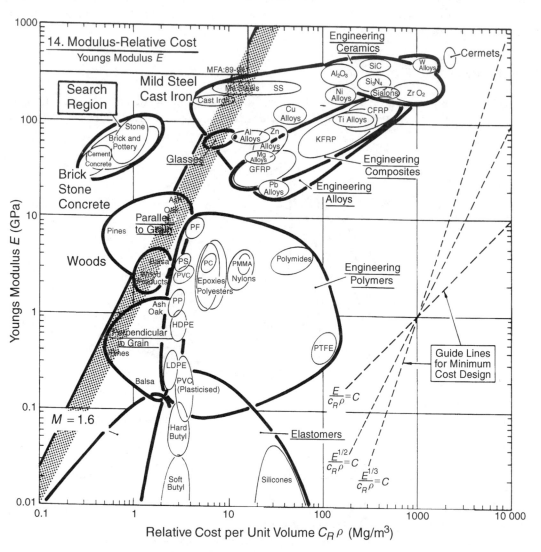

Fig. 6.8 The selection of cheap, stiff materials for the structural frames of buildings.

where, as always, E is Young's modulus, σ_f is the failure strength, ρ is the density and C_m material cost.

The selection

Cost appears in two of the charts. Figure 6.8 shows the first of them: modulus against relative cost per unit volume. The shaded band has the appropriate slope; it isolates concrete, stone, brick, softwoods, cast irons and the cheaper steels. The second, strength against relative cost, is shown in Figure 6.9. The shaded band — M_2 this time — gives almost the same selection. They are listed, with values, in the table. They are exactly the materials of which buildings have been, and are, made.

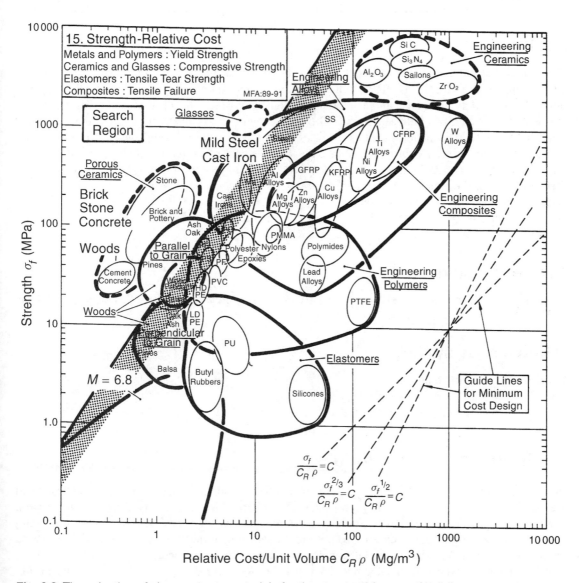

Fig. 6.9 The selection of cheap, strong materials for the structural frames of buildings.

Table 6.8 Structural materials for buildings

Material	M_1 $(GPa^{1/2}/(k\$/m^3))$	M_2 $(MPa^{2/3}/(k\$/m^3))$	Comment
Concrete	40	80	Use in compression only
Brick	20	45	
Stone	15	45	
Woods	15	80	Tension and compression, with
Cast iron	5	20	freedom of section shape
Steel	3	21	
Reinforced concrete	20	60	

Postscript

It is sometimes suggested that architects live in the past; that in the late 20th century they should be building with fibreglass (GFRP), aluminium alloys and stainless steel. Occasionally they do, but the last two figures give an idea of the penalty involved: the cost of achieving the same stiffness and strength is between 5 and 10 times greater. Civil construction (buildings, bridges, roads and the like) is materials-intensive: the cost of the material dominates the product cost, and the quantity used is enormous. Then only the cheapest of materials qualify, and the design must be adapted to use them. Concrete, stone and brick have strength only in compression; the form of the building must use them in this way (columns, arches). Wood, steel and reinforced concrete have strength both in tension and compression, and steel, additionally, can be given efficient shapes (I-sections, box sections, tubes); the form of the building made from these has much greater freedom.

Further reading

Cowan, H.J. and Smith, P.R. (1988) *The Science and Technology of Building Materials*, Van Nostrand-Reinhold, New York.

Related case studies

Case Study 6.2: Materials for oars
Case Study 6.4: Materials for table legs
Case Study 8.4: Floor joists: wood or steel?

6.6 Materials for flywheels

Flywheels store energy. Small ones — the sort found in children's toys — are made of lead. Old steam engines have flywheels; they are made of cast iron. More recently flywheels have been proposed for power storage and regenerative braking systems for vehicles; a few have been built, some of high-strength steel, some of composites. Lead, cast iron, steel, composites — there is a strange diversity here. What *is* the best choice of material for a flywheel?

An efficient flywheel stores as much *energy per unit weight* as possible, without *failing*. Failure (were it to occur) is caused by centrifugal loading: if the centrifugal stress exceeds the tensile strength (or fatigue strength) the flywheel flies apart. One constraint is that this should not occur.

The flywheel of a child's toy is not efficient in this sense. Its velocity is limited by the pulling-power of the child, and never remotely approaches the burst velocity. In this case, and for the flywheel of an automobile engine — we wish to maximize the *energy stored per unit volume* at a constant (specified) *angular velocity*. There is also a constraint on the outer radius, R, of the flywheel so that it will fit into a confined space.

The answer therefore depends on the application. The strategy for optimizing flywheels for efficient energy-storing systems differs from that for children's toys. The two alternative sets of design requirements are listed in Tables 6.9(a) and (b).

The model

An efficient flywheel of the first type stores as much energy per unit weight as possible, without failing. Think of it as a solid disc of radius R and thickness t, rotating with angular velocity ω (Figure 6.10). The energy U stored in the flywheel is

$$U = \frac{1}{2}J\omega^2 \tag{6.10}$$

Table 6.9(a) Design requirements for maximum-energy flywheel

Function	Flywheel for energy storage
Objective	Maximize kinetic energy per unit mass
Constraints	(a) Must not burst
	(b) Adequate toughness to give crack-tolerance

Table 6.9(b) Design requirements for limited-velocity flywheel

Function	Flywheel for child's toy
Objective	Maximize kinetic energy per unit volume
Constraints	Outer radius fixed

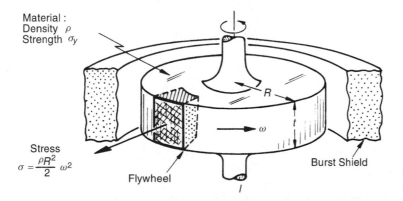

Fig. 6.10 A flywheel. The maximum kinetic energy it can store is limited by its strength.

Here $J = \frac{\pi}{2}\rho R^4 t$ is the polar moment of inertia of the disc and ρ the density of the material of which it is made, giving

$$U = \frac{\pi}{4}\rho R^4 t\omega^2 \qquad (6.11)$$

The mass of the disc is

$$m = \pi R^4 t\rho \qquad (6.12)$$

The quantity to be maximized is the kinetic energy per unit mass, which is the ratio of the last two equations:

$$\frac{U}{m} = \frac{1}{4}R^2\omega^2 \qquad (6.13)$$

As the flywheel is spun up, the energy stored in it increases, but so does the centrifugal stress. The maximum principal stress in a spinning disc of uniform thickness is

$$\sigma_{\text{max}} = \left(\frac{3+\nu}{8}\right)\rho R^2\omega^2 \qquad (6.14)$$

where ν is Poisson's ratio. This stress must not exceed the failure stress σ_f (with an appropriate factor of safety, here omitted). This sets an upper limit to the angular velocity, ω, and disc radius, R (the free variables). Eliminating $R\omega$ between the last two equations gives

$$\frac{U}{m} = \left(\frac{2}{(3+\nu)}\right)\left(\frac{\sigma_f}{\rho}\right) \qquad (6.15)$$

Poissons's ratio, ν, is roughly 1/3 for solids; we can treat it as a constant. The best materials for high-performance flywheels are those with high values of the material index

$$\boxed{M = \frac{\sigma_f}{\rho}} \qquad (6.16)$$

It has units of kJ/kg.

But what of the other sort of flywheel — that of the child's toy? Here we seek the material which stores the most energy per unit volume V at constant velocity. The energy per unit volume at a given ω is (from equation (6.2)):

$$\frac{U}{V} = \frac{1}{4}\rho R^2\omega^2$$

Both R and ω are fixed by the design, so the best material is now that with the greatest value of

$$\boxed{M_2 = \rho} \qquad (6.17)$$

The selection

Figure 6.11 shows Chart 2: strength against density. Values of M correspond to a grid of lines of slope 1. One such line is shown at the value $M = 100$ kJ/kg. Candidate materials with high values

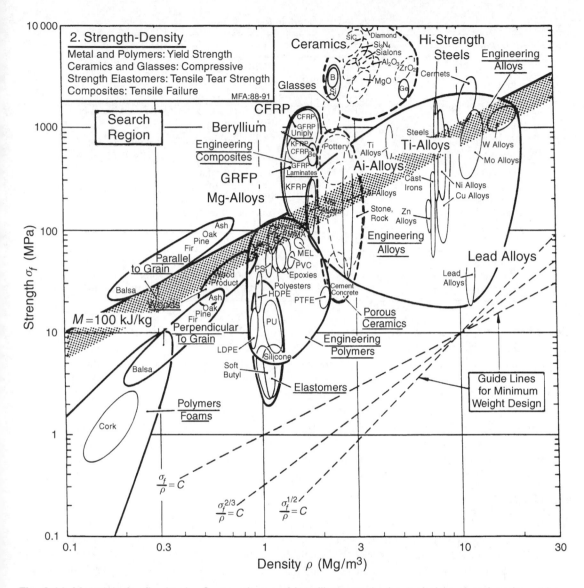

Fig. 6.11 Materials for flywheels. Composites and beryllium are the best choices. Lead and cast iron, traditional for flywheels, are good when performance is limited by rotational velocity, not strength.

of M lie in the search region towards the top left. They are listed in the upper part of Table 6.10. The best choices are unexpected ones: beryllium and composites, particularly glass-fibre reinforced polymers. Recent designs use a filament-wound glass-fibre reinforced rotor, able to store around 150 kJ/kg; a 20 kg rotor then stores 3 MJ or 800 kWh. A lead flywheel, by contrast, can store only 3 kJ/kg before disintegration; a cast-iron flywheel, about 10. All these are small compared with the energy density in gasoline: roughly 20 000 kJ/kg.

Even so, the energy density in the flywheel is considerable; its sudden release in a failure could be catastrophic. The disc must be surrounded by a burst-shield and precise quality control in manufacture is essential to avoid out-of-balance forces. This has been achieved in a number of

Table 6.10 Materials for flywheels

Material	M (kJ/kg)	Comment
Ceramics	200–2000 (compression only)	Brittle and weak in tension — eliminate.
Composites: CFRP	200–500	The best performance — a good choice.
GFRP	100–400	Almost as good as CFRP and cheaper. Excellent choice.
Beryllium	300	Good but expensive, difficult to work and toxic.
High-strength steel	100–200	All about equal
High-strength Al alloys	100–200	in performance. Steel and Al alloys
High-strength Mg alloys	100–200	cheaper than Mg and Ti alloys.
Ti alloys	100–200	
Lead alloys	3	High density makes these a good (and
Cast iron	8–10	traditional) selection when performance is velocity-limited, not strength-limited.

glass-fibre energy-storage flywheels intended for use in trucks and buses, and as an energy reservoir for smoothing wind-power generation.

But what of the lead flywheels of children's toys? There could hardly be two more different materials than GFRP and lead: the one, strong and light, the other, soft and heavy. Why lead? It is because, in the child's toy, the constraint is different. Even a super-child cannot spin the flywheel of his toy up to its burst velocity. The angular velocity ω is limited, instead, by the drive mechanism (pull-string, friction drive). Then, as we have seen, the best material is that with the largest density (Table 6.10, bottom section). Lead is good. Cast iron is less good, but cheaper. Gold, platinum and uranium are better, but may be thought unsuitable for other reasons.

Postscript

And now a digression: the electric car. By the turn of the century electric cars will be on the roads, powered by a souped-up version of the lead-acid battery. But batteries have their problems: the energy density they can contain is low (see Table 6.11); their weight limits both the range and the performance of the car. It is practical to build flywheels with an energy density of roughly five times that of the battery. Serious consideration is now being given to a flywheel for electric cars. A pair of counter-rotating CFRP discs are housed in a steel burst-shield. Magnets embedded in the discs pass near coils in the housing, inducing a current and allowing power to be drawn to the electric motor which drives the wheels. Such a flywheel could, it is estimated, give an electric car a range of 600 km, at a cost competitive with the gasoline engine.

Further reading

Christensen, R.M. (1979) *Mechanics of Composite Materials*, Wiley Interscience, New York, p. 213 *et seq.*
Lewis, G. (1990) *Selection of Engineering Materials*, Prentice Hall, Englewood Cliffs, NJ, Part 1, p. 1.
Medlicott, P.A.C. and Potter, K.D. (1986) The development of a composite flywheel for vehicle applications, in *High Tech — the Way into the Nineties*, edited by Brunsch, K., Golden, H-D., and Horkert, C-M. Elsevier, Amsterdam, p. 29.

Table 6.11 Energy density of power sources

Source	Energy density kJ/kg	Comment
Gasoline	20 000	Oxidation of hydrocarbon — mass of oxygen not included.
Rocket fuel	5000	Less than hydrocarbons because oxidizing agent forms part of fuel.
Flywheels	Up to 350	Attractive, but not yet proven.
Lead–acid battery	40–50	Large weight for acceptable range.
Springs rubber bands	Up to 5	Much less efficient method of energy storage than flywheel.

Related case studies

Case Study 6.7: Materials for high-flow fans
Case Study 6.15: Safe pressure vessels

6.7 Materials for high-flow fans

Automobile engines have a fan which cools the radiator when the forward motion of the car is insufficient to do the job. Commonly, the fan is driven by a belt from the main drive-shaft of the engine. The blades of the fan are subjected both to centrifugal forces and to bending moments caused by sudden acceleration of the motor. At least one fatality has been caused by the disintegration of a fan when an engine which had been reluctant to start suddenly sprang to life and was violently raced while a helper leaned over it. What criteria should one adopt in selecting materials to avoid this? The material chosen for the fan must be cheap. Any automaker who has survived to the present day has cut costs relentlessly on every component. But safety comes first.

The radius, R, of the fan is determined by design considerations: flow rate of air, and the space into which it must fit. The fan must not fail. The design requirements, then, are those of Table 6.12.

The model

A blade (Figure 6.12) has mean section area A and length αR, where α is the fraction of the fan radius R which is blade (the rest is hub). Its volume is αRA and the angular acceleration is $\omega^2 R$, so

Table 6.12 Design requirements for the fan

Function	Cooling fan
Objective	Maximum angular velocity without failure
Constraints	(a) Radius R specified
	(b) Must be cheap and easy to form

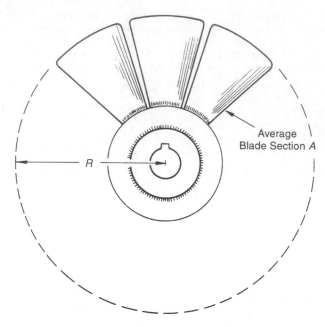

Fig. 6.12 A fan. The flow-rate of gas through the fan is related to its rotation speed, which is ultimately limited by its strength.

the centrifugal force at the blade root is

$$F = \rho(\alpha R A)\omega^2 R \tag{6.18}$$

The force is carried by the section A, so the stress at the root of the blade is

$$\sigma = \frac{F}{A} = \alpha\rho\omega^2 R^2 \tag{6.19}$$

This stress must not exceed the failure stress Σ_f divided by a safety factor (typically about 3) which does not affect the analysis and can be ignored. Thus for safety:

$$\omega < \frac{1}{\sqrt{\alpha}R} \left(\frac{\sigma_f}{\rho}\right)^{1/2}$$

The length R is fixed, as is α. The safe rotational velocity ω is maximized by selecting materials with large values of

$$M = \frac{\sigma_f}{\rho} \tag{6.21}$$

The selection

Figure 6.13 shows strength σ_f plotted against density, ρ. The materials above the selection line (slope = 1) have high values of M. This selection must be balanced against the cost. Low cost fans can be made by die-casting a metal, or by injection-moulding a polymer (Table 6.13).

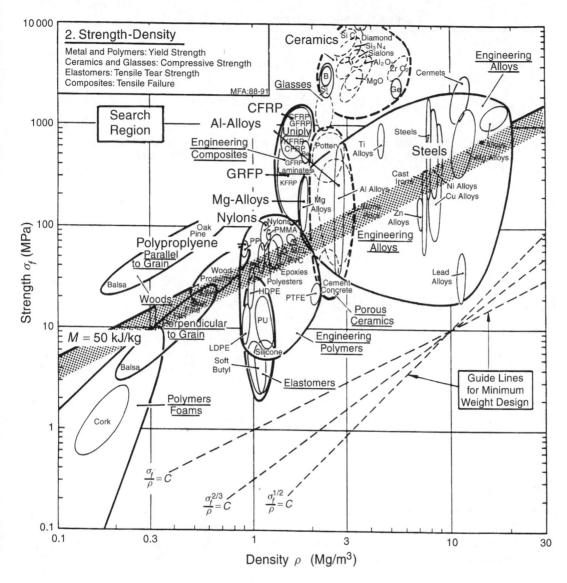

Fig. 6.13 Materials for cheap high-flow fans. Polymers — nylons and polypropylenes — are good; so are die-cast aluminium and magnesium alloys. Composites are better, but more difficult to fabricate.

Postscript

To an auto-maker additional cost is anathema, but the risk of a penal law suit is worse. Here (as elsewhere) it is possible to 'design' a way out of the problem. The problem is not really the fan; it is the undisciplined speed-changes of the engine which drives it. The solution (now we put it this way) is obvious: decouple the two. Increasingly, the cooling fans of automobiles are driven, not by the engine, but by an electric motor (cost: about that of a fan-belt) which limits it to speeds which are safe — and gives additional benefits in allowing independent control and more freedom in where the fan is placed.

Table 6.13 Candidate materials for a high-flow fan

Material	Comment
Cast iron	Cheap and easy to cast but poor σ_f/ρ.
Cast Al alloys	Can be die-cast to final shape.
High density polyethylene (HDPE)	Mouldable and cheap.
Nylons	
Rigid PVCs	
GFRP (chopped fibre)	Lay-up methods too expensive and slow. Press
CFRP (chopped fibre)	from chopped-fibre moulding material.

Related case studies

Case Study 6.6:   Materials for flywheels
Case Study 12.2: Forming a fan
Case Study 14.3: A non-ferrous alloy: Al–Si die casting alloys

6.8 Golf-ball print heads

Mass is important when inertial forces are large, as they are in high-speed machinery. The golf-ball typewriter is an example: fast positioning of the golf-ball requires large accelerations and decelerations. Years before they came on the market, both the golf-ball and the daisy-wheel design had been considered and rejected: in those days print heads could only be made of heavy type-metal, and had too much inertia. The design became practical when it was realized that a polymer (density, $1\,\text{Mg/m}^3$) could be moulded to carry the type, replacing the lead-based type-metal (density, about $10\,\text{Mg/m}^3$). The same idea has contributed to other high-speed processes, which include printing, textile manufacture, and packaging.

The model

A golf-ball print head is a thin-walled shell with the type faces moulded on its outer surface (Figure 6.14). Its outer radius, R, is fixed by the requirement that it carry the usual 88 standard characters; the other requirements are summarized in Table 6.14. The time to reposition it varies as the square root of its mass, m, where

$$m \cong 4\pi R^2 t \rho \tag{6.22}$$

and t is the wall thickness and ρ the density of the material of which it is made. We wish to minimize this mass. The wall thickness must be sufficient to bear the strike force: a force F, distributed over

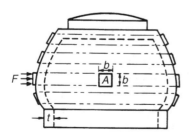

Fig. 6.14 A golf-ball print head. It must be strong yet light, to minimize inertial forces during rapid repositioning.

Table 6.14 Design requirements for golf-ball print heads

Function	Rapidly positioned print head
Objective	Minimize the mass (and thus inertia)
Constraints	(a) Outer radius R fixed
	(b) Adequate strength; must not fail under striking loads
	(c) Adequate stiffness
	(d) Can be moulded or cast to give sharply defined type-faces

an area of roughly b^2, where b is the average linear dimension of a character. When golf-ball print heads fail, they do so by cracking through the shell wall. We therefore require as a constraint that the through-thickness shear stress, $F/4bt$, be less than the failure strength, which, for shear, we approximate by $\sigma_f/2$:

$$\frac{F}{4bt} \leq \frac{\sigma_f}{2} \tag{6.23}$$

The free variable is the wall thickness, t. Solving for t and substituting into the equation (6.22) gives

$$m = F\left(\frac{2\pi R^2}{b}\right)\left(\frac{\rho}{\sigma_f}\right) \tag{6.24}$$

The repositioning time is minimized by choosing a material with the largest possible value of

$$\boxed{M = \frac{\sigma_f}{\rho}}$$

The material must also be mouldable or castable.

The selection

Materials for golf-balls require high σ_f/ρ; then Chart 2 is the appropriate one. It is reproduced in Figure 6.15, with appropriate selection lines constructed on it. It isolates two viable classes of candidate materials: metals, in the form of aluminium or magnesium casting alloys (which can be pressure die-cast) and the stronger polymers (which can be moulded to shape). Both classes, potentially, can meet the design requirements at a weight which is 15 to 20 times less than lead-based alloys which are traditional for type. We reject ceramics which are strong in compression but not in bending, and composites which cannot be moulded to give fine detail.

Data for the candidates are listed in Table 6.15, allowing a more detailed comparison. The final choice is an economic one: achieving high character-definition requires high-pressure moulding techniques which cost less, per unit, for polymers than for metals. High-modulus, high-strength polymers become the primary choice for the design.

Postscript

Printers are big business: long before computers were invented, IBM was already a large company made prosperous by selling typewriters. The scale of the market has led to sophisticated designs. Golf-balls and daisy-wheels are made of polymers, for the reasons given above; but not just one polymer. A modern daisy-wheel uses at least two: one for the type-face, which must resist wear

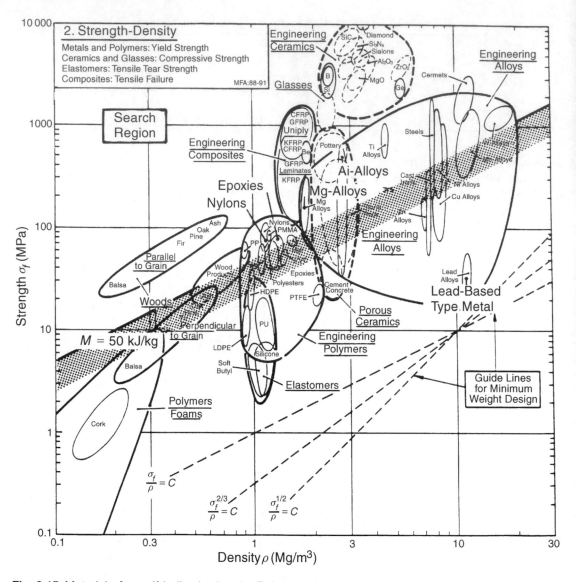

Fig. 6.15 Materials for golf-ball print heads. Polymers, because of their low density, are better than type-metal, which is mostly lead, and therefore has high inertia.

and impact, and a second for the fingers, which act as the return springs. Golf-balls have a surface coating for wear resistance, or simply to make the polymer look like a metal. Their days, however, are numbered. Laser and bubble-jet technologies have already largely displaced them. These, too, present problems in material selection, but of a different kind.

Related case studies

Case Study 6.6: Materials for flywheels
Case Study 6.7: Materials for high-flow fans

Table 6.15 Materials for golf-ball and daisy-wheel print heads

Material	$M = \dfrac{\sigma_f}{\rho}$ $(MPa/(Mg/m^3))$	Comment
Nylons	80	Mouldable thermoplastic.
Epoxy	75	Castable thermoset.
Cast Mg alloys	60	Character definition poor.
Cast Al alloys	60	Character definition poor.
Type metal (Pb-5% Sn-10% Sb)	4	15 to 20 times heavier than the above for the same strength.

6.9 Materials for springs

Springs come in many shapes (Figure 6.16) and have many purposes: one thinks of axial springs (a rubber band, for example), leaf springs, helical springs, spiral springs, torsion bars. Regardless of their shape or use, the best material for a spring of minimum volume is that with the greatest value of σ_f^2/E, and for minimum weight it is that with the greatest value of $\sigma_f^2/E\rho$ (derived below). We use them as a way of introducing two of the most useful of the charts: Young's modulus E plotted against strength σ_f (Chart 4), and specific modulus, E/ρ, plotted against specific strength σ_f/ρ (Chart 5).

The model

The primary function of a spring is that of storing elastic energy and — when required — releasing it again (Table 6.16). The elastic energy stored per unit volume in a block of material stressed

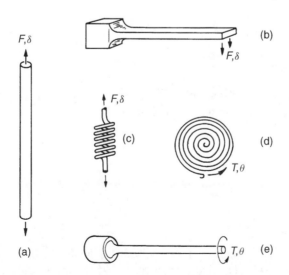

Fig. 6.16 Springs store energy. The best material for any spring, regardless of its shape or the way in which it is loaded, is that with the highest value of σ_f^2/E, or, if weight is important, $\sigma_f^2/E\rho$.

Table 6.16 Design requirements for springs

Function	Elastic spring
Objectives	(a) Maximum stored elastic energy per unit volume
	(b) Maximum stored elastic energy per unit mass
Constraints	(a) No failure by yield, fracture or fatigue (whichever is the most restrictive), meaning $\sigma < \sigma_f$ everywhere in the spring
	(b) Adequate toughness: $G_c > 1 \, \text{kJ/m}^2$

uniformly to a stress σ is

$$W_v = \frac{1}{2} \frac{\sigma^2}{E}$$

where E is Young's modulus. It is this W_v that we wish to maximize. The spring will be damaged if the stress σ exceeds the yield stress or failure stress σ_f; the constraint is $\sigma \leq \sigma_f$. So the maximum energy density is

$$W_v = \frac{1}{2} \frac{\sigma_f^2}{E} \tag{6.25}$$

Torsion bars and leaf springs are less efficient than axial springs because much of the material is not fully loaded: the material at the neutral axis, for instance, is not loaded at all. For torsion bars

$$W_v = \frac{1}{3} \frac{\sigma_f^2}{E}$$

and for leaf springs

$$W_v = \frac{1}{4} \frac{\sigma_f^2}{E}$$

But — as these results show — this has no influence on the choice of material. The best material for springs is that with the biggest value of

$$M_1 = \frac{\sigma_f^2}{E} \tag{6.26}$$

If weight, rather than volume, matters, we must divide this by the density ρ (giving energy stored per unit weight), and seek materials with high values of

$$M_2 = \frac{\sigma_f^2}{\rho E} \tag{6.27}$$

The selection

The choice of materials for springs of minimum volume is shown in Figure 6.17. A family lines of slope 1/2 link materials with equal values of $M_1 = \sigma_f^2/E$; those with the highest values of M_1

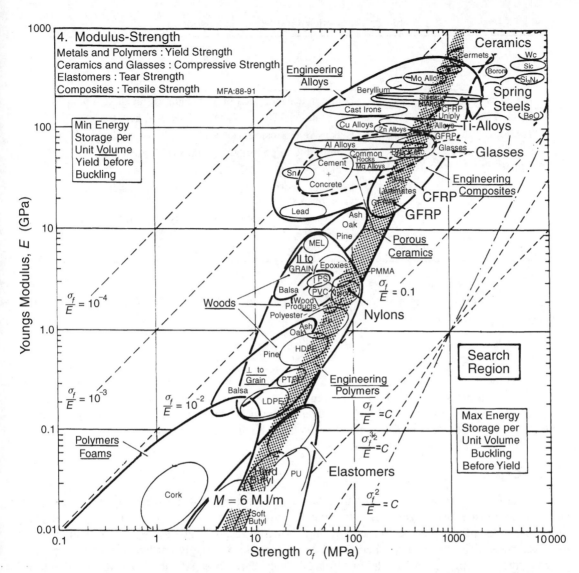

Fig. 6.17 Materials for small springs. High strength ('spring') steel is good. Glass, CFRP and GFRP all, under the right circumstances, make good springs. Elastomers are excellent. Ceramics are eliminated by their low tensile strength.

lie towards the bottom right. The heavy line is one of the family; it is positioned so that a subset of materials is left exposed. The best choices are a *high-strength steel* ((spring steel, in fact) lying near the top end of the line, and, at the other end, *rubber*. But certain other materials are suggested too: *GFRP* (now used for truck springs), *titanium alloys* (good but expensive), *glass* (used in galvanometers) and *nylon* (children's toys often have nylon springs). Note how the procedure has identified a candidate from almost every class of material: metals, glasses, polymers, elastomers and composites. They are listed, with commentary, in Table 6.17.

Table 6.17 Materials for efficient small springs

Material	$M_1 = \dfrac{\sigma_f^2}{E}$ (MJ/m^3)	Comment
Ceramics	(10–100)	Brittle in tension; good only in compression.
Spring steel	15–25	The traditional choice: easily formed and heat treated.
Ti alloys	15–20	Expensive, corrosion-resistant.
CFRP	15–20	Comparable in performance with steel; expensive.
GFRP	10–12	Almost as good as CFRP and much cheaper.
Glass (fibres)	30–60	Brittle in torsion, but excellent if protected against damage; very low loss factor.
Nylon	1.5–2.5	The least good; but cheap and easily shaped, but high loss factor.
Rubber	20–50	Better than spring steel; but high loss factor.

Materials selection for light springs is shown in Figure 6.18. A family of lines of slope 2 link materials with equal values of

$$M_2 = \left(\frac{\sigma_f}{\rho}\right)^2 \Big/ \left(\frac{E}{\rho}\right) = \frac{\sigma_f^2}{E\rho}$$

One is shown at the value $M_2 = 2\,kJ/kg$. Metals, because of their high density, are less good than composites, and much less good than elastomers. (You can store roughly eight times more elastic energy, per unit weight, in a rubber band than in the best spring steel.) Candidates are listed in Table 6.18. Wood, the traditional material for archery bows, now appears.

Postscript

Many additional considerations enter the choice of a material for a spring. Springs for vehicle suspensions must resist fatigue and corrosion; IC valve springs must cope with elevated temperatures. A subtler property is the loss coefficient, shown in Chart 7. Polymers have a relatively high loss factor and dissipate energy when they vibrate; metals, if strongly hardened, do not. Polymers, because they creep, are unsuitable for springs which carry a steady load, though they are still perfectly good for catches and locating-springs which spend most of their time unstressed.

Further reading

Boiton, R.G. (1963) The mechanics of instrumentation, *Proc. I. Mech. E.*, Vol. 177, No. 10, 269–288.
Hayes, M. (1990) Materials update 2: springs, *Engineering*, May, p. 42.

Related case studies

Case Study 6.10: Elastic hinges
Case Study 6.12: Diaphragms for pressure actuators
Case Study 8.6:   Ultra-efficient springs

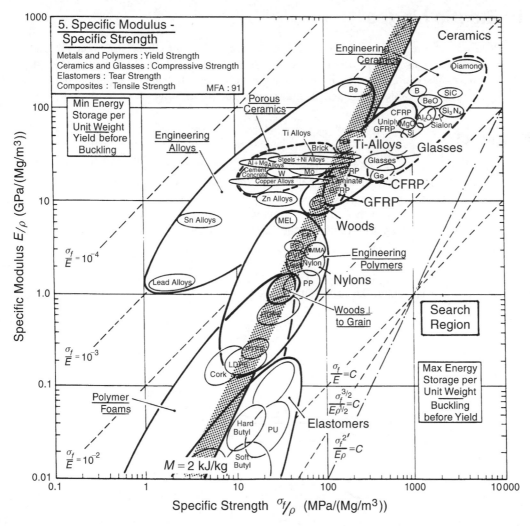

Fig. 6.18 Materials for light springs. Metals are disadvantaged by their high densities. Composites are good; so is wood. Elastomers are excellent.

Table 6.18 Materials for efficient light springs

Material	$M_2 = \dfrac{\sigma_f^2}{E\rho}$ (kJ/kg)	Comment
Ceramics	(5–40)	Brittle in tension; good only in compression.
Spring steel	2–3	Poor, because of high density.
Ti alloys	2–3	Better than steel; corrosion-resistant; expensive.
CFRP	4–8	Better than steel; expensive.
GFRP	3–5	Better than steel; less expensive than CFRP.
Glass (fibres)	10–30	Brittle in torsion, but excellent if protected.
Wood	1–2	On a weight basis, wood makes good springs.
Nylon	1.5–2	As good as steel, but with a high loss factor.
Rubber	20–50	Outstanding; 10 times better than steel, but with high loss factor.

6.10 Elastic hinges

Nature makes much use of elastic hinges: skin, muscle, cartilage all allow large, recoverable deflections. Man, too, designs with *flexure and torsion hinges*: devices which connect or transmit load between components while allowing limited relative movement between them by deflecting elastically (Figure 6.19 and Table 6.19). Which materials make good hinges?

The model

Consider the hinge for the lid of a box. The box, lid and hinge are to be moulded in one operation. The hinge is a thin ligament of material which flexes elastically as the box is closed, as in the figure, but it carries no significant axial loads. Then the best material is the one which (for given ligament dimensions) bends to the smallest radius without yielding or failing. When a ligament of thickness t is bent elastically to a radius R, the surface strain is

$$\varepsilon = \frac{t}{2R} \tag{6.28}$$

and, since the hinge is elastic, the maximum stress is

$$\sigma \geq E \frac{t}{2R} \tag{6.29}$$

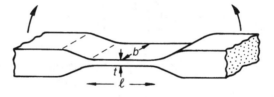

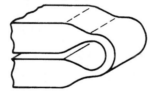

Fig. 6.19 Elastic or 'flexure' hinges. The ligaments must bend repeatedly without failing. The cap of a shampoo bottle is an example; elastic hinges are used in high performance applications too, and are found widely in nature.

Table 6.19 Design requirements for elastic hinges

Function	Elastic hinge (possibly with additional axial load)
Objective	Maximize elastic flexure or twisting
Constraints	No failure by yield, fracture or fatigue (whichever is the most restrictive) (a) with no axial load (b) with additional axial load

This must not exceed the yield or failure strength σ_f. Thus the radius to which the ligament can be bent without damage is

$$R \leq \frac{t}{2}\left[\frac{E}{\sigma_f}\right] \tag{6.30}$$

The best material is the one that can be bent to the smallest radius, that is, the one with the greatest value of the index

$$M_1 = \frac{\sigma_f}{E}$$

We have assumed thus far that the hinge thickness, t, is dictated by the way the hinge is made. But in normal use, the hinge may also carry repeated axial (tensile) forces, F, due to handling or to the weight of the box and its contents. This sets a minimum value for the thickness, t, which is found by requiring that the tensile stress, F/tw (where w is the hinge width) does not exceed the strength limit σ_f:

$$t^* = \frac{F}{\sigma_f w}$$

Substituting this value of t into equation (6.30) gives

$$R \leq \frac{F}{2w}\left[\frac{E}{\sigma_f^2}\right]$$

and the second index

$$M_2 = \frac{\sigma_f^2}{E}$$

The selection

The criteria both involve ratios of σ_f and E; we need Chart 4 (Figure 6.20). Candidates are identified by using the guide line of slope 1; a line is shown at the position $M = \sigma_y/E = 3 \times 10^{-2}$. The best choices for the hinge are all polymeric materials. The shortlist (Table 6.20) includes polyethylenes, polypropylene, nylon and, best of all, elastomers, though these may be too flexible for the body of the box itself. Cheap products with this sort of elastic hinge are generally moulded from polyethylene, polypropylene or nylon. Spring steel and other metallic spring materials (like phosphor bronze) are possibilities: they combine usable σ_f/E with high E, giving flexibility with good positional stability (as in the suspensions of relays). The tables gives further details.

Postscript

Polymers give more design-freedom than metals. The elastic hinge is one example of this, reducing the box, hinge and lid (three components plus the fasteners needed to join them) to a single box-hinge-lid, moulded in one operation. Their spring-like properties allow snap-together, easily-joined

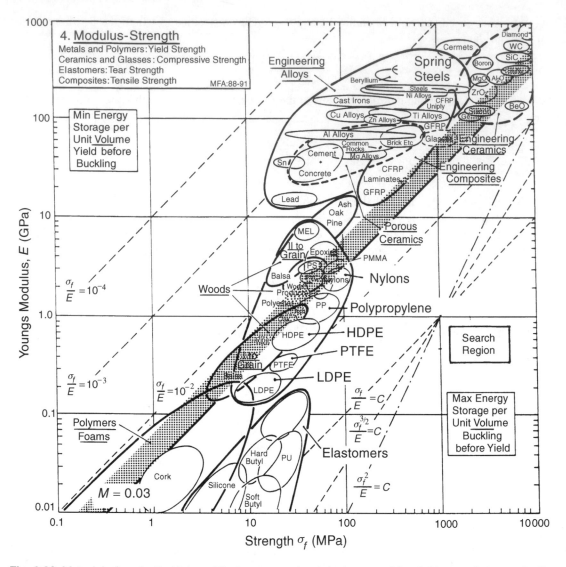

Fig. 6.20 Materials for elastic hinges. Elastomers are best, but may not be rigid enough to meet other design needs. Then polymers such as nylon, PTFE and PE are better. Spring steel is less good, but much stronger.

parts. Another is the elastomeric coupling — a flexible universal joint, allowing an exceptionally high angular, parallel and axial flexibility with good shock absorption characteristics. Elastomeric hinges offer many more opportunities, to be exploited in engineering design.

Related case studies

Case Study 6.9:   Materials for springs
Case Study 6.11: Materials for seals
Case Study 6.12: Diaphragms for pressure actuators

Table 6.20 Materials for elastic hinges

Material	M_1 $(\times 10^{-3})$	M_2 (MJ/m^3)	Comment
Polyethylenes	30–45	1.6–1.8	Widely used for cheap hinged bottle caps, etc.
Polypropylene	30	1.6–1.7	Stiffer than PEs. Easily moulded.
Nylon	30	2–2.1	Stiffer than PEs. Easily moulded.
PTFE	35	2–2.1	Very durable; more expensive than PE, PP, etc.
Elastomers	100–300	10–20	Outstanding, but low modulus.
Beryllium-copper	5–10	8–12	M_1 less good than polymers. Use when high stiffness required.
Spring steel	5–10	10–20	M_1 less good than polymers. Use when high stiffness required.

6.11 Materials for seals

A reusable elastic seal consists of a cylinder of material compressed between two flat surfaces (Figure 6.21). The seal must form the largest possible contact width, b, while keeping the contact stress, σ sufficiently low that it does not damage the flat surfaces; and the seal itself must remain elastic so that it can be reused many times. What materials make good seals? Elastomers — everyone knows that. But let us do the job properly; there may be more to be learnt. We build the selection around the requirements of Table 6.21.

The model

A cylinder of diameter $2R$ and modulus E, pressed on to a rigid flat surface by a force f per unit length, forms an elastic contact of width b (Appendix A: 'Useful Solutions') where

$$b \approx 2 \left(\frac{fR}{E} \right)^{1/2} \tag{6.31}$$

This is the quantity to be maximized: the objective function. The contact stress, both in the seal and in the surface, is adequately approximated (Appendix A again) by

$$\sigma = 0.6 \left(\frac{fE}{R} \right)^{1/2} \tag{6.32}$$

The constraint: the seal must remain elastic, that is, σ must be less than the yield or failure strength, σ_f, of the material of which it is made. Combining the last two equations with this condition gives

$$b \le 3.3R \left(\frac{\sigma_f}{E} \right) \tag{6.33}$$

The contact width is maximized by maximizing the index

$$\boxed{M_1 = \frac{\sigma_f}{E}}$$

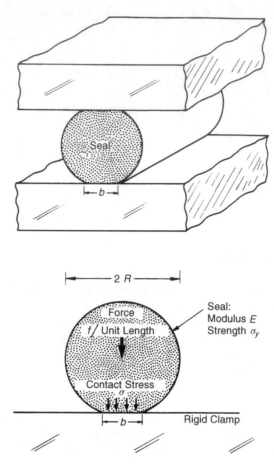

Fig. 6.21 An elastic seal. A good seal gives a large conforming contact area without imposing damaging loads on itself or on the surfaces with which it mates.

Table 6.21 Design requirements for the elastic seals

Function	Elastic seal
Objective	Maximum conformability
Constraints	(a) Limit on contact pressure
	(b) low cost

It is also required that the contact stress σ be kept low to avoid damage to the flat surfaces. Its value when the maximum contact force is applied (to give the biggest width) is simply σ_f, the failure strength of the seal. Suppose the flat surfaces are damaged by a stress of greater than 100 MPa. The contact pressure is kept below this by requiring that

$$M_2 = \sigma_f \leq 100\,\text{MPa}$$

The selection

The two indices are plotted on the $\sigma_f - E$ chart in Figure 6.22 isolating elastomers, foams and cork. The candidates are listed in Table 6.22 with commentary. The value of $M_2 = 100$ MPa admits all elastomers as candidates. If M_2 were reduced to 10 MPa, all but the most compliant elastomers are eliminated, and foamed polymers become the best bet.

Postscript

The analysis highlights the functions that seals must perform: large contact area, limited contact pressure, environmental stability. Elastomers maximize the contact area; foams and cork minimize

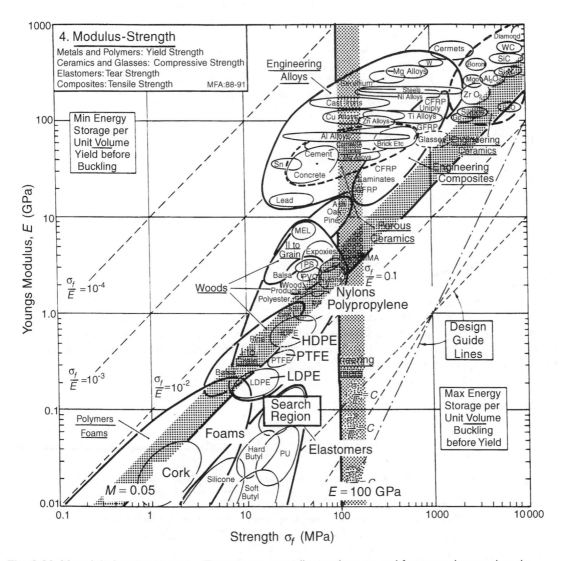

Fig. 6.22 Materials for elastic seals. Elastomers, compliant polymers and foams make good seals.

Table 6.22 Materials for reusable seals

Material	$M_1 = \dfrac{\sigma_f}{E}$	Comment
Butyl rubbers	1–3	The natural choice; poor resistance to heat and to some solvents.
Polyurethanes	0.5–4.5	Widely used for seals.
Silicone rubbers	0.1–0.8	Higher temperature capability than carbon-chain elastomers, chemically inert.
PTFE	0.1	Expensive but chemically stable and with high temperature capability.
Polyethylenes	0.05–0.2	Cheap.
Polypropylenes	0.1	Cheap.
Nylons	0.05	Near upper limit on contact pressure.
Cork	0.1	Low contact stress, chemically stable.
Polymer foams	up to 0.5	Very low contact pressure; delicate seals.

the contact pressure; PTFE and silicone rubbers best resist heat and organic solvents. The final choice depends on the conditions under which the seal will be used.

Related case studies

Case Study 6.9: Materials for springs
Case Study 6.10: Elastic hinges
Case Study 6.12: Diaphragms for pressure actuators
Case Study 6.13: Knife edges and pivots

6.12 Diaphragms for pressure actuators

A barometer is a pressure actuator. Changes in atmospheric pressure, acting on one side of a diaphragm, cause it to deflect; the deflection is transmitted through mechanical linkage or electro-magnetic sensor to a read-out. Similar diaphragms form the active component of altimeters, pressure gauges, and gas-flow controls for diving equipment. Which materials best meet the requirements for diaphragms, summarized in Table 6.23?

The model

Figure 6.23 shows a diaphragm of radius a and thickness t. A pressure difference $\Delta p = p_1 - p_2$ acts across it. We wish to maximize the deflection of the centre of the diaphragm, subject to the

Table 6.23 Design requirements for diaphragms

Function	Diaphragm for pressure sensing
Objective	Maximize displacement for given pressure difference
Constraints	(a) Must remain elastic (no yield or fracture)
	(b) No creep
	(c) Low damping for quick, accurate response

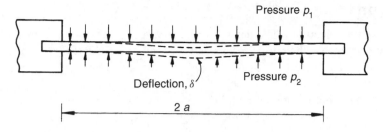

Fig. 6.23 A diaphragm. Its deflection under a pressure difference is used to sense and actuate.

constraint that it remain elastic — that is, that the stresses in it are everywhere less than the yield or fracture stress, σ_f, of the material of which it is made. The deflection δ of a diaphragm caused by Δp (Appendix A: 'Useful Solutions') depends on whether its edges are clamped or free:

$$\delta = \frac{C_1 \Delta p a^4 (1 - v^2)}{E t^3} \tag{6.34}$$

with
$$C_1 = \frac{3}{16} \qquad \text{(clamped edges)}$$

or
$$C_1 \approx \frac{9}{8} \qquad \text{(free edges)}$$

Here E is Young's modulus, and v is Poisson's ratio. The maximum stress in the diaphragm (Appendix A again) is

$$\sigma_{\max} = C_2 \Delta p \frac{a^2}{t^2} \tag{6.35}$$

with
$$C_1 \approx \frac{1}{2} \qquad \text{(clamped edges)}$$

or
$$C_2 \approx \frac{3}{2} \qquad \text{(free edges)}$$

This stress must not exceed the yield or failure stress, σ_f.

The radius of the diaphragm is determined by the design; the thickness t is free. Eliminating t between the two equations gives

$$\delta = \frac{C_1}{C_2^{3/2}} \left(\frac{a}{\Delta p^{1/2}} \right) \left(\frac{\sigma_f^{3/2} (1 - v^2)}{E} \right) \tag{6.36}$$

The material properties are grouped in the last brackets. The quantity $(1 - v^2)$ is close to 1 for all solids. The best material for the diaphragm is that with the largest value of

$$\boxed{M = \frac{\sigma_f^{3/2}}{E}} \tag{6.37}$$

The selection

Figure 6.24 shows the selection. Candidates with large values of M are listed in Table 6.24 together with approximate values of their loss coefficients, η read from Chart 8. Ceramics are eliminated because the stresses of equation (6.35) are tensile. Metals make good diaphragms, notably spring steel, and high-strength titanium alloys. Certain polymers are possible — nylon, polypropylene and PTFE — but they have high damping and they creep. So do elastomers: both natural and artificial rubbers acquire a permanent set under static loads.

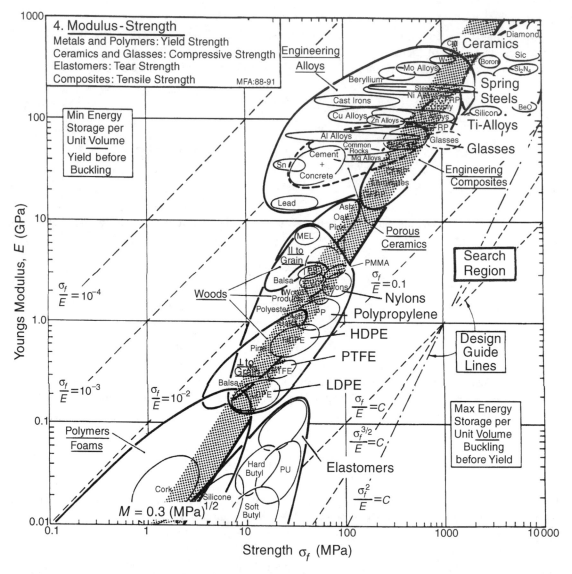

Fig. 6.24 Materials for elastic diaphragms. Elastomers, polymers, metals and even ceramics can be used; the final selection depends on details of the design.

Table 6.24 Materials for diaphragms

Material	$M = \dfrac{\sigma_f^{3/2}}{E}$ $(MPa)^{1/2}$	Loss coefficient η	Comment
Ceramics	0.3–3	$<10^{-4}$	Weak in tension. Eliminate.
Glasses	0.5	$\approx 10^{-4}$	Possible if protected from damage.
Spring Steel	0.3	$\approx 10^{-4}$	The standard choice. Low loss coefficient gives rapid response.
Ti-Alloys	0.3	$\approx 3 \times 10^{-4}$	As good as steel, corrosion resistant, expensive.
Nylons	0.3	$\approx 2 \times 10^{-2}$	Polymers creep and
Polypropylene	0.3	$\approx 5 \times 10^{-2}$	have high loss coefficients,
HDPE	0.3	$\approx 10^{-1}$	giving an actuator with
PTFE	0.3	$\approx 10^{-1}$	poor reproducibility.
Elastomers	0.5–10	$\approx 10^{-1}–1$	Excellent M value, giving large elastic deflection, but high loss coefficient limits response time.

Postscript

As always, application of the primary design criterion (large δ without failure) leads to a subset of materials to which further criteria are now applied. Elastomers have the best values of M, but they have high loss coefficients, are easily punctured, and may be permeable to certain gases or liquids. If corrosive liquids (sea water, cleaning fluids) may contact the diaphragm, then stainless steel or bronze may be preferable to a high-carbon steel, even though they have smaller values of M. This can be overcome by design: crimping the diaphragm or shaping it like a bellows magnifies deflection without increase in stress, but adding manufacturing cost.

Related case studies

Case Study 6.9: Materials for springs
Case Study 6.10: Elastic hinges
Case Study 6.11: Materials for seals
Case Study 6.13: Knife edges and pivots
Case Study 6.16: High damping materials for shaker tables

6.13 Knife edges and pivots

Middle-aged readers may remember the words '17 Sapphires' printed on the face of a watch, roughly where the word 'Quartz' now appears. A really expensive watch had, not sapphires, but diamonds. They are examples of good materials for knife edges and pivots. These are bearings in which two members are loaded together in nominal line or point contact, and can tilt relative to one another, or rotate freely about the load axis (Figure 6.25). The essential material properties, arising directly from the design requirements of Table 6.25, are high hardness (to carry the contact pressures) and high modulus (to give positional precision and to minimize frictional losses). But in what combination? And which materials have them?

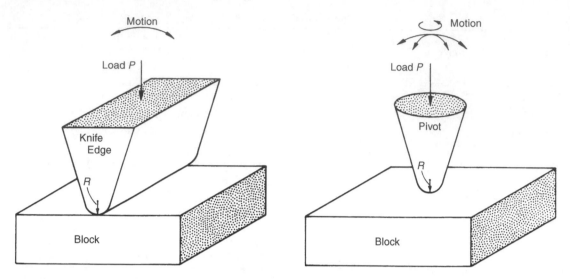

Fig. 6.25 A knife edge and a pivot. Good performance requires a high strength (to prevent plastic indentation or fracture) and a high modulus (to minimize elastic flattening at the contact which leads to frictional losses).

Table 6.25 Design requirements for knife edges and pivots

Function	Knife edges and pivots
Objective	(a) Maximize positional precision for given load, or
	(b) Maximize load capacity for given geometry
Constraints	(a) Contact stress must not damage either surface
	(b) Low thermal expansion (precision pivots)
	(c) High toughness (pivots exposed to shock loading)

The model

The first design goal is to maximize the load P that the contact can support, subject to the constraint that both faces of the bearing remain elastic. The contact pressure p at an elastic, non-conforming, contact (one which appears to touch at a point or along a line) is proportional to $(PE^2/R^2)^{1/3}$, where P is the load and R the radius of the knife-edge or pivot (Appendix A: 'Useful Solutions'). Check the dimensions: they are those of stress, MPa. Young's modulus, E, appears on the top because the elastic contact area decreases if E is large, and this increases the contact pressure. The knife or pivot will indent the block, or deform itself, if the contact pressure exceeds the hardness, H; and H is proportional to the strength, σ_f. The constraint is described by:

$$\left[\frac{PE^2}{R^2}\right]^{1/3} \leq C\sigma_f \tag{6.38}$$

where C is a constant (approximately 3.2). Thus, for a given geometry, the maximum bearing load is

$$P = C^3 R^2 \left[\frac{\sigma_f^3}{E^2}\right] \tag{6.39}$$

The subset of materials which maximizes the permitted bearing load is that with the greatest values of

$$M_1 = \frac{\sigma_f^3}{E^2}$$

The second constraint is that of low total contact area. The contact area A of any non-conforming contact has the form (Appendix A again)

$$A = C \left[\frac{PR}{E} \right]^{2/3} \tag{6.40}$$

where C is another constant (roughly 1). For any value of P less than that given by equation (6.39), this constraint is met by selecting from the subset those with the highest values of

$$M_2 = E$$

The selection

Once again, the material indices involve σ_f and E only. Chart 4 is shown in Figure 6.26. The two requirements isolate the top corner of the diagram and this time the loading is compressive, so ceramics are usable. Glasses, high-carbon steels and ceramics are all good choices. Table 6.26 gives more details: note the superiority of diamond.

Postscript

The final choice depends on the details of its application. In sensitive force balances and other measuring equipment, very low friction is important: then we need the exceptionally high modulus of sapphire or diamond. In high load-capacity devices (weigh bridges, mechanical testing equipment),

Table 6.26 Materials selection for knife edges and pivots

Material	$M = \frac{\sigma_f^3}{E^2}$ (MPa)	$M_2 = E$ (GPa)	Comment
Quartz	0.5	70	Good M_1 but brittle — poor impact resistance.
High-Carbon Steel	0.2	210	Some ductility, giving impact
Tool Steel	0.3	210	resistance; poor corrosion resistance.
Silicon	1	120	Good M_1, but brittle. Readily available in large quantities.
Sapphire, Al_2O_3	0.9	380	
Silicon Carbide, SiC	1	410	Excellent M_1 and M_2 with good corrosion
Silicon Nitride, Si_3N_4	1.1	310	resistance, but damaged by impact because of low toughness.
Tungsten Carbide, WC	1	580	
Diamond	2	1000	Outstanding on all counts except cost.

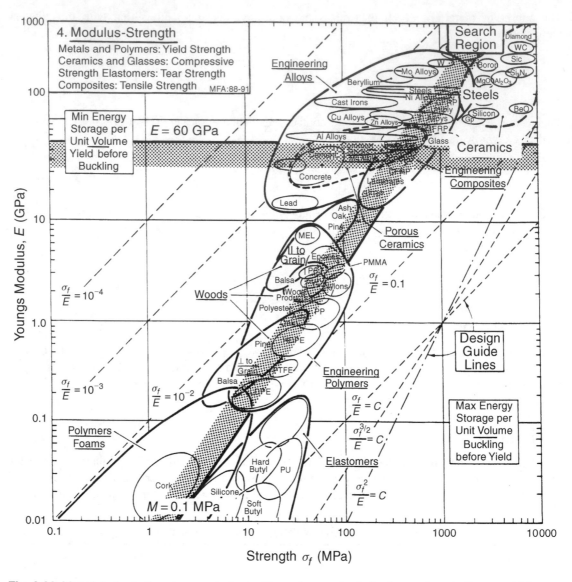

Fig. 6.26 Materials for knife edges and pivots. Ceramics, particularly diamond and silicon carbide, are good; fully hardened steel is a good choice too.

some ability to absorb overloads by limited plasticity is an advantage, and hardened steel is a good choice. If the environment is a potentially corrosive one — and this includes ordinary damp air — glass or a ceramic may be best. Note how the primary design criteria — high σ_f^3/E^2 and E — identify a subset from which, by considering further requirements, a single choice can be made.

Related case studies

Case Study 6.9: Materials for springs
Case Study 6.10: Elastic hinges

6.14 Deflection-limited design with brittle polymers

Among mechanical engineers there is a rule-of-thumb: avoid materials with fracture toughnesses K_{Ic} less than 15 MPa m$^{1/2}$. Almost all metals pass: they have values of K_{Ic} in the range of 20–100 in these units. White cast iron, and a few powder metallurgy products fail; they have values around 10 MPa m$^{1/2}$. Ordinary engineering ceramics have fracture toughnesses in the range 1–6 MPa m$^{1/2}$; mechanical engineers view them with deep suspicion. But engineering polymers are even less tough, with K_{Ic} values in the range 0.5–3 MPa m$^{1/2}$, and yet engineers use them all the time. What is going on here?

When a brittle material is deformed, it deflects elastically until it fractures. The stress at which this happens is

$$\sigma_f = \frac{CK_c}{\sqrt{\pi a_c}} \tag{6.41}$$

where K_c is an appropriate fracture toughness, a_c is the length of the largest crack contained in the material and C is a constant which depends on geometry, but is usually about 1. In a *load-limited* design — a tension member of a bridge, say — the part will fail in a brittle way if the stress exceeds that given by equation (6.41). Here, obviously, we want materials with high values of K_c.

But not all designs are load limited; some are *energy limited*, others are *deflection limited*. Then the criterion for selection changes. Consider, then, the three scenarios created by the three alternative constraints of Table 6.27.

The model

In *load-limited* design the component must carry a specified load or pressure without fracturing. Then the local stress must not exceed that specified by equation (6.41) and, for minimum volume, the best choice of materials are those with high values of

$$\boxed{M_1 = K_c} \tag{6.42}$$

Table 6.27 Design requirements for

Function	Resist brittle fracture
Objective	Minimize volume (mass, cost...)
Constraints	(a) Design load specified or
	(b) Design energy specified or
	(c) Design deflection specified

It is usual to identify K_c with the plane-strain fracture toughness, corresponding to the most highly constrained cracking conditions, because this is conservative. For load-limited design using thin sheet, a plane-stress fracture toughness may be more appropriate; and for multi-layer materials, it may be an interface fracture toughness that matters. The point, though, is clear enough: the best materials for load-limited design are those with large values of appropriate K_c.

But, as we have said, not all design is load limited. Springs, and containment systems for turbines and flywheels are *energy* limited. Take the spring (Figure 6.16) as an example. The elastic energy per unit volume stored in the spring is the integral over the volume of

$$U_e = \frac{1}{2}\sigma\varepsilon = \frac{1}{2}\frac{\sigma^2}{E}$$

The stress is limited by the fracture stress of equation (6.41) so that — if 'failure' means 'fracture' — the maximum energy the spring can store is

$$U_e^{max} = \frac{C^2}{2\pi a_c}\left(\frac{K_{Ic}^2}{E}\right)$$

For a given initial flaw size, energy is maximized by choosing materials with large values of

$$\boxed{M_2 = \frac{K_{Ic}^2}{E} \approx J_c} \tag{6.43}$$

where J_c is the toughness (usual units: kJ/m^2).

There is a third scenario: that of *displacement*-limited design (Figure 6.27). Snap-on bottle tops, snap together fasteners and such like are displacement limited: they must allow sufficient elastic displacement to permit the snap-action without failure, requiring a large failure strain ε_f. The strain is related to the stress by Hooke's law

$$\varepsilon = \frac{\sigma}{E}$$

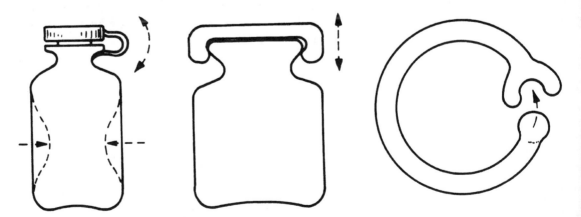

Fig. 6.27 Load and deflection-limited design. Polymers, having low moduli, frequently require deflection-limited design methods.

and the stress is limited by the fracture equation (6.41). Thus the failure strain is

$$\varepsilon_f = \frac{CK_{Ic}}{\sqrt{\pi a_c}E}$$

The best materials for displacement-limited design are those with large values of

$$M_3 = \frac{K_{Ic}}{E}$$

The selection

Figure 6.28 shows a chart of fracture toughness, K_{Ic}, plotted against modulus E. It allows materials to be compared by values of fracture toughness, M_1, by toughness, M_2, and by values of the deflection-limited index M_3. As the engineer's rule-of-thumb demands, almost all metals have values of K_{Ic} which lie above the $15\,\text{MPa}\,\text{m}^{1/2}$ acceptance level for load-limited design. Polymers and ceramics do not.

The line showing M_2 on Figure 6.28 is placed at the value $1\,\text{kJ/m}^2$. Materials with values of M_2 greater than this have a degree of shock-resistance with which engineers feel comfortable (another rule-of-thumb). Metals, composites and some polymers qualify (Table 6.28); ceramics do not. When we come to deflection-limited design, the picture changes again. The line shows the index $M_3 = K_{Ic}/E$ at the value $10^{-3}\,\text{m}^{1/2}$. It illustrates why polymers find such wide application: when the design is deflection limited, polymers — particularly nylons, polycarbonates and polystyrene — are as good as the best metals.

Postscript

The figure gives further insights. The mechanical engineers' love of metals (and, more recently, of composites) is inspired not merely by the appeal of their K_{Ic} values. They are good by all three criteria (K_{Ic}, K_{Ic}^2/E and K_{Ic}/E). Polymers have good values of K_{Ic}/E but not the other two. Ceramics are poor by all three criteria. Herein lie the deeper roots of the engineers' distrust of ceramics.

Further reading

Background in fracture mechanics and safety criteria can be found in these books:

Brock, D. (1984) *Elementary Engineering Fracture Mechanics*, Martinus Nijoff, Boston.
Hellan, K. (1985) *Introduction to Fracture Mechanics*, McGraw-Hill.
Hertzberg, R.W. (1989) *Deformation and Fracture Mechanics of Engineering Materials*, Wiley, New York.

Related case studies

Case Study 6.9: Materials for springs
Case Study 6.10: Elastic hinges and couplings
Case Study 6.15: Safe pressure vessels

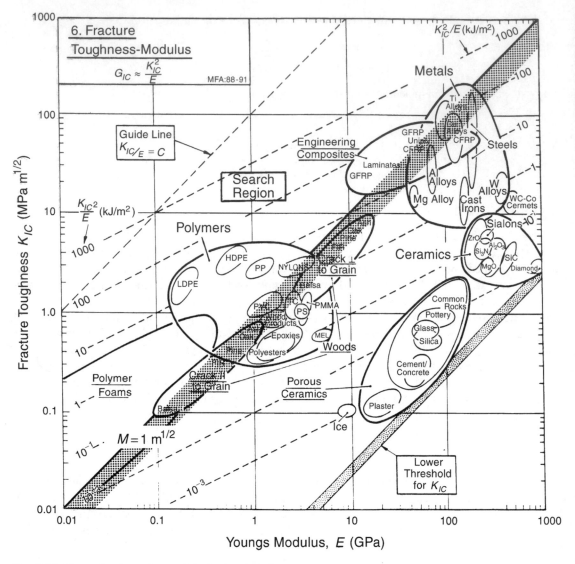

Fig. 6.28 The selection of materials for load, deflection and energy-limited design. In deflection-limited design, polymers are as good as metals, despite having very low values of fracture toughness.

Table 6.28 Materials for fracture-limited design

Design type, and rule-of-thumb	Material
Load-limited design $K_{Ic} > 15\,\mathrm{MPa\,m^{1/2}}$	Metals, polymer-matrix composites.
Energy-limited design $J_c > 1\,\mathrm{kJ/m^2}$	Metals, composites and some polymers.
Displacement-limited design $K_{Ic}/E > 10^{-3}\,\mathrm{m^{1/2}}$	Polymers, elastomers and some metals.

6.15 Safe pressure vessels

Pressure vessels, from the simplest aerosol-can to the biggest boiler, are designed, for safety, to yield or leak before they break. The details of this design method vary. Small pressure vessels are usually designed to allow general yield at a pressure still too low to cause any crack the vessel may contain to propagate ('yield before break'); the distortion caused by yielding is easy to detect and the pressure can be released safely. With large pressure vessels this may not be possible. Instead, safe design is achieved by ensuring that the smallest crack that will propagate unstably has a length greater than the thickness of the vessel wall ('leak before break'); the leak is easily detected, and it releases pressure gradually and thus safely (Table 6.29). The two criteria lead to different material indices. What are they?

The model

The stress in the wall of a thin-walled spherical pressure vessel of radius R (Figure 6.29) is

$$\sigma = \frac{pR}{2t} \tag{6.45}$$

In pressure vessel design, the wall thickness, t, is chosen so that, at the working pressure p, this stress is less than the yield strength, σ_f, of the wall. A small pressure vessel can be examined

Table 6.29 Design requirements for safe pressure vessels

Function	Pressure vessel = contain pressure, p
Objective	Maximum safety
Constraints	(a) Must yield before break or
	(b) Must leak before break
	(c) Wall thickness small to reduce mass and cost

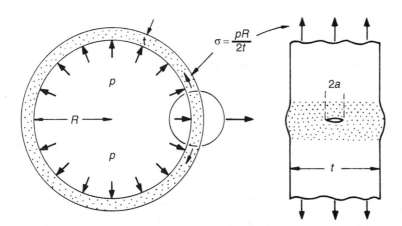

Fig. 6.29 A pressure vessel containing a flaw. Safe design of small pressure vessels requires that they yield before they break; that of large pressure vessels may require, instead, that they leak before they break.

ultrasonically, or by X-ray methods, or proof tested, to establish that it contains no crack or flaw of diameter greater than $2a_c$; then the stress required to make the crack propagate* is

$$\sigma = \frac{CK_{Ic}}{\sqrt{\pi a_c}} \tag{6.46}$$

where C is a constant near unity. Safety can be achieved by ensuring that the working stress is less than this; but greater security is obtained by requiring that the crack will not propagate even if the stress reaches the general yield stress — for then the vessel will deform stably in a way which can be detected. This condition is expressed by setting σ equal to the yield stress, σ_f, giving

$$\pi a_c \leq C^2 \left[\frac{K_{Ic}}{\sigma_f} \right]^2$$

The tolerable crack size is maximized by choosing a material with the largest value of

$$\boxed{M = \frac{K_{Ic}}{\sigma_f}}$$

Large pressure vessels cannot always be X-rayed or sonically tested; and proof testing them may be impractical. Further, cracks can grow slowly because of corrosion or cyclic loading, so that a single examination at the beginning of service life is not sufficient. Then safety can be ensured by arranging that a crack just large enough to penetrate both the inner and the outer surface of the vessel is still stable, because the leak caused by the crack can be detected. This is achieved if the stress is always less than or equal to

$$\sigma = \frac{CK_{Ic}}{\sqrt{\pi t/2}} \tag{6.47}$$

The wall thickness t of the pressure vessel was, of course, designed to contain the pressure p without yielding. From equation (6.45), this means that

$$t \geq \frac{pR}{2\sigma_f} \tag{6.48}$$

Substituting this into the previous equation (with $\sigma = \sigma_f$) gives

$$C^2 \frac{\pi pR}{4} = \left[\frac{K_{Ic}^2}{\sigma_f} \right] \tag{6.49}$$

The maximum pressure is carried most safely by the material with the greatest value of

$$\boxed{M_2 = \frac{K_{Ic}^2}{\sigma_f}}$$

---

* If the wall is sufficiently thin, and close to general yield, it will fail in a plane-stress mode. Then the relevant fracture toughness is that for plane stress, not the smaller value for plane strain.

Both M_1 and M_2 could be made large by making the yield strength of the wall, σ_f, very small: lead, for instance, has high values of both, but you would not choose it for a pressure vessel. That is because the vessel wall must also be as thin as possible, both for economy of material, and to keep it light. The thinnest wall, from equation (6.48), is that with the largest yield strength, σ_f. Thus we wish also to maximize

$$\boxed{M_3 = \sigma_f}$$

narrowing further the choice of material.

The selection

These selection criteria are applied by using the chart shown in Figure 6.30: the fracture toughness, K_{Ic}, plotted against strength σ_f. The three criteria appear as lines of slope 1, 1/2 and as lines that are vertical. Take 'yield before break' as an example. A diagonal line corresponding to $M = K_{Ic}/\sigma_f = C$ links materials with equal performance; those above the line are better. The line shown in the figure at $M_1 = 0.6\,\mathrm{m}^{1/2}$ excludes everything but the toughest steels, copper and aluminium alloys, though some polymers nearly make it (pressurized lemonade and beer containers are made of these polymers). A second selection line at $M_3 = 100\,\mathrm{MPa}$ eliminates aluminium alloys. Details are given in Table 6.30.

Large pressure vessels are always made of steel. Those for models (a model steam engine, for instance) are copper; it is favoured in the small-scale application because of its greater resistance to corrosion. The reader may wish to confirm that the alternative criterion

$$\boxed{M_2 = \frac{K_{Ic}^2}{\sigma_f}}$$

favours steel more strongly, but does not greatly change the conclusions.

Postscript

Boiler failures used to be common place — there are even songs about it. Now they are rare, though when safety margins are pared to a minimum (rockets, new aircraft designs) pressure vessels still

Table 6.30 Materials for safe pressure vessels

Material	$M_1 = \dfrac{K_{Ic}}{\sigma_f}$ $(m^{1/2})$	$M_3 = \sigma_f$ (MPa)	Comment
Tough steels	>0.6	300	These are the pressure-vessel steels, standard in this application.
Tough copper alloys	>0.6	120	OFHC Hard drawn copper.
Tough Al-alloys	>0.6	80	1000 and 3000 series Al-alloys.
Ti-alloys	02	700	High yield but low
High-strength Al-alloys	0.1	500	safety margin. Good for light
GFRP/CFRP	0.1	500	pressure vessels.

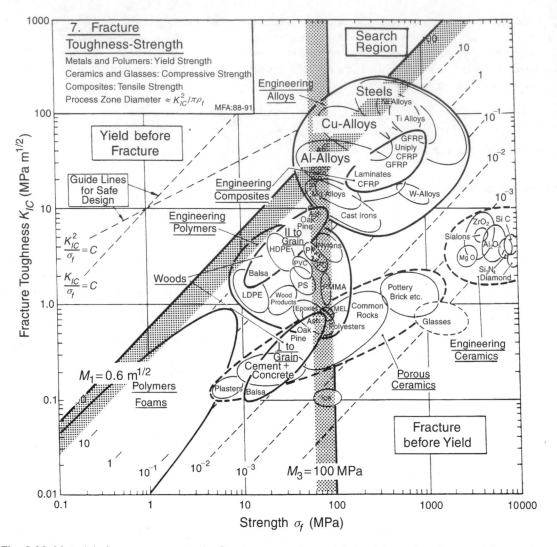

Fig. 6.30 Materials for pressure vessels. Steel, copper alloys and aluminium alloys best satisfy the 'yield before break' criterion. In addition, a high yield strength allows a high working pressure. The materials in the 'search area' triangle are the best choice. The leak-before-break criterion leads to essentially the same selection.

occasionally fail. This (relative) success is one of the major contributions of fracture mechanics to engineering practice.

Further reading

Background in fracture mechanics and safety criteria can be found in these books:

Brock, D. (1984) *Elementary Engineering Fracture Mechanics*, Martinus Nijoff, Boston.
Hellan, K. (1985) *Introduction to Fracture Mechanics*, McGraw-Hill.
Hertzberg, R.W. (1989) *Deformation and Fracture Mechanics of Engineering Materials*, Wiley, New York.

Related case studies

Case Study 6.6:   Materials for flywheels
Case Study 6.14: Deflection-limited design with brittle polymers

6.16 Stiff, high damping materials for shaker tables

Shakers, if you live in Pennsylvania, are the members of an obscure and declining religious sect, noted for their austere wooden furniture. To those who live elsewhere they are devices for vibration testing. This second sort of shaker consists of an electromagnetic actuator driving a table, at frequencies up to 1000 Hz, to which the test-object (a space probe, an automobile, an aircraft component or the like) is clamped (Figure 6.31). The shaker applies a spectrum of vibration frequencies, f, and amplitudes, A, to the test-object to explore its response.

   A big table operating at high frequency dissipates a great deal of power. The primary objective is to minimize this, but subject to a number of constraints itemized in Table 6.31. What materials make good shaker tables?

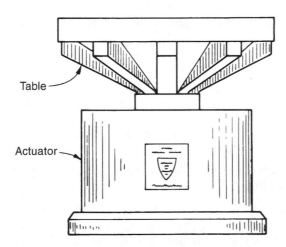

Fig. 6.31 A shaker table. It is required to be stiff, but have high intrinsic 'damping' or loss coefficient.

Table 6.31 Design requirements for shaker tables

Function	Table for vibration tester (shaker table) .
Objective	Minimize power consumption
Constraints	(a) Radius, R, specified
	(b) Must be stiff enough to avoid distortion by clamping forces
	(c) Natural frequencies above maximum operating frequency (to avoid resonance)
	(d) High damping to minimize stray vibrations
	(e) Tough enough to withstand mishandling and shock

The model

The power p (watts) consumed by a dissipative vibrating system with a sinusoidal input is equal to

$$p = C_1 m A^2 \omega^3$$

where m is the mass of the table, A is the amplitude of vibration, ω is the frequency (rad/s) and C_1 is a constant. Provided the operating frequency ω is significantly less than the resonant frequency of the table, then $C_1 \approx 1$. The amplitude A and the frequency ω are prescribed. To minimize the power lost in shaking the table itself, we must minimize its mass m. We idealize the table as a disc of given radius, R. Its thickness, t, is a free variable which we may choose. Its mass is

$$m = \pi R^2 t \rho \tag{6.49}$$

where ρ is the density of the material of which it is made. The thickness influences the bending-stiffness of the table — and this is important both to prevent the table flexing too much under clamping loads, and because it determines its lowest natural vibration frequency. The bending stiffness, S, is

$$S = \frac{C_2 EI}{R^3},$$

where C_2 is a constant. The second moment of the section, I, is proportional to $t^3 R$. Thus, for a given stiffness S and radius R,

$$t = C_3 \left(\frac{SR^2}{E} \right)^{1/3}$$

where C_3 is another constant. Inserting this into equation (6.49) we obtain

$$m = C_3 \pi R^{8/3} S^{1/3} \frac{\rho}{E^{1/3}}$$

The mass of the table, for a given stiffness and minimum vibration frequency, is therefore minimized by selecting materials with high values of

$$\boxed{M_1 = \frac{E^{1/3}}{\rho}}$$

There are three further requirements. The first is that of high mechanical damping η. The second that the fracture toughness K_{Ic} of the table be sufficient to withstand mishandling and clamping forces. And the third is that the material should not cost too much.

The selection

Figure 6.32 shows Chart 8: loss coefficient η plotted against modulus E. The vertical line shows the constraint $E \geq 30\,\text{GPa}$, the horizontal one, the constraint $\eta > 0.01$. The search region contains several suitable materials, notably magnesium, cast iron, various composites and concrete (Table 6.32). Of these, magnesium and composites have high values of $E^{1/3}/\rho$, and both have low densities. Among metals, magnesium is the best choice; otherwise GFRP.

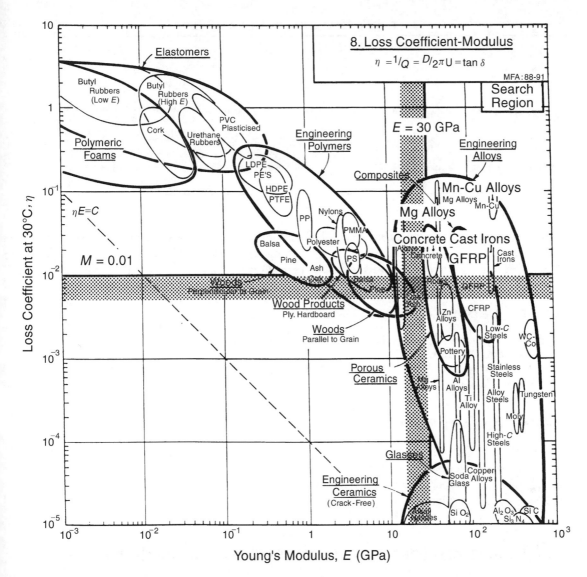

Fig. 6.32 Selection of materials for the shaker table. Magnesium alloys, cast irons, GFRP, concrete and the special high-damping Mn–Cu alloys are candidates.

Postscript

Stiffness, high natural frequencies and damping are qualities often sought in engineering design. The shaker table found its solution (in real life as well as this case study) in the choice of a cast magnesium alloy.

Sometimes, a solution is possible by combining materials. The loss coefficient chart shows that polymers and elastomers have high damping. Sheet steel panels, prone to lightly-damped vibration, can be damped by coating one surface with a polymer, a technique exploited in automobiles, typewriters and machine tools. Aluminium structures can be stiffened (raising natural frequencies) by bonding carbon fibre to them: an approach sometimes used in aircraft design. And structures

Table 6.32 Materials for shaker tables

Material	Loss coeff., η	$M = E^{1/3}/\rho$	ρ $(Mg/m)^3$	Comment
Mg-alloys	$10^{-2}-10^{-1}$		1.75	The best combination of properties.
Mn–Cu alloys	10^{-1}		8.0	Good damping but heavy.
KFRP/GFRP	2×10^{-2}		1.8	Less damping than Mg-alloys, but possible.
Cast irons	2×10^{-2}		7.8	Good damping but heavy.
Concrete	2×10^{-2}		2.5	Less damping than Mg-alloys, but possible for a large table.

loaded in bending or torsion can be made lighter, for the same stiffness (again increasing natural frequencies), by shaping them efficiently: by attaching ribs to their underside, for instance. Shaker tables — even the austere wooden tables of the Pennsylvania Shakers — exploit shape in this way.

Further reading

Tustin, W. and Mercado, R. (1984) *Random Vibrations in Perspective*. Tustin Institute of Technology Inc, Santa Barbara, CA, USA.
Cebon, D. and Ashby, M.F. (1994) Materials selection for precision instruments, *Meas. Sci. and Technol.*, Vol. 5, pp. 296–306.

Related case studies

Case Study 6.4:   Materials for table legs
Case Study 6.9:   Materials for springs
Case Study 6.12: Diaphragms for pressure actuators
Case Study 6.20: Minimizing distortion in precision devices

6.17 Insulation for short-term isothermal containers

Each member of the crew of a military aircraft carries, for emergencies, a radio beacon. If forced to eject, the crew member could find himself in trying circumstances — in water at 4°C, for example (much of the earth's surface is ocean with a mean temperature of roughly this). The beacon guides friendly rescue services, minimizing exposure time.

But microelectronic metabolisms (like those of humans) are upset by low temperatures. In the case of the beacon, it is its transmission frequency which starts to drift. The design specification for the egg-shaped package containing the electronics (Figure 6.33) requires that, when the temperature of the outer surface is changed by 30°C, the temperature of the inner surface should not change significantly for an hour. To keep the device small, the wall thickness is limited to a thickness w of 20 mm. What is the best material for the package? A dewar system is out — it is too fragile.

A foam of some sort, you might think. But here is a case in which intuition leads you astray. So let us formulate the design requirements (Table 6.33) and do the job properly.

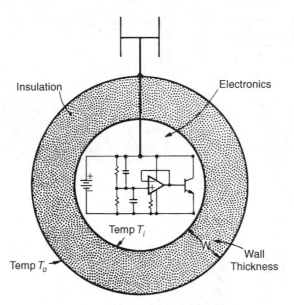

Fig. 6.33 An isothermal container. It is designed to maximize the time before the inside temperature changes after the outside temperature has suddenly changed.

Table 6.33 Design requirements for short-term insulation

Function	Short-term thermal insulation
Objective	Maximize time t before internal temperature of container falls appreciably when external temperature suddenly drops
Constraints	Wall thickness must not exceed w

The model

We model the container as a wall of thickness w, thermal conductivity λ. The heat flux J through the wall, once a steady-state has been established, is

$$J = \lambda \frac{(T_i - T_o)}{w} \tag{6.50}$$

where T_o is the temperature of the outer surface and T_i that of the inner one (Figure 6.33). The only free variable here is the thermal conductivity, λ. The flux is minimized by choosing a wall material with the lowest possible value of λ. Chart 9 (Figure 6.34) shows that this is, indeed, a foam.

But we have answered the wrong question. The design brief was not to minimize the heat flux, but the *time* before the temperature of the inner wall changed appreciably. When the surface temperature of a body is suddenly changed, a temperature wave, so to speak, propagates inwards. The distance x it penetrates in time t is approximately $\sqrt{2at}$. Here a is the thermal diffusivity, defined by $a = \lambda/\rho C_p$, where ρ is the density and C_p is the specific heat (Appendix A: 'Useful Solutions'). Equating this to the wall thickness w gives

$$t \approx \frac{w^2}{2a} \tag{6.51}$$

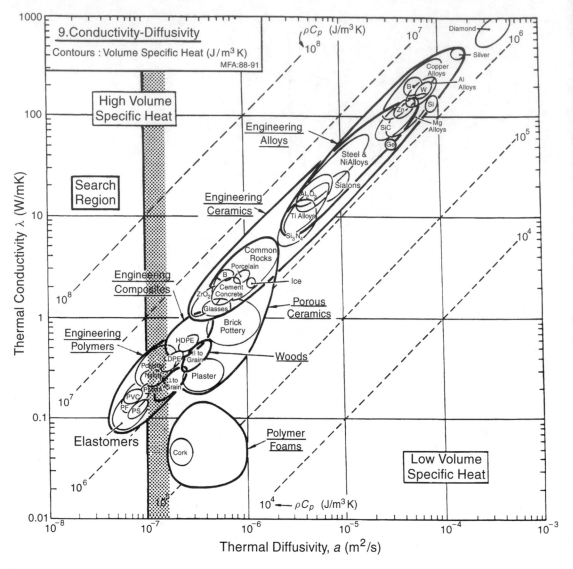

Fig. 6.34 Materials for short-term isothermal containers. Elastomers are good; foams are not.

The time is maximized by choosing the smallest value of the thermal diffusivity, a, not the conductivity λ.

The selection

Chart 9 (Figure 6.34) shows that the thermal diffusivities of foams are not particularly low; it is because they have so little mass, and thus heat capacity. The diffusivity of heat in a solid polymer or elastomer is much lower because they have specific heats which are particularly large. A package made of solid rubber, polystyrene or nylon, would — if of the same thickness — give the beacon a transmission life 10 times greater than one made of (say) a polystyrene foam, although of course

Table 6.34 Materials for short-term thermal insulation

Material	Comment
Elastomers: Butyl rubber (BR), Polychloroprene (CR), and Chlorosulfinated polyethylene (CSM) are examples	Best choice for short-term insulation.
Commodity polymers: Polyethylenes and Polypropylenes	Cheaper than elastomers, but somewhat less good for short-term insulation.
Polymer foams	Much less good than elastomers for short-term insulation; best choice for long-term insulation at steady state.

it would be heavier. The reader can confirm that 22 mm of a solid elastomer ($a = 7 \times 10^{-8}\,\mathrm{m^2/s}$, read from Chart 9) will allow a time interval of 1 hour after an external temperature change before the internal temperature shifts much. Table 6.34 summarizes the results of materials selection.

Postscript

One can do better than this. The trick is to exploit other ways of absorbing heat. If a liquid — a low-melting wax, for instance — can be found that solidifies at a temperature equal to the minimum desired operating temperature for the transmitter (T_i), it can be used as a 'latent-heat sink'. Channels in the package are filled with the liquid; the inner temperature can only fall below the desired operating temperature when all the liquid has solidified. The latent heat of solidification must be supplied to do this, giving the package a large (apparent) specific heat, and thus an exceptionally low diffusivity for heat at the temperature T_i. The same idea is, in reverse, used in 'freezer packs' which solidify when placed in the freezer compartment of a refrigerator and remain cold (by melting, at 4°C) when packed around warm beer cans in a portable cooler.

Further reading

Holman, J.P. (1981) *Heat Transfer*, 5th edition, McGraw-Hill, New York.

Related case studies

Case Study 6.18: Energy-efficient kiln walls
Case Study 6.19: Materials for heat-storing walls

6.18 Energy-efficient kiln walls

The energy cost of one firing cycle of a large pottery kiln (Figure 6.35) is considerable. Part is the cost of the energy which is lost by conduction through the kiln walls; it is reduced by choosing a wall material with a low conductivity, and by making the wall thick. The rest is the cost of the energy used to raise the kiln to its operating temperature; it is reduced by choosing a wall material with a low heat capacity, and by making the wall thin. Is there a material index which captures these apparently conflicting design goals? And if so, what is a good choice of material for kiln walls? The choice is based on the requirements of Table 6.35.

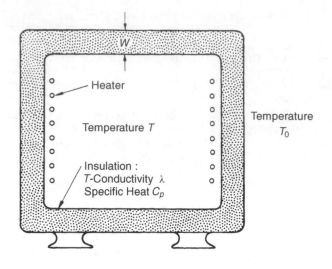

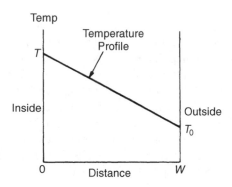

Fig. 6.35 A kiln. On firing, the kiln wall is first heated to the operating temperature, then held at this temperature. A linear gradient is then expected through the kiln wall.

Table 6.35 Design requirements for kiln walls

Function	Thermal insulation for kiln (cyclic heating and cooling)
Objective	Minimized energy consumed in firing cycle
Constraints	(a) Maximum operating temperature 1000 K
	(b) Possible limit on kiln-wall thickness for space reasons

The model

When a kiln is fired, the temperature rises quickly from ambient, T_o, to the operating temperature, T_i, where it is held for the firing time t. The energy consumed in the firing time has, as we have said, two contributions. The first is the heat conducted out: at steady state the heat loss by conduction, Q_1, per unit area, is given by the first law of heat flow. If held for time t it is

$$Q_1 = -\lambda \frac{dT}{dx} t = \lambda \frac{(T_i - T_o)}{w} t \tag{6.52}$$

Here λ is the thermal conductivity, dT/dx is the temperature gradient and w is the insulation wall-thickness. The second contribution is the heat absorbed by the kiln wall in raising it to T_i, and this can be considerable. Per unit area, it is

$$Q_2 = C_p \rho w \left(\frac{T_i - T_o}{2} \right) \tag{6.53}$$

where C_p is the specific heat of the wall material and ρ is its density. The total energy consumed per unit area is the sum of these two:

$$Q = Q_1 + Q_2 = \frac{\lambda(T_i - T_o)t}{w} + \frac{C_p \rho w(T_i - T_o)}{2} \tag{6.54}$$

A wall which is too thin loses much energy by conduction, but absorbs little energy in heating the wall itself. One which is too thick does the opposite. There is an optimum thickness, which we find by differentiating equation (6.54) with respect to wall thickness w and equating the result to zero, giving:

$$w = \left(\frac{2\lambda t}{C_p \rho} \right)^{1/2} = (2at)^{1/2} \tag{6.55}$$

where $a = \lambda/C_p\rho$ is the thermal diffusivity. The quantity $(2at)^{1/2}$ has dimensions of length and is a measure of the distance heat can diffuse in time t. Equation (6.55) says that the most energy-efficient kiln wall is one that only starts to get really hot on the outside as the firing cycle approaches completion. Substituting equation (6.55) back into equation (6.54) to eliminate w gives:

$$Q = (T_i - T_o)(2t)^{1/2}(\lambda C_p \rho)^{1/2}$$

Q is minimized by choosing a material with a low value of the quantity $(\lambda C_p\rho)^{1/2}$, that is, by maximizing

$$\boxed{M_1 = (\lambda C_p \rho)^{-1/2} = \frac{a^{1/2}}{\lambda}} \tag{6.56}$$

But, by eliminating the wall thickness w we have lost track of it. It could, for some materials, be excessively large. We must limit it. A given firing time, t, and wall thickness, w, defines, via equation (6.55), an upper limit for the thermal diffusivity, a:

$$a \leq \frac{w^2}{2t}$$

Selecting materials which maximize equation (6.56) with the constraint on a defined by the last equation minimizes the energy consumed per firing cycle.

The selection

Figure 6.36 shows the λ–a chart with a selection line corresponding to $M = a^{1/2}/\lambda$ plotted on it. Polymer foams, cork and solid polymers are good, but only if the internal temperature is less than $100°C$. Real kilns operate near $1000°C$. Porous ceramics are the obvious choice (Table 6.36). Having chosen a material, the acceptable wall thickness is calculated from equation (6.55). It is listed, for a firing time of 3 hours (approximately 10^4 seconds) in Table 6.35.

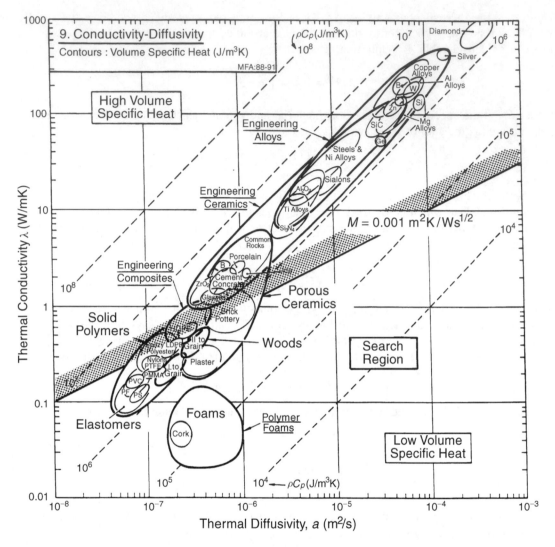

Fig. 6.36 Materials for kiln walls. Low density, porous or foam-like ceramics are the best choice.

Postscript

It is not generally appreciated that, in an efficiently-designed kiln, as much energy goes in heating up the kiln itself as is lost by thermal conduction to the outside environment. It is a mistake to make kiln walls too thick; a little is saved in reduced conduction-loss, but more is lost in the greater heat capacity of the kiln itself.

That, too is the reason that foams are good: they have a low thermal conductivity *and* a low heat capacity. Centrally heated houses in which the heat is turned off at night suffer a cycle like that of the kiln. Here (because T_{max} is lower) the best choice is a polymeric foam, cork or fibreglass (which has thermal properties like those of foams). But as this case study shows — turning the heat off at night doesn't save you as much as you think, because you have to supply the heat capacity of the walls in the morning.

Table 6.36 Materials for energy-efficient kilns

Material	$M = a^{1/2}/\lambda$ $(m^2 K/W s^{1/2})$	Thickness, t (m)	Comment
Porous ceramics	$3 \times 10^{-4} - 3 \times 10^{-3}$	0.1	The obvious choice: the lower the density, the better the performance.
Solid elastomers	$10^{-3} - 3 \times 10^{-3}$	0.05	Good values of material index. Useful if the wall must be very thin.
Solid polymers	10^{-3}		Limited to temperatures below 150°C.
Polymer foam, Cork[*]	$3 \times 10^{-3} - 3 \times 10^{-2}$	0.09	The highest value of M — hence their use in house insulation. Limited to temperatures below 150°C.
Woods	3×10^{-3}	0.07	The boiler of Stevenson's 'Rocket' was insulated with wood.
Fibreglass	10^{-2}	0.1	Thermal properties comparable with polymer foams; usable to 200°C.

Further reading

Holman, J.P. (1981) *Heat Transfer* 5th edition, McGraw-Hill, New York.

Related case studies

Case Study 6.17: Insulation for short-term isothermal containers
Case Study 6.19: Materials for passive solar heating

6.19 Materials for passive solar heating

There are a number of schemes for capturing solar energy for home heating: solar cells, liquid filled heat exchangers, and solid heat reservoirs. The simplest of these is the heat-storing wall: a thick wall, the outer surface of which is heated by exposure to direct sunshine during the day, and from which heat is extracted at night by blowing air over its inner surface (Figure 6.37). An essential of such a scheme is that the time-constant for heat flow through the wall be about 12 hours; then the wall first warms on the inner surface roughly 12 hours after the sun first warms the outer one, giving out at night what it took in during the day. We will suppose that, for architectural reasons, the wall must not be more than 0.5 m thick. What materials maximize the thermal energy captured by the wall while retaining a heat-diffusion time of up to 12 hours? Table 6.37 summarizes the requirements.

The model

The heat content, Q, per unit area of wall, when heated through a temperature interval ΔT gives the objective function

$$Q = w\rho C_p \Delta T \tag{6.57}$$

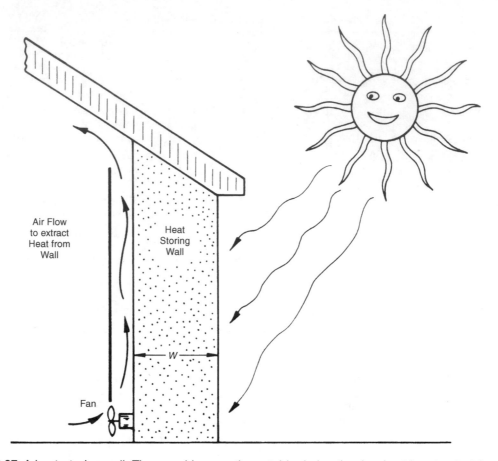

Fig. 6.37 A heat-storing wall. The sun shines on the outside during the day; heat is extracted from the inside at night. The heat diffusion-time through the wall must be about 12 hours.

Table 6.37 Design requirements for passive solar heating

Function	Heat-storing medium
Objective	Maximize thermal energy stored per unit material cost
Constraints	(a) Heat diffusion time through wall $t \approx 12$ hours
	(b) Wall thickness ≤ 0.5 m
	(c) Adequate working temperature $T_{max} > 100°C$

where w is the wall thickness, and ρC_p is the volumetric specific heat (the density ρ times the specific heat C_p). The 12-hour time constant is a constraint. It is adequately estimated by the approximation (see Appendix A, 'Useful Solutions')

$$w = \sqrt{2at} \tag{6.58}$$

where a is the thermal diffusivity and t the time. Eliminating the free variable w gives

$$Q = \sqrt{2t}\Delta T a^{1/2} \rho C_p \tag{6.59}$$

or, using the fact that $a = \lambda/\rho C_p$ where λ is the thermal conductivity,

$$Q = \sqrt{2t}\Delta T\lambda/a^{1/2}$$

The heat capacity of the wall is maximized by choosing material with a high value of

$$M_1 = \frac{\lambda}{a^{1/2}} \qquad (6.60)$$

— it is the inverse of the index of Case Study 6.17. The restriction on thickness w requires (from equation 6.58) that

$$a \le \frac{w^2}{2t}$$

with $w \le 0.5\,\text{m}$ and $t = 12\,\text{hours}$ $(4 \times 10^4\,\text{s})$, we obtain a material limit

$$M_2 = a \le 3 \times 10^{-6}\,\text{m}^2/\text{s}$$

The selection

Figure 6.38 shows Chart 9 (thermal conductivity plotted against thermal diffusivity) with M_1 and M_2 plotted on it. It identifies the group of materials, listed in Table 6.38: they maximize M_1 while meeting the constraint expressed by M_2. Solids are good; porous materials and foams (often used in walls) are not.

Postscript

All this is fine, but what of cost? If this scheme is to be used for housing, cost is an important consideration. The relative costs per unit volume, read from Chart 14 (Figure 4.15), are listed in Table 6.38 — it points to the selection of cement, concrete and brick.

Table 6.38 Materials for passive solar heat storage

Material	$M_1 = \lambda/a^{1/2}$ $(W\,s^{1/2}/m^2 K)$	Relative Cost (Mg/m^3)	Comment
Cement		0.5	The right choice
Concrete	3×10^{-3}	0.35	depending on availability
Common rocks		1.0	and cost.
Glass	3×10^3	10	Good M_1; transmits visible radiation.
Brick	10^3	0.8	Less good than concrete.
HDPE	10^3	3	Too expensive.
Ice	3×10^3	0.1	Attractive value of M; pity it melts at 0°C.

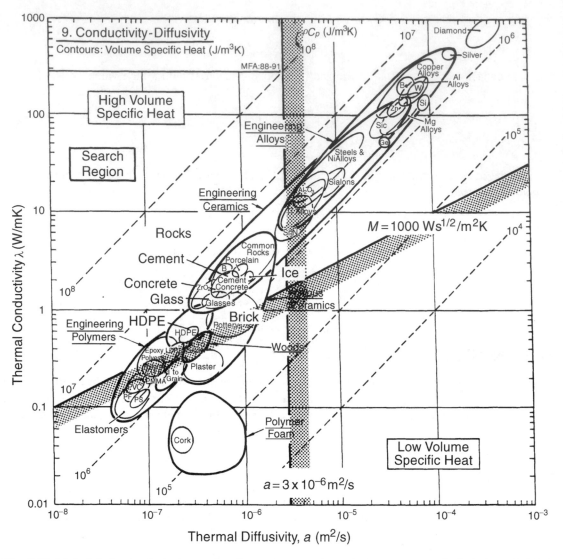

Fig. 6.38 Materials for heat-storing walls. Cement, concrete and stone are practical choices; brick is less good.

If minimizing cost, rather than maximizing Q, were the primary design goal, the model changes. The cost per unit area, C, of the wall is

$$C = w\rho C_m$$

where C_m is the cost per kg of the wall material. The requirement of the 12-hour time-constant remains the same as before (equation (6.58)). Eliminating w gives

$$C = (t)^{1/2}(a^{1/2}\rho C_m)$$

We now wish to maximize

$$M_3 = (a^{1/2}\rho C_m)^{-1} \tag{6.61}$$

This is a new index, one not contained in Figure 6.38, and there is no chart for making the selection. Software, described in Chapter 5, allows a chart to be constructed for use with any material index. Running this software identifies cement, concrete and ice as the cheapest candidates.

Ice appears in both selections. Here is an example of a forgotten constraint. If a material is to be used in a given temperature range, its maximum use temperature, T_{max}, must lie above it. Restricting the selection to materials with $T_{max} > 100°C$ eliminates ice.

Related case studies

Case Study 6.17: Insulation for short-term isothermal containers
Case Study 6.18: Energy-efficient kiln walls

6.20 Materials to minimize thermal distortion in precision devices

The precision of a measuring device, like a sub-micrometer displacement gauge, is limited by its stiffness and by the dimensional change caused by temperature gradients. Compensation for elastic deflection can be arranged; and corrections to cope with thermal expansion are possible too — provided the device is at a uniform temperature. *Thermal gradients* are the real problem: they cause a change of shape — that is, a distortion of the device — for which compensation is not possible. Sensitivity to *vibration* is also a problem: natural excitation introduces noise and thus imprecision into the measurement. So it is permissible to allow expansion in precision instrument design, provided distortion does not occur (Chetwynd, 1987). Elastic deflection is allowed, provided natural vibration frequencies are high.

What, then, are good materials for precision devices? Table 6.39 lists the requirements.

The model

Figure 6.39 shows, schematically, such a device: it consists of a force loop, an actuator and a sensor. We aim to choose a material for the force loop. It will, in general, support heat sources: the fingers of the operator of the device in the figure, or, more usually, electrical components which generate heat. The relevant material index is found by considering the simple case of one-dimensional heat flow through a rod insulated except at its ends, one of which is at ambient and the other connected

Table 6.39 Design requirements for precision devices

Function	Force loop (frame) for precision device
Objective	Maximize positional accuracy (minimize distortion)
Constraints	(a) Must tolerate heat flux
	(b) Must tolerate vibration

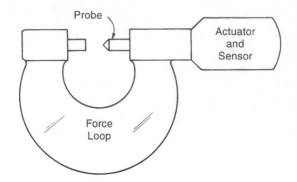

Fig. 6.39 A schematic of a precision measuring device. Super-accurate dimension-sensing devices include the atomic-force microscope and the scanning tunnelling microscope.

to the heat source. In the steady state, Fourier's law is

$$q = -\lambda \frac{dT}{dx} \qquad (6.67)$$

where q is heat input per unit area, λ is the thermal conductivity and dT/dx is the resulting temperature gradient. The strain is related to temperature by

$$\varepsilon = \alpha(T_o - T) \qquad (6.68)$$

where α is the thermal conductivity and T_o is ambient temperature. The distortion is proportional to the gradient of the strain:

$$\frac{d\varepsilon}{dx} = \frac{\alpha \, dT}{dx} = \left(\frac{\alpha}{\lambda}\right) q \qquad (6.69)$$

Thus for a given geometry and heat flow, the distortion $d\varepsilon/dx$ is minimized by selecting materials with large values of the index

$$\boxed{M_1 = \frac{\lambda}{\alpha}}$$

The other problem is vibration. The sensitivity to external excitation is minimized by making the natural frequencies of the device as high as possible. The flexural vibrations have the lowest frequencies; they are proportional to

$$\boxed{M_2 = \frac{E^{1/2}}{\rho}}$$

A high value of this index will minimize the problem. Finally, of course, the device must not cost too much.

The selection

Chart 10 (Figure 6.40) shows the expansion coefficient, α, plotted against the thermal conductivity, λ. Contours show constant values of the quantity λ/α. A search region is isolated by the line $\lambda/\alpha = 10^7$ W/m, giving the shortlist of Table 6.40. Values of $M_2 = E^{1/2}/\rho$ read from Chart 1 (Figure 4.2) are included in the table. Diamond is outstanding, but practical only for very small devices. The metals, except for beryllium, are disadvantaged by having high densities and thus poor values of M_2. The best choice is silicon, available in large sections, with high purity. Silicon carbide is an alternative.

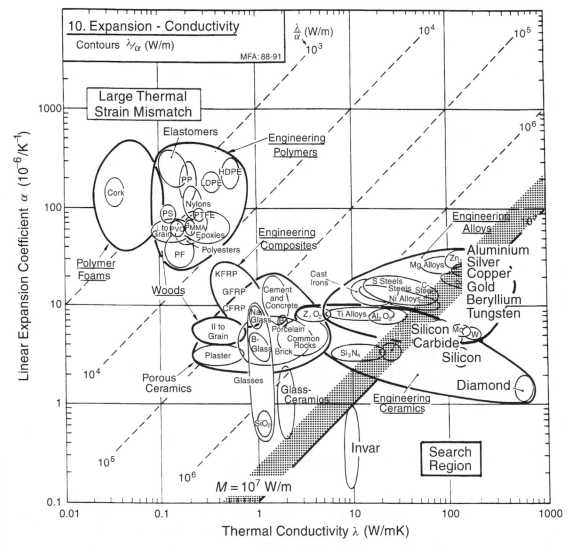

Fig. 6.40 Materials for precision measuring devices. Metals are less good than ceramics because they have lower vibration frequencies. Silicon may be the best choice.

Table 6.40 Materials to minimize thermal distortion

Material	$M_1 = \lambda/\alpha$ (W/m)	$M_2 = E^{1/2}/\rho$ ($GPa^{1/2}/(Mg/m^3)$)	Comment
Diamond	5×10^8	8.6	Outstanding M_1 and M_2; expensive.
Silicon	4×10^7	6.0	Excellent M_1 and M_2; cheap.
Silicon carbide	2×10^7	6.2	Excellent M_1 and M_2; potentially cheap.
Beryllium	10^7	9	Less good than silicon or SiC.
Aluminium	10^7	3.1	Poor M_1, but very cheap.
Silver	2×10^7	1.0	High density
Copper	2×10^7	1.3	gives poor
Gold	2×10^7	0.6	value of M_2.
Tungsten	3×10^7	1.1	Better than copper, silver or
Molybdenum	2×10^7	1.3	gold, but less good than
Invar	3×10^7	1.4	silicon, SiC, diamond.

Postscript

Nano-scale measuring and imaging systems present the problem analysed here. The atomic-force microscope and the scanning-tunnelling microscope both support a probe on a force loop, typically with a piezo-electric actuator and electronics to sense the proximity of the probe to the test surface. Closer to home, the mechanism of a video recorder and that of a hard disk drive qualify as precision instruments; both have an actuator moving a sensor (the read head) attached, with associated electronics, to a force loop. The materials identified in this case study are the best choice for force loop.

Further reading

Chetwynd, D.G. (1987) *Precision Engineering*, **9**(1), 3.
Cebon, D. and Ashby, M.F. (1994) *Meas. Sci. and Technol.*, **5**, 296.

Related case studies

Case Study 6.3:   Mirrors for large telescopes
Case Study 6.17: Insulation for short-term isothermal containers
Case Study 6.21: Ceramic valves for taps

6.21 Ceramic valves for taps

Few things are more irritating than a dripping tap. Taps drip because the rubber washer is worn, or the brass seat is pitted by corrosion, or both. Could an alternative choice of materials overcome the problem? Ceramics wear well, and they have excellent corrosion resistance in both pure and salt water. How about a tap with a ceramic valve and seat?

Figure 6.41 shows a possible arrangement. Two identical ceramic discs are mounted one above the other, spring-loaded so that their faces, polished to a tolerance of 0.5 µm, are in contact. The

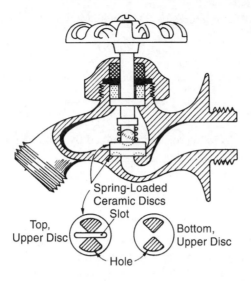

Fig. 6.41 A design for a ceramic valve: two ceramic discs, spring loaded, have holes which align when the tap is turned on.

outer face of each has a slot which registers it, and allows the upper disc to be rotated through 90° (1/4 turn). In the 'off' position the holes in the upper disc are blanked off by the solid part of the lower one; in the 'on' position the holes are aligned. Normal working loads should give negligible wear in the expected lifetime of the tap. Taps with vitreous alumina valves are now available. The manufacturers claim that they do not need any servicing and that neither sediment nor hard water can damage them.

But do they live up to expectation? As cold-water taps they perform well. But as hot-water taps, there is a problem: the discs sometimes crack. The cracking appears to be caused by thermal shock or by thermal mismatch between disc and tap body when the local temperature suddenly changes (as it does when the tap is turned on). Would another ceramic be better? Table 6.41 lists the requirements.

The model

When the water flowing over the ceramic disc suddenly changes in temperature (as it does when you run the tap) the surface temperature of the disc changes suddenly by ΔT. The thermal strain of the surface is proportional to $\alpha \Delta T$ where α is the linear expansion coefficient; the constraint

Table 6.41 Design requirements for ceramic valves for taps

Function	Ceramic valve
Objective	Maximize life
Constraints	(a) Must withstand thermal shock
	(b) High hardness to resist wear
	(c) No corrosion in tap water

exerted by the interior of the disc generates a thermal stress

$$\sigma \approx E\alpha\Delta T \qquad (6.72)$$

If this exceeds the tensile strength of the ceramic, fracture will result. We require, for damage-free operation, that

$$\sigma \leq \sigma_t$$

The safe temperature interval ΔT is therefore maximized by choosing materials with large values of

$$M_1 = \frac{\sigma_t}{E\alpha}$$

This self-induced stress is one possible origin for valve failures. Another is the expansion mismatch between the valve and the metal components with which it mates. The model for this is almost the same; it is simply necessary to replace the thermal expansion coefficient of the ceramic, α, by the difference, $\Delta\alpha$, between the ceramic and the metal.

The selection

The thermal shock resistance of materials is summarized by Chart 12, reproduced as Figure 6.42. From it we see that alumina ceramics (particularly those containing a high proportion of glassy phases) have poor thermal shock resistance: a sudden temperature change of 80°C can crack them, and mechanical loading makes this worse.

   The answer is to select a ceramic with a greater resistance to thermal shock. Almost any engineering ceramic is better — notably zirconia, silicon nitride, silicon carbide or sialon (Table 6.42).

Postscript

So ceramic valves for taps appear to be viable. The gain is in service life: the superior wear and corrosion resistance of the ceramic reduce both to a negligible level. But the use of ceramics and metals together raises problems of matching which require careful redesign, and informed material selection procedures.

Related case studies

Case Study 6.20: Minimizing distortion in precision devices

Table 6.42 Materials for ceramic valves

Material	Comment
Aluminas, Al_2O_3 with glass	Cheap, but poor thermal shock resistance.
Zirconia, ZrO_2 Silicon carbides, SiC Silicon nitrides, Si_3N_4 Sialons Mullites	All are hard, corrosion resistant in water and most aqueous solutions, and have better thermal shock resistance than aluminas.

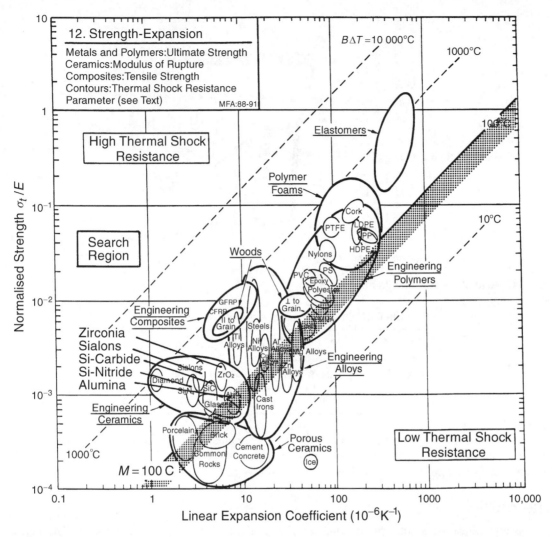

Fig. 6.42 The selection of a material for the ceramic valve of a tap. A ceramic with good thermal shock resistance is desirable.

6.22 Nylon bearings for ships' rudders

Rudder bearings of ships (Figure 6.43) operate under the most unpleasant conditions. The sliding speed is low, but the bearing pressure is high and adequate lubrication is often difficult to maintain. The rudder lies in the wake of the propeller, which generates severe vibration and consequent fretting. Sand and wear debris tend to get trapped between the bearing surfaces. Add to this the environment — aerated salt water — and you can see that bearing design is something of a challenge (Table 6.43).

Ship bearings are traditionally made of bronze. The wear resistance of bronzes is good, and the maximum bearing pressure (important here) is high. But, in sea water, galvanic cells are set up

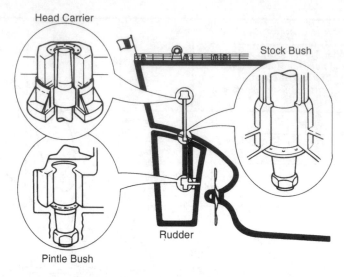

Fig. 6.43 A ship's rudder and its bearings.

Table 6.43 Design requirements for rudder bearings

Function	Sliding bearing
Objective	Maximize life
Constraints	(a) Wear resistant with water lubrication
	(b) Resist corrosion in sea water
	(c) High damping desirable

between the bronze and any other metal to which it is attached by a conducting path (no matter how remote), and in a ship such connections are inevitable. So galvanic corrosion, as well as abrasion by sand, is a problem. Is there a better choice than bronze?

The model

We assume (reasonably) that the bearing *force* F is fixed by the design of the ship. The bearing *pressure*, P, can be controlled by changing the area A of the bearing surface:

$$P \propto \frac{F}{A}$$

This means that we are free to choose a material with a lower maximum bearing pressure provided the length of the bearing itself is increased to compensate. With this thought in mind, we seek a bearing material which will not corrode in salt water and can function without full lubrication.

The selection

Figure 6.44 shows Chart 16, the chart of wear-rate constant, k_a, and hardness, H. The wear-rate, W, is given by equation (4.29), which, repeated, is

$$\Omega = k_a P = C \left(\frac{P}{P_{\max}} \right) k_a H$$

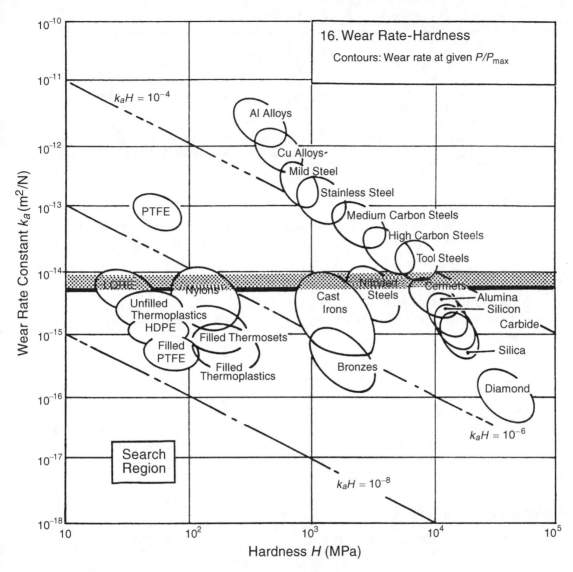

Fig. 6.44 Materials for rudder bearings. Wear is very complex, so the chart gives qualitative guidance only. It suggests that polymers such as nylon or filled or reinforced polymers might be an alternative to bronze provided the bearing area is increased appropriately.

where C is a constant, P is the bearing pressure, P_{max} the maximum allowable bearing pressure for the material, and H is its hardness. If the bearing is not re-sized when a new material is used, the bearing pressure P is unchanged and the material with the lowest wear-rate is simply that with the smallest value of k_a. Bronze performs well, but filled thermoplastics are nearly as good and have superior corrosion resistance in salt water. If, on the other hand, the bearing is re-sized so that it operates at a set fraction of P_{max} (0.5, say), the material with the lowest wear-rate is that with the smallest value of k_aH. Here polymers are clearly superior. Table 6.44 summarizes the conclusions.

Table 6.44 Materials for rudder bearings

Material	Comment
PTFE, polyethylenes polypropylenes	Low friction and good wear resistance at low bearing pressures.
Glass-reinforced PTFE, polyethylenes and polypropylenes	Excellent wear and corrosion resistance in sea water. A viable alternative to bronze if bearing pressures are not too large.
Silica, alumina, magnesia	Good wear and corrosion resistance but poor impact properties and very low damping.

Postscript

Recently, at least one manufacturer of marine bearings has started to supply cast nylon 6 bearings for large ship rudders. The makers claim just the advantages we would expect from this case study:

(a) wear and abrasion resistance with water lubrication is improved;
(b) deliberate lubrication is unnecessary;
(c) corrosion resistance is excellent;
(d) the elastic and damping properties of nylon 6 protect the rudder from shocks (see Chart 7: Damping/modulus);
(e) there is no fretting.

Further, the material is easy to handle and install, and is inexpensive to machine.

Figure 6.44 suggests that a filled polymer or composite might be even better. Carbon-fibre filled nylon has better wear resistance than straight nylon, but it is less tough and flexible, and it does not damp vibration as effectively. As in all such problems, the best material is the one which comes closest to meeting *all* the demands made on it, not just the primary design criterion (in this case, wear resistance). The suggestion of the chart is a useful one, worth a try. It would take sea-tests to tell whether it should be adopted.

Related case studies

Case Study 6.21: Ceramic valves for taps

6.23 Summary and conclusions

The case studies of this chapter illustrate how the choice of material is narrowed from the initial, broad, menu to a small subset which can be tried, tested, and examined further. Most designs make certain non-negotiable demands on a material: it must withstand a temperature greater than T, it must resist corrosive fluid F, and so forth. These constraints narrow the choice to a few broad classes of material. The choice is narrowed further by seeking the combination of properties which maximize performance (combinations like $E^{1/2}/\rho$) or maximize safety (combinations like K_{Ic}/σ_f). These, plus economics, isolate a small subset of materials for further consideration.

The final choice between these will depend on more detailed information on their properties, considerations of manufacture, economics and aesthetics. These are discussed in the chapters which follow.

6.24 Further reading

Compilations of case studies starting with the full materials menu

A large compilation of case studies, including many of those given here but with more sophisticated, computer-based selections, is to be found in

Ashby, M.F. and Cebon, D. (1996) *Case Studies in Materials Selection*, published by Granta Design, Trumpington Mews, 40B High Street, Trumpington CB2 2LS, UK.

General texts

The texts listed below give detailed case studies of materials selection. They generally assume that a shortlist of candidates is already known and argue their relative merits, rather than starting with a clean slate, as we do here.

Charles, J.A., Crane, F.A.A. and Furness J.A.G. (1987) *Selection and Use of Engineering Materials*, 3rd edition, Butterworth-Heinemann, Oxford.
Dieter, G.E. (1991) *Engineering Design, A Materials and Processing Approach*, 2nd edition, McGraw-Hill, New York.
Lewis, G. (1990) *Selection of Engineering Materials*, Prentice-Hall, Englewood Cliffs, NJ.

Chapter 7

Selection of material and shape

7.1 Introduction and synopsis

Shaped sections carry bending, torsional and axial-compressive loads more efficiently than solid sections do. By 'shaped' we mean that the cross-section is formed to a tube, a box-section, an I-section or the like. By 'efficient' we mean that, for given loading conditions, the section uses as little material, and is therefore as light, as possible. Tubes, boxes and I-sections will be referred to as 'simple shapes'. Even greater efficiencies are possible with sandwich panels (thin load-bearing skins bonded to a foam or honeycomb interior) and with structures (the Warren truss, for instance).

This chapter extends the concept of indices so as to include shape (Figure 7.1). Often it is not necessary to do so: in the case studies of Chapter 6, shape either did not enter at all, or, when it did, it was not a variable (that is, we compared materials with the same shape). But when two materials are available with different section shapes and the design is one in which shape matters (a beam in bending, for example), the more general problem arises: how to choose, from among the vast range of materials and the section shapes in which they are available — or could, potentially, be made — the one which maximizes the performance. Take the example of a bicycle: its forks are loaded in bending. It could, say, be made of steel or of wood — early bikes *were* made of wood. But steel is available as thin-walled tube, whereas the wood is not; wood, usually, has a solid section. A solid wood bicycle is certainly lighter and stiffer than a solid steel one, but is it better than one made of steel tubing? Might a magnesium I-section be better still? What about a webbed polymer moulding? How, in short, is one to choose the best combination of material and shape?

A procedure for answering these and related questions is outlined in this chapter. It involves the definition of *shape factors*: simple numbers which characterize the efficiency of shaped sections. These allow the definition of material indices which are closely related to those of Chapter 5, but which now include shape. When shape is constant, the indices reduce exactly to those of Chapter 5; but when shape is a variable, the shape factor appears in the expressions for the indices.

The ideas in this chapter are a little more difficult than those of Chapter 5; their importance lies in the connection they make between materials selection and the designs of load-bearing structures. A feel for the method can be had by reading the following section and the final section alone; these, plus the results listed in Tables 7.1 and 7.2, should be enough to allow the case studies of Chapter 8 (which apply the method) to be understood. The reader who wishes to grasp how the results arise will have to read the whole thing.

7.2 Shape factors

As explained in Chapter 5, the loading on a component is generally axial, bending or torsional: *ties* carry tensile loads; *beams* carry bending moments; *shafts* carry torques; *columns* carry compressive

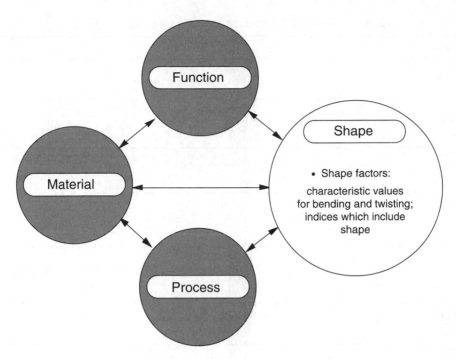

Fig. 7.1 Section shape is important for certain modes of loading. When shape is a variable a new term, the shape factor, appears in some of the material indices: they then allow optimum selection of material and shape.

axial loads. Figure 7.2 shows these modes of loading, applied to shapes that resist them well. The point it makes is that the best material-and-shape combination depends on the mode of loading. In what follows, we separate the modes, dealing with each separately.

In axial tension, the area of the cross-section is important but its shape is not: all sections with the same area will carry the same load. Not so in bending: beams with hollow-box or I-sections are better than solid sections of the same cross-sectional area. Torsion too, has its 'best' shapes: circular tubes, for instance, are better than either solid sections or I-sections. To deal with this, we define a *shape factor* (symbol ϕ) which measures, for each mode of loading, the efficiency of a shaped section. We need four of them, which we now define.

A *material* can be thought of as having properties but no shape; a *component* or a *structure* is a material made into a shape (Figure 7.3). A *shape factor* is a dimensionless number which characterizes the efficiency of the shape, regardless of its scale, in a given mode of loading. Thus there is a shape factor, ϕ_B^e, for elastic bending of beams, and another, ϕ_T^e, for elastic twisting of shafts (the superscript e means elastic). These are the appropriate shape factors when design is based on stiffness; when, instead, it is based on strength (that is, on the first onset of plastic yielding or on fracture) two more shape factors are needed: ϕ_B^f and ϕ_T^f (the superscript f meaning failure). All four shape factors are defined so that they are equal to 1 for a solid bar with a circular cross-section.

Elastic extension (Figure 7.2(a))

The elastic extension or shortening of a tie or strut under a given load (Figure 7.2(a)) depends on the area A of its section, but not on its shape. No shape factor is needed.

Table 7.1 Moments of areas of sections for common shapes

Section Shape	$A\,(m^2)$	$I_{xx}\,(m^4)$	$K\,(m^4)$	$Z\,(m^3)$	$Q\,(m^3)$
circle, radius r	πr^2	$\dfrac{\pi}{4}r^4$	$\dfrac{\pi}{2}r^4$	$\dfrac{\pi}{4}r^3$	$\dfrac{\pi}{2}r^3$
square, side b	b^2	$\dfrac{b^4}{12}$	$0.14b^4$	$\dfrac{b^3}{6}$	$0.21b^3$
ellipse $2a \times 2b$	πab	$\dfrac{\pi}{4}a^3b$	$\dfrac{\pi a^3 b^3}{(a^2+b^2)}$	$\dfrac{\pi}{4}a^2b$	$\dfrac{\pi a^2 b}{2}$ $(a<b)$
rectangle $b \times h$	bh	$\dfrac{bh^3}{12}$	$\dfrac{b^3h}{3}\left(1-0.58\dfrac{b}{h}\right)$ $(h>b)$	$\dfrac{bh^2}{6}$	$\dfrac{b^2h^2}{3h+1.8b}$ $(h>b)$
triangle, side a	$\dfrac{\sqrt{3}}{4}a^2$	$\dfrac{a^4}{32\sqrt{3}}$	$\dfrac{a^4\sqrt{3}}{80}$	$\dfrac{a^3}{32}$	$\dfrac{a^3}{20}$
tube, radii $r_o,\,r_i$	$\pi(r_o^2-r_i^2)$ $\approx 2\pi rt$	$\dfrac{\pi}{4}(r_o^4-r_i^4)$ $\approx \pi r^3 t$	$\dfrac{\pi}{2}(r_o^4-r_i^4)$ $\approx 2\pi r^3 t$	$\dfrac{\pi}{4r_o}(r_o^4-r_i^4)$ $\approx \pi r^2 t$	$\dfrac{\pi}{2r_o}(r_o^4-r_i^4)$ $\approx 2\pi r^2 t$
square tube	$4bt$	$\dfrac{2}{3}b^3t$	$b^3t\left(1-\dfrac{t}{b}\right)^4$	$\dfrac{4}{3}b^2t$	$2b^2t\left(1-\dfrac{t}{b}\right)^2$
thin ellipse $2a\times 2b$	$\pi(a+b)t$	$\dfrac{\pi}{4}a^3t\left(1+\dfrac{3b}{a}\right)$	$\dfrac{4\pi(ab)^{5/2}t}{(a^2+b^2)}$	$\dfrac{\pi a^2 t}{4}\left(1+\dfrac{3b}{a}\right)$	$2\pi t(a^3b)^{1/2}$ $(b>a)$
I / channel (hollow)	$b(h_o-h_i)$ $\approx 2bt$	$\dfrac{b}{12}(h_o^3-h_i^3)$ $\approx \dfrac{1}{2}bth_o^2$	—	$\dfrac{b}{6h_o}(h_o^3-h_i^3)$ $\approx bth_o$	—
box / I-beam	$2t(h+b)$	$\dfrac{1}{6}h^3t\left(1+\dfrac{3b}{h}\right)$	$\mathrm{I}:\ \approx\dfrac{2tb^2h^2}{h+b}$; $\Box:\ \dfrac{2}{3}bt^3\left(1+\dfrac{4h}{b}\right)$	$\dfrac{h^2t}{3}\left(1+\dfrac{3b}{h}\right)$	$\mathrm{I}:\ 2tbh$; $\Box:\ \dfrac{2}{3}bt^2\left(1+\dfrac{4h}{b}\right)$
H / T sections	$2t(h+b)$	$\dfrac{t}{6}(h^3+4bt^2)$	$\mathrm{H}:\ \dfrac{t^3}{3}(8b+h)$; $\vdash:\ \dfrac{2}{3}ht^3\left(1+\dfrac{4b}{h}\right)$	$\dfrac{t}{3h}(h^3+4bt^2)$	$\mathrm{H}:\ \dfrac{t^2}{3}(8b+h)$; $\vdash:\ \dfrac{2}{3}ht^2\left(1+\dfrac{4b}{h}\right)$
corrugated sheet	$t\lambda\left(1+\dfrac{\pi^2d^2}{4\lambda^2}\right)$	$\dfrac{t\lambda d^2}{8}$	—	$\dfrac{t\lambda d}{4}$	—

Table 7.2 Values for the four shape factors

Section shape	Stiffness		Strength	
	ϕ_B^e	ϕ_T^e	ϕ_B^f	ϕ_T^f
circle ($2r_o$)	1	1	1	1
square ($b \times b$)	$\dfrac{\pi}{3}=1.05$	0.88	$\dfrac{2}{3}\sqrt{\pi}=1.18$	0.74
ellipse ($2a \times 2b$)	$\dfrac{a}{b}$	$\dfrac{2ab}{(a^2+b^2)}$	$\sqrt{\dfrac{a}{b}}$	$\sqrt{\dfrac{a}{b}}$ $(a<b)$
rectangle ($h \times b$)	$\dfrac{\pi h}{3 b}$	$\dfrac{2\pi}{3}\dfrac{b}{h}\left(1-0.58\dfrac{b}{h}\right)$ $(h>b)$	$\dfrac{2}{3}\sqrt{\pi}\left(\dfrac{h}{b}\right)^{1/2}$	$\dfrac{2}{3}\sqrt{\pi}\dfrac{(b/h)^{1/2}}{(1+0.6b/h)}$ $(h>b)$
triangle	$\dfrac{2\pi}{3\sqrt{3}}=1.21$	$\dfrac{2\pi}{5\sqrt{3}}=0.73$	0.77	0.62
tube ($2r_o$, t)	$\dfrac{r}{t}$	$\dfrac{r}{t}$	$\left(\dfrac{2r}{t}\right)^{1/2}$	$\left(\dfrac{2r}{t}\right)^{1/2}$

(continued overleaf)

Table 7.2 (continued)

Section shape	Stiffness		Strength	
	ϕ_B^e	ϕ_T^e	ϕ_B^f	ϕ_T^f
(square tube: b, t)	$\dfrac{\pi}{6}\dfrac{b}{t}$	$\dfrac{\pi}{8}\dfrac{b}{t}\left(1-\dfrac{t}{b}\right)^4$	$\dfrac{2}{3}\sqrt{\pi}\left(\dfrac{b}{t}\right)^{1/2}$	$\dfrac{\sqrt{\pi}}{2}\left(\dfrac{b}{t}\right)^{1/2}\left(1-\dfrac{t}{b}\right)^2$
(ellipse: $2a$, $2b$, t)	$\dfrac{a}{t}\dfrac{(1+3b/a)}{(1+b/a)^2}$	$\dfrac{8(ab)^{5/2}}{t(a^2+b^2)(a+b)^2}$	$\left(\dfrac{a}{t}\right)^{1/2}\dfrac{(1+3b/a)}{(1+b/a)^{3/2}}$	$\dfrac{4a^{1/2}}{t^{1/2}(1+a/b)^{3/2}}$
(two plates: h_0, h_i, t, b)	$\dfrac{\pi}{2}\dfrac{h^2}{bt}$	—	$\sqrt{2\pi}\,\dfrac{h}{(bt)^{1/2}}$	—
(box)	$\dfrac{\pi}{6}\dfrac{h}{t}\dfrac{(1+3b/h)}{(1+b/h)^2}$	$\dfrac{\pi b^2 h^2}{t(h+b)^3}$	$\dfrac{\sqrt{2\pi}}{3}\left(\dfrac{h}{t}\right)^{1/2}\dfrac{(1+3b/h)}{(1+b/h)^{3/2}}$	$\dfrac{\sqrt{2\pi}\,h}{(bt)^{1/2}(1+h/b)^{3/2}}$
(I)		$\dfrac{\pi}{3}\dfrac{t}{b}\dfrac{(1+4h/b)}{(1+h/b)^2}$	$\dfrac{\sqrt{2\pi}}{3}\left(\dfrac{t}{b}\right)^{1/2}\dfrac{(1+4h/b)}{(1+h/b)^{3/2}}$	$\dfrac{\sqrt{2\pi}}{3}\left(\dfrac{t}{b}\right)^{1/2}\dfrac{(1+4h/b)}{(1+h/b)^{3/2}}$
(T)		$\dfrac{\pi}{6}\dfrac{t}{h}\dfrac{(1+8b/h)}{(1+b/h)^2}$	$\dfrac{\sqrt{\pi}}{2}\left(\dfrac{h}{t}\right)^{1/2}\dfrac{(1+4b^2/h^3)}{(1+b/h)^{3/2}}$	$\dfrac{\pi}{18}\left(\dfrac{t}{h}\right)^{1/2}\dfrac{(1+8b/h)}{(1+b/h)^{3/2}}$
(H)	$\dfrac{\pi}{6}\dfrac{h}{t}\dfrac{(1+4b^2/h^3)}{(1+b/h)^2}$	$\dfrac{\pi}{3}\dfrac{t}{h}\dfrac{(1+4b/h)}{(1+b/h)^2}$		$\dfrac{\sqrt{2\pi}}{3}\left(\dfrac{t}{h}\right)^{1/2}\dfrac{(1+4b/h)}{(1+b/h)^{3/2}}$
(curved plate: d, λ)	$\dfrac{\pi d^2}{2\,t\lambda}$	—	$\sqrt{\pi}\,\dfrac{d}{(t\lambda)^{1/2}}$	—

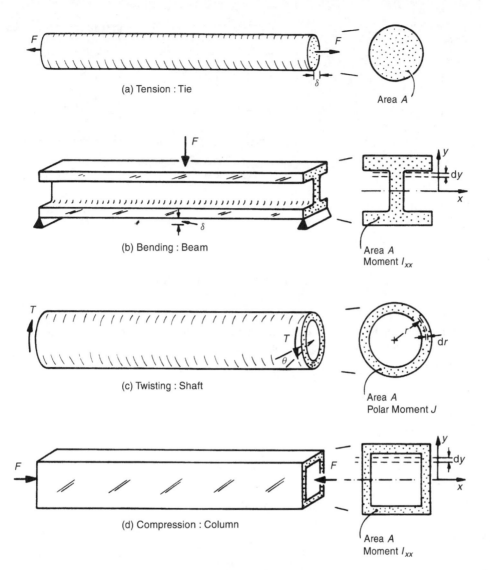

Fig. 7.2 Common modes of loading: (a) axial tension; (b) bending; (c) torsion; and (d) axial compression, which can lead to buckling.

Elastic bending and twisting (Figure 7.2(b) and (c))

If, in a beam of length ℓ, made of a material with Young's modulus E, shear is negligible, then its bending stiffness (a force per unit displacement) is

$$S_B = \frac{C_1 EI}{\ell^3} \tag{7.1}$$

where C_1 is a constant which depends on the details of the loading (values are given in Appendix A, Section A3). Shape enters through the second moment of area, I, about the axis of bending

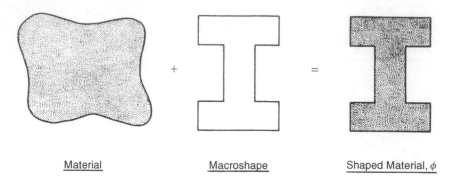

Material   Macroshape   Shaped Material, ϕ

Fig. 7.3 Mechanical efficiency is obtained by combining material with macroscopic shape. The shape is characterized by a dimensionless shape factor, ϕ. The schematic is suggested by Parkhouse (1987).

(the x axis):

$$I = \int_{\text{section}} y^2 \, dA \tag{7.2}$$

where y is measured normal to the bending axis and dA is the differential element of area at y. Values of I and of the area A for common sections are listed in Table 7.1. Those for the more complex shapes are approximate, but completely adequate for present needs.

The first shape factor — that for elastic bending — is defined as the ratio of the stiffness S_B of the shaped beam to that, S_B^o, of a solid circular section (second moment I^o) with the same cross-section A, and thus the mass per unit length. Using equation (7.1) we find

$$\phi_B^e = \frac{S_B}{S_B^o} = \frac{I}{I^o}$$

Now I^o for a solid circular section of area A (Table 7.1) is just

$$I^o = \pi r^4 = \frac{A^2}{4\pi} \tag{7.3}$$

from which

$$\boxed{\phi_B^e = \frac{4\pi I}{A^2}} \tag{7.4}$$

Note that it is dimensionless — I has dimensions of (length)^4 and so does A^2. It depends only on shape: big and small beams have the same value of ϕ_B^e if their section shapes are the same. This is shown in Figure 7.4: the three rectangular wood sections all have the same shape factor ($\phi_B^e = 2$); the three I-sections also have the same shape factor ($\phi_B^e = 10$). In each group the scale changes but the shape does not — each is a magnified or shrunken version of its neighbour. Shape factors ϕ_B^e for common shapes, calculated from the expressions for A and I in Table 7.1, are listed in the first column of Table 7.2. Solid equiaxed sections (circles, squares, hexagons, octagons) all have values very close to 1 — for practical purposes they can be set equal to 1. But if the section is elongated, or hollow, or of I-section, or corrugated, things change: a thin-walled tube or a slender I-beam can have a value of ϕ_B^e of 50 or more. Such a shape is efficient in that it uses less material (and thus

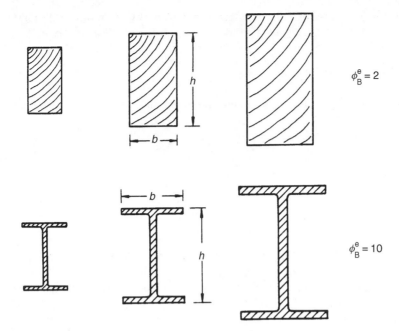

Fig. 7.4 A set of rectangular sections with $\phi_B^e = 2$, and a set of I-sections with $\phi_B^e = 10$. Members of a set differ in size but not in shape.

less mass) to achieve the same bending stiffness* A beam with $\phi_B^e = 50$ is 50 times stiffer than a solid beam of the same weight.

Shapes which resist bending well may not be so good when twisted. The stiffness of a shaft — the torque T divided by the angle of twist θ (Figure 7.2(c)) — is given by

$$S_T = \frac{KG}{\ell} \tag{7.5}$$

where G is the shear modulus. Shape enters this time through the torsional moment of area, K. For circular sections it is identical with the polar moment of area, J:

$$J = \int_{\text{section}} r^2 \, \mathrm{d}A \tag{7.6}$$

where $\mathrm{d}A$ is the differential element of area at the radial distance r, measured from the centre of the section. For non-circular sections, K is less than J; it is defined (Young, 1989) such that the angle of twist θ is related to the torque T by

$$\theta = \frac{T\ell}{KG} \tag{7.7}$$

where ℓ is length of the shaft and G the shear modulus of the material of which it is made. Approximate expressions for K are listed in Table 7.1.

---

* This shape factor is related to the radius of gyration, R_g, by $\phi_B^e = 4\pi R_g^2/A$. It is related to the 'shape parameter', k_1, of Shanley (1960) by $\phi_B^e = 4\pi k_1$. Finally, it is related to the 'aspect ratio' α and 'sparsity ratio' i of Parkhouse (1984, 1987) by $\phi_B^e = i\alpha$.

The shape factor for elastic twisting is defined, as before, by the ratio of the torsional stiffness of the shaped section, S_T, to that, S_T^o, of a solid circular shaft of the same length ℓ and cross-section A, which, using equation (7.5), is

$$\phi_T^e = \frac{S_T}{S_T^o} = \frac{K}{K^o}$$

The torsional constant K^o for a solid cylinder (Table 7.1) is

$$K^o = \frac{\pi}{2} r^4 = \frac{A^2}{2\pi}$$

giving

$$\phi_T^e = \frac{2\pi K}{A^2} \qquad (7.8)$$

It, too, has the value 1 for a solid circular cylinder, and values near 1 for any solid, equiaxed section; but for thin-walled shapes, particularly tubes, it can be large. As before, sets of sections with the same value of ϕ_T^e differ in size but not shape. Values, derived from the expressions for K and A in Table 7.1, are listed in Table 7.2.

Failure in bending and twisting*

Plasticity starts when the stress, somewhere, first reaches the yield strength, σ_y; fracture occurs when this stress first exceeds the fracture strength, σ_{fr}; fatigue failure if it exceeds the endurance limit σ_e. Any one of these constitutes failure. As in earlier chapters, we use the symbol σ_f for the failure stress, meaning 'the local stress which will first cause yielding or fracture or fatigue failure.' One shape factor covers all three.

In bending, the stress is largest at the point y_m in the surface of the beam which lies furthest from the neutral axis; it is:

$$\sigma = \frac{M y_m}{I} = \frac{M}{Z} \qquad (7.9)$$

where M is the bending moment. Thus, in problems of failure of beams, shape enters through the section modulus, $Z = I/y_m$. If this stress exceeds σ_f the beam will fail, giving the failure moment

$$M_f = Z\sigma_f \qquad (7.10)$$

The shape factor for failure in bending, ϕ_B^f, is defined as the ratio of the failure moment M_f (or equivalent failure load F_f) of the shaped section to that of a solid circular section with the same cross-sectional area A:

$$\phi_B^f = \frac{M_f}{M_f^o} = \frac{Z}{Z^o}$$

The quantity Z^o for the solid cylinder (Table 7.1) is

$$Z^o = \frac{\pi}{4} r^3 = \frac{A^{3/2}}{4\sqrt{\pi}}$$

---

* The definitions of ϕ_B^f and of ϕ_T^f differ from those in the first edition of this book; each is the square root of the old one. The new definitions allow simplification.

giving

$$\phi_B^f = \frac{4\sqrt{\pi}Z}{A^{3/2}} \tag{7.11}$$

Like the other shape factors, it is dimensionless, and therefore independent of scale; and its value for a beam with a solid circular section is 1. Table 7.2 gives expressions for other shapes, derived from the values of the section modulus Z which can be found in Table 7.1.

In torsion, the problem is more complicated. For circular tubes or cylinders subjected to a torque T (as in Figure 7.2c) the shear stress τ is a maximum at the outer surface, at the radial distance r_m from the axis of bending:

$$\tau = \frac{Tr_m}{J} \tag{7.12}$$

The quantity J/r_m in twisting has the same character as $Z = I/y_m$ in bending. For non-circular sections with ends that are free to warp, the maximum surface stress is given instead by

$$\tau = \frac{T}{Q} \tag{7.13}$$

where Q, with units of m^3, now plays the role of J/r_m or Z (details in Young, 1989). This allows the definition of a shape factor, ϕ_T^f for failure in torsion, following the same pattern as before:

$$\phi_T^f = \frac{T_f}{T_f^o} = \frac{Q}{Q^o} = \frac{2\sqrt{\pi}Q}{A^{3/2}} \tag{7.14}$$

Values of Q and ϕ_T^f are listed in Tables 7.1 and 7.2. Shafts with solid equiaxed sections all have values of ϕ_T^f close to 1.

Fully plastic bending or twisting (such that the yield strength is exceeded throughout the section) involve a further pair of shape factors. But, generally speaking, shapes which resist the onset of plasticity well are resistant to full plasticity also. New shape factors for these are not, at this stage, necessary.

Axial loading and column buckling

A column, loaded in compression, buckles elastically when the load exceeds the Euler load

$$F_c = \frac{n^2\pi^2E\,I_{min}}{\ell^2} \tag{7.15}$$

where n is a constant which depends on the end-constraints. The resistance to buckling, then, depends on the smallest second moment of area, I_{min}, and the appropriate shape factor (ϕ_B^e) is the same as that for elastic bending (equation (7.4)) with I replaced by I_{min}.

A beam or shaft with an elastic shape factor of 50 is 50 times stiffer than a solid circular section of the same mass per unit length; one with a failure shape factor of 20 is 20 times stronger. If you wish to make stiff, strong structures which are efficient (using as little material as possible) then making the shape factors as large as possible is the way to do it. It would seem, then, that the bigger the value of ϕ the better. True, but there are limits. We examine them next.

7.3 The efficiency of standard sections

There are practical limits for the thinness of sections, and these determine, for a given material, the maximum attainable efficiency. These limits may be imposed by manufacturing constraints: the difficulty or expense of making an efficient shape may simply be too great. More often they are imposed by the properties of the material itself because these determine the failure mode of the section. Here we explore the ultimate limits for shape efficiency. This we do in two ways. The first (this section) is empirical: by examining the shapes in which real materials — steel, aluminium, etc. — are actually made, recording the limiting efficiency of available sections. The second is by the analysis of the mechanical stability of shaped sections, explored in the following section.

Standard sections for beams, shafts, and columns are generally *prismatic*; prismatic shapes are easily made by rolling, extrusion, drawing, pultrusion or sawing. Figure 7.5 shows the taxonomy of the kingdom of prismatic shapes. The section may be solid, closed-hollow (like a tube or box) or open-hollow (an I-, U- or L-section, for instance). Each class of shape can be made in a range of materials. Those for which standard, off-the-shelf, sections are available are listed on the figure: steel, aluminium, GFRP and wood. Each section has a set of *attributes*: they are the parameters used in structural or mechanical design. They include its dimensions and its section properties (the 'moments' I, K and the 'section moduli' Z and Q) defined in the previous section.

These are what we need to allow the limits of shape to be explored. Figures 7.6 show I, K, Z and Q plotted against A, on logarithmic scales for standard steel sections. Consider the first, Figure 7.6(a). It shows $\log(I)$ plotted against $\log(A)$. Taking logarithms of the equation for the first shape factor ($\phi_B^e = 4\pi I/A^2$) gives, after rearrangement,

$$\log I = 2\log A + \log \frac{\phi_B^e}{4\pi}$$

meaning that values of ϕ_B^e appear as a family of parallel lines, all with slope 2, on the figure. The data are bracketed by the values $\phi_B^e = 1$ (solid circular sections) and $\phi_B^e = 65$, the *empirical upper limit* for the shape factor characterizing stiffness in bending for simple structural steel sections. An analogous construction for torsional stiffness (involving $\phi_T^e = 2\pi K/A^2$), shown in Figure 7.6(b), gives a measure of the upper limits for this shape factor; they are listed in the first row of Table 7.3. Here the closed sections group into the upper band of high ϕ_T^e; the open sections group into a band with a much lower ϕ_T^e because they have poor torsional stiffness, and shape factors which are less than 1.

The shape factors for strength are explored in a similar way. Taking logs of that for failure in bending (using $\phi_B^f = 4\sqrt{\pi}Z/A^{3/2}$) gives

$$\log Z = \frac{3}{2}\log A + \log \frac{\phi_B^f}{4\sqrt{\pi}}$$

Values of ϕ_B^f appear as lines of slope 3/2 on Figure 7.6(c), which shows that, for steel, real sections have values of this shape factor with an upper limit of about 13. The analogous construction for torsion (using $\phi_T^f = 2\sqrt{\pi}Q/A^{3/2}$), shown in Figure 7.6(d), gives the results at the end of the first row of Table 7.3. Here, again, the open sections cluster in a lower band than the closed ones because they are poor in torsion.

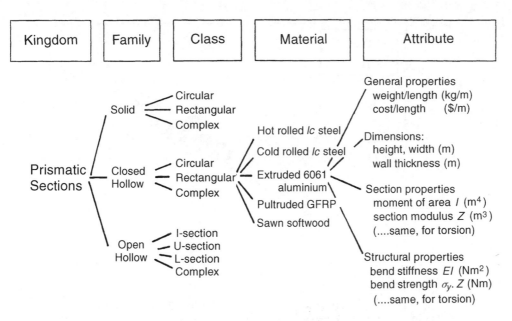

Fig. 7.5 A taxonomy of prismatic shapes, illustrating the attributes of a shaped section.

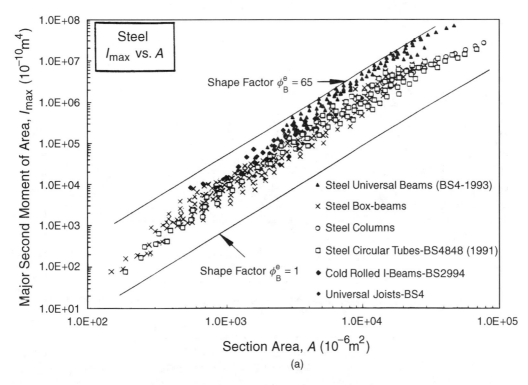

Fig. 7.6 Empirical upper limits for shape factors for steel sections: (a) log(I) plotted against log(A); (b) log(Z) plotted against log(A); (c) log(K) plotted against log(A); (d) log(Q) plotted against log(A).

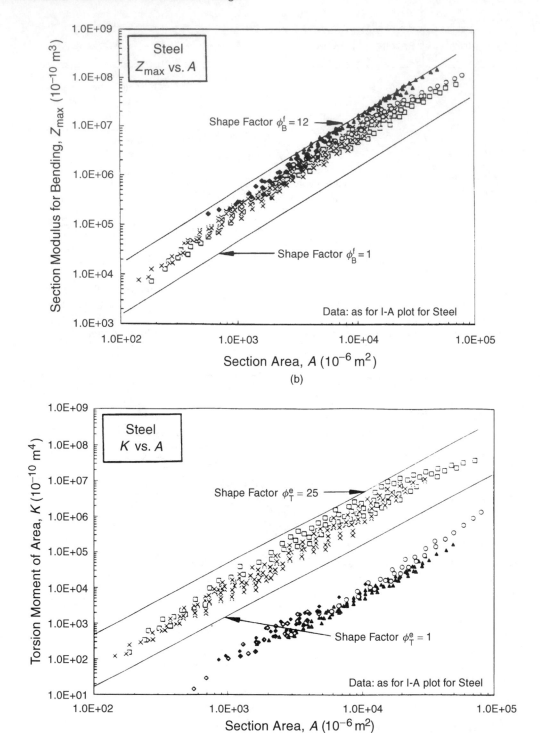

Fig. 7.6 (*continued*)

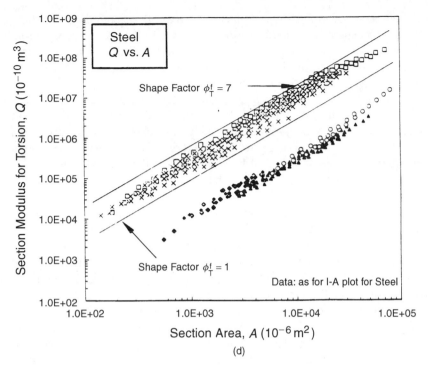

Fig. 7.6 (*continued*)

Table 7.3 Upper limits for the shape factors ϕ_B^e, ϕ_T^e, ϕ_B^f and ϕ_T^f

Material	$(\phi_B^e)_{max}$	$(\phi_T^e)_{max}$	$(\phi_B^f)_{max}$	$(\phi_T^f)_{max}$
Structural steels	65	25	13	7
Aluminium alloys	44	31	10	8
GFRP and CFRP	39	26	9	7
Polymers (e.g. nylons)	12	8	5	4
Woods (solid sections)	5	1	3	1
Elastomers	<6	3	—	—

Similar plots for extruded aluminium, pultruded GFRP, wood, nylon and rubber give the results shown in the other rows of the table. It is clear that the *upper-limiting shape factor for simple shapes depends on material.*

The upper limits for shape efficiency are important. They are central to the design of lightweight structures, and structures in which, for other reasons (cost, perhaps) the material content should be minimized. Three questions then arise. What sets the upper limit on shape efficiency of Table 7.3? Why does the limit depend on material? And what, in a given application where efficiency is sought, is the best combination of material and shape? We address these questions in turn.

7.4 Material limits for shape factors

The range of shape factor for a given material is limited either by manufacturing constraints, or by local buckling. Steel, for example, can be drawn to thin-walled tubes or formed (by rolling, folding or welding) into efficient I-sections; shape factors as high as 50 are common. Wood cannot so easily be shaped; ply-wood technology could, in principle, be used to make thin tubes or I-sections, but in practice, shapes with values of ϕ greater than 5 are uncommon. That is a manufacturing constraint. Composites, too, can be limited by the present difficulty in making them into thin-walled shapes, although the technology for doing this now exists.

When efficient shapes *can* be fabricated, the limits of the efficiency derive from the competition between failure modes. Inefficient sections fail in a simple way: they yield, they fracture, or they suffer large-scale buckling. In seeking efficiency, a shape is chosen which raises the load required for the simple failure modes, but in doing so the structure is pushed nearer the load at which other modes — particularly those involving local buckling — become dominant. It is a characteristic of shapes which approach their limiting efficiency that two or more failure modes occur at almost the same load.

Why? Here is a simple-minded explanation. If failure by one mechanism occurs at a lower load than all others, the section shape can be adjusted to suppress it; but this pushes the load upwards until another mechanism becomes dominant. If the shape is described by a single variable (ϕ) then when two mechanisms occur at the same load you have to stop — no further shape adjustment can improve things. Adding webs, ribs or other stiffeners, gives further variables, allowing shape to be optimized further, but we shall not pursue that here.

The best way to illustrate this is with an example. We take that of a tubular column. The column (Figure 7.7) is progressively loaded in compression. If sufficiently long and thin, it will first fail by general *elastic (Euler) buckling*. The buckling load is increased with no change in mass if the diameter of the tube is increased and the wall thickness correspondingly reduced. But there is a limit to how far this can go because new failure modes appear: if the load rises too far, the tube will *yield plastically*, and if the tube wall is made too thin, it will fail by *local buckling*. Thus there are three competing failure modes: general buckling, local buckling (both influenced by the modulus of the material and the section shape) and plastic collapse (dependent on the yield strength of the material and — for axial loading — dependent on the area of the cross-section but not on its shape). The most efficient shape for a given material is the one which, for a given load, uses the least material. It is derived as follows.

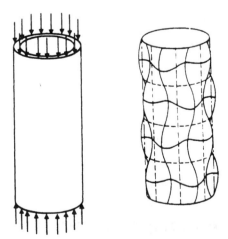

Fig. 7.7 A tube loaded in compression. The upper limit on shape is determined by a balance between failure mechanisms, of which one — local ('chessboard') buckling — is shown in the right-hand figure.

General buckling of a column of height ℓ, radius r, wall thickness t and cross-sectional area $A = 2\pi rt$ with ends which are free to rotate, occurs at the load

$$F = \frac{\pi^2 EI}{\ell^2} \tag{7.16}$$

where, for thin-walled tubes, $I = \pi r^3 t$, and E is the value of Young's modulus for the material of which the column is made. Dividing equation (7.16) by A^2, substituting for I/A^2 from

$$\phi = \frac{4\pi I}{A^2} = \frac{r}{t} \tag{7.17}$$

where we use the short-hand ϕ for ϕ_B^e. Writing $F/A = \sigma$ where σ is the axial stress in the tube wall, we obtain an expression for the value of the stress σ_1 at the onset of general buckling:

$$(\textit{mechanism 1}) \qquad \sigma_1 = \left(\frac{\pi}{4} E\phi \frac{F}{\ell^2} \right)^{1/2} \tag{7.18}$$

Local buckling is characterized by the 'chessboard' pattern of Figure 7.7. This second failure mode occurs in a thin-walled tube when the axial stress exceeds, approximately, the value (Young, 1989, p. 262–263)

$$(\textit{mechanism 2}) \qquad \sigma_2 = 0.6\alpha E \frac{t}{r} = 0.6\alpha \frac{E}{\phi} \tag{7.19}$$

(using equation (7.17) to introduce ϕ). This expression contains an empirical knockdown factor, α, which Young (1989) takes to equal 0.5 to allow for the interaction of different buckling modes.

The final failure mode is that of general yield. It occurs when the wall-stress exceeds the value

$$(\textit{mechanism 3}) \qquad \sigma_3 = \sigma_y \tag{7.20}$$

where σ_y is the yield strength of the material of the tube.

We now have the stresses at which each failure mechanism first occurs. The one which is dominant is the one that cuts in first — that is, it has the lowest failure stress. Mechanism 1 is dominant when the value of σ_1 is lower than either σ_2 or σ_3, mechanism 2 when σ_2 is the least, and so on. The boundaries between the three fields of dominance are found by equating the equations for σ_1, σ_2 and σ_3 (equations (7.18), (7.19) and (7.20)) taken in pairs, giving

$$(\textit{1--2 boundary}) \qquad \frac{F}{\sigma_y \ell^2} = \frac{1.44\alpha^2}{\pi} \left(\frac{E}{\sigma_y} \right) \frac{1}{\phi^3} \tag{7.21a}$$

$$(\textit{1--3 boundary}) \qquad \frac{F}{\sigma_y \ell^2} = \frac{4}{\pi} \left(\frac{\sigma_y}{E} \right) \frac{1}{\phi} \tag{7.21b}$$

$$(\textit{2--3 boundary}) \qquad \phi = 0.6\alpha \left(\frac{E}{\sigma_y} \right) \tag{7.21c}$$

Here we have arranged the variables into dimensionless groups. There are just three: the first is the *load factor* $F/\sigma_y \ell^2$, the second is the *yield strain* σ_y/E and the last is the shape factor ϕ. This allows a simple presentation of the failure-mechanism boundaries, and the associated fields of dominance, as shown in Figure 7.8. The axes are the load factor $F/\sigma_y \ell^2$ and the shape factor ϕ. The diagram is constructed for a specific value of the yield strain σ_y/E of 3×10^{-3}. Changing σ_y/E moves the boundaries a little, but leaves the general picture unchanged.

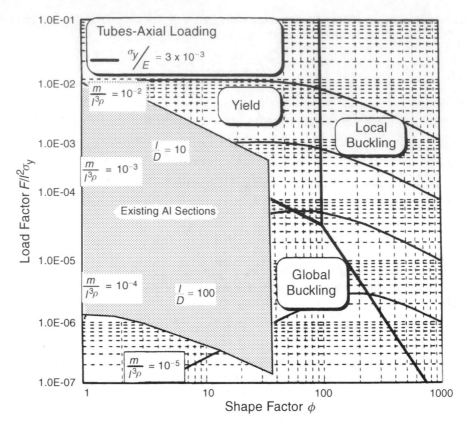

Fig. 7.8 A plot of the load factor $F/\sigma_y \ell^2$ against shape factor ϕ for $\sigma_y/E = 3 \times 10^{-3}$ for axially loaded tubes. The grey area shows where standard sections lie. The upper limit falls just below the boundary between yield and local buckling.

To explore *efficiency* we need one more step. According to the simple-minded argument, above, maximum efficiency is found when two failure modes occur at the same load. Let us be more precise, and see whether simple-mindedness is justified. To do this we calculate the mass of the column which will just not fail by any one of the mechanisms, and then seek a way of minimizing this with respect to ϕ. The mass, m, of the column is

$$m = A\ell\rho \tag{7.22}$$

where A is the area of its cross-section and ρ is the density of the material of which it is made. Within the general-buckling regime 1, the minimum section area A which will just support F is

$$A_1 = \frac{F}{\sigma_1}$$

Inserting this into equation (7.22) and replacing σ_1 by equation (7.18) gives for regime 1:

$$(\textit{mass in regime 1}) \quad \left(\frac{m}{\ell^3\rho}\right) = \left(\frac{4}{\pi}\left(\frac{1}{\phi}\right)\left(\frac{F}{\sigma_y\ell^2}\right)\left(\frac{\sigma_y}{E}\right)\right)^{1/2} \tag{7.23a}$$

Within the local buckling regime 2, equation (7.19) for σ_2 dominates and we find instead

$$(mass\ in\ regime\ 2)\qquad \left(\frac{m}{\ell^3\rho}\right) = \left(\frac{\phi}{0.6\alpha}\right)\left(\frac{F}{\sigma_y\ell^2}\right)\left(\frac{\sigma_y}{E}\right) \qquad\qquad (7.23b)$$

and for the yield regime 3, using equation (7.20) for σ_3:

$$(mass\ in\ regime\ 3)\qquad \left(\frac{m}{\ell^3\rho}\right) = \left(\frac{F}{\sigma_y\ell^2}\right) \qquad\qquad (7.23c)$$

As before, the variables have been assembled into dimensionless groups; there is one new one: the mass is described by the group $(m/\ell^3\rho)$. For a chosen value of this quantity and of the yield strain σ_y/E, each equation becomes a relation between the load factor, $F/\sigma_y\ell^2$, and the shape factor, ϕ, allowing contours of mass to be plotted on the diagram, as shown in Figure 7.8.

We can now approach the question: what is the most efficient shape, measured by ϕ, for the cross-section of the column? Tracking across Figure 7.8 from left to right at a given value of the load factor, the mass at first falls and then rises again. In the lower half of the diagram the minimum mass lies at or near the 1–2 boundary; higher up it lies slightly to the left of the 2–3 boundary. So, like all good simple-minded explanations, this one is almost right — right enough to be useful.

If the column is designed for a *specific* value of the load factor, the optimum ϕ can be read from the diagram. But if the column is intended as a *general-purpose component*, the load factor is not known, though all reasonable values lie well within the range shown in the vertical axis of Figure 7.8. Then the safest choice is a value of ϕ a little to the left of the 2–3 (yield-local buckling) boundary, since this ensures that, if the column were to fail, it would fail by yield rather than the more catastrophic local buckling. This boundary lies at the position given by equation (7.21c). Allowing a margin of reserve of 1.5 (by reducing ϕ by a factor of 2/3) we find the optimal shape factor for the tubular column to be

$$\phi_{opt} = \left(\frac{r}{t}\right)_{opt} \approx 0.4\alpha\frac{E}{\sigma_y}$$

which for $\alpha = 0.5$ is

$$\phi_{opt} = \left(\frac{r}{t}\right)_{opt} \approx 0.2\frac{E}{\sigma_y} \qquad\qquad (7.24)$$

This is a single example of how competing failure mechanisms determine shape efficiencies. Other modes of loading (bending, torsion) and other classes of shape (box-sections, I-sections) each require analysis, and this is a painfully tedious process, best left to others. Others have done it* and find that all combinations of loading and shape lead to diagrams which resemble Figure 7.8. The limiting efficiency depends to some extent on details of loading and class of shape, but not much. The broad conclusion: the ultimate limit for simple shapes (tubes, box-sections, I-sections) is set by material properties, and is approximated by equation (7.4).

Much higher efficiencies are possible when precise loading conditions are known, allowing customized application of stiffeners and webs to suppress local buckling. This allows a further increase in the ϕs until failure or new, localized, buckling modes appear. These, too, can be suppressed by a further hierarchy of structuring; ultimately, the ϕs are limited only by manufacturing constraints. But for a general selection of material and shape, this is getting too sophisticated, and equation (7.24) above is the best approximation.

---

* See, for example, the Weaver and Ashby (1998).

7.5 Material indices which include shape

The performance-maximizing combination of material and section shape, for a given mode of loading, is found as follows. The method follows that of Chapter 5, with one extra step to bring in the shape.

Axial tension of ties

The ability of a tie to carry a load F without deflecting excessively or failing depends only on the area of its section, but not on its shape. The material index for stiffness at minimum weight, E/ρ, holds for all section shapes. This, as we have said, is not true of bending or twisting, or when columns buckle.

Elastic bending of beams and twisting of shafts

Consider the selection of a material for a beam of specified stiffness S_B and length ℓ, and it is to have minimum mass, m. The selection must allow for the fact that the available candidate-materials have section shapes which differ. The mass m of a beam of length ℓ and section area A is given by equation (7.22). Its bending stiffness is given by equation (7.1). Replacing I by ϕ_B^e using equation (7.4) gives

$$S_B = \frac{C_1}{4\pi} \frac{E}{\ell^3} \phi_B^e A^2 \tag{7.25}$$

Using this to eliminate A in equation (7.25) gives the mass of the beam:

$$m = \left[\frac{4\pi S_B}{C_1 \ell}\right]^{1/2} \ell^3 \left[\frac{\rho^2}{\phi_B^e E}\right]^{1/2} \tag{7.26}$$

For beams with the same shape, for which ϕ_B^e is constant, the best choice for the lightest beam is the material with the greatest value of $E^{1/2}/\rho$ — the result derived in Chapter 5 (note that this applies to material selection for all self-similar shapes, not just solid ones). But if we wish to select a material–shape combination for a light stiff beam, the best choice is that with the greatest value of the index

$$M_1 = \frac{[E\phi_B^e]^{1/2}}{\rho} \tag{7.27}$$

Exactly the same result holds for the general elastic buckling of an axially loaded column.

The procedure for elastic twisting of shafts is similar. A shaft of section A and length ℓ is subjected to a torque T. It twists through an angle θ. It is required that the torsional stiffness, T/θ, meet a specified target S_T, at minimum mass. The mass of the shaft is given, as before, by equation (7.24). Its torsional stiffness is

$$S_T = \frac{KG}{\ell}$$

where G is the shear modulus, and K was defined earlier. Replacing K by ϕ_T^e using equation (7.8) gives

$$S_T = \frac{G}{2\pi\ell} \phi_T^e A^2 \tag{7.28}$$

Using this to eliminate A in equation (7.24) gives

$$m = \left[2\pi \frac{S_T}{\ell^3}\right]^{1/2} \ell^3 \left[\frac{\rho^2}{\phi_T^e G}\right]^{1/2}$$

The best material-and-shape combination is that with the greatest value of $[\phi_T^e G]^{1/2}/\rho$. The shear modulus, G, is closely related to Young's modulus E. For the practical purposes we approximate G by $3/8E$; then the index becomes

$$M_2 = \frac{[\phi_T^e E]^{1/2}}{\rho} \tag{7.29}$$

For shafts of the same shape, this reduces to $E^{1/2}/\rho$ again. When shafts differ in both material and shape, the material index (7.29) is the one to use.

Failure of beams and shafts

A beam, loaded in bending, must support a specified load F without failing. The mass of the beam is to be minimized. When shape is not a consideration, the best choice (Chapter 5) is that of the material with the greatest value of $\sigma_f^{2/3}/\rho$ where σ_f is the failure strength of the material. When section-shape is a variable, the best choice is found as follows.

Failure occurs if the load exceeds the failure moment

$$M_f = Z\sigma_f$$

Replacing Z by the appropriate shape-factor ϕ_B^f via equation (7.11) gives

$$M_f = \frac{\sigma_f}{4\sqrt{\pi}} \phi_B^f A^{3/2} \tag{7.30}$$

Substituting this into equation (7.22) for the mass of the beam gives

$$m = \left(4\sqrt{\pi}\frac{M_f}{\ell^3}\right)^{2/3} \ell^3 \left[\frac{\rho^{3/2}}{\phi_B^f \sigma_f}\right]^{2/3} \tag{7.31}$$

The best material-and-shape combination is that with the greatest value of the index

$$M_3 = \frac{(\phi_B^f \sigma_f)^{2/3}}{\rho} \tag{7.32}$$

At constant shape the index reduces to the familiar $\sigma_f^{2/3}/\rho$ of Chapter 5; but when shape as well as material can be chosen, the full index must be used.

The twisting of shafts is treated in the same way. A shaft must carry a torque T without failing. This requires that T not exceed the failure torque T_f, where, from equation (7.13),

$$T_f = Q\tau_f$$

Replacing Q by ϕ_T^f with equation (7.14) gives

$$T_f = \frac{\sigma_f}{4\sqrt{\pi}} \phi_T^f A^{3/2} \tag{7.33}$$

where τ_f, the shear-failure strength has been replaced by $\sigma_f/2$, the tensile failure strength. Using this to eliminate the area A in equation (7.34) for the mass of the shaft gives

$$m = \left(4\sqrt{\pi}\frac{T_f}{\ell^3} \right)^{2/3} \ell^3 \left[\frac{\rho^{3/2}}{\phi_T^f \sigma_f} \right]^{2/3} \tag{7.34}$$

Performance is maximized by the selection which has the greatest value of

$$\boxed{M_4 = \frac{(\phi_T^f \sigma_f)^{2/3}}{\rho}} \tag{7.35}$$

Constrained shapes

The geometry of a design sometimes imposes constraints on shape. Panels, for example, usually have a fixed width but a thickness which is 'free', meaning that it can be chosen to give a desired bending stiffness; the shape of the section, too, is free: it could, for example, be a honeycomb. Beams, too, may be constrained in either height or width. When there is a dimensional constraint, the definition of the shape factor changes. Material indices for constrained shapes are discussed in the Appendix to this chapter.

7.6 The microscopic or micro-structural shape factor

Microscopic shape

The sections listed in Tables 7.1 and 7.2 achieve efficiency through their *macroscopic* shape. Efficiency can be achieved in another way: through shape on a small scale; *microscopic* or 'micro-structural' shape. Wood is an example. The solid component of wood (a composite of cellulose, lignin and other polymers) is shaped into little prismatic cells, dispersing the solid further from the axis of bending or twisting of the branch or trunk of the tree. This gives wood a greater bending and torsional stiffness than the solid of which it is made. The added efficiency (Figure 7.9) is character-ized by a set of *microscopic shape factors*, ψ, with definitions and characteristics exactly like those of ϕ. The characteristic of microscopic shape is that the structure repeats itself: it is *extensive*. The micro-structured solid can be thought of as a 'material' in its own right: it has a modulus, a density, a strength, and so forth. Shapes can be cut from it which — provided they are large compared with the size of the cells — inherit its properties. It is possible, for instance, to fabricate an I-section out of wood, and such a section has macroscopic shape (as defined earlier) as well as microscopic shape (Figure 7.10). It is shown in a moment that the total shape factor for a wooden I-beam is the product of the shape factor for the wood structure and that for the I-beam; and this can be large.

Many natural materials have microscopic shape. Wood is just one example. Bone, stalk and cuttle all have structures which give high stiffness at low weight. It is harder to think of man-made examples, although it would appear possible to make them. Figure 7.11 shows four extensive

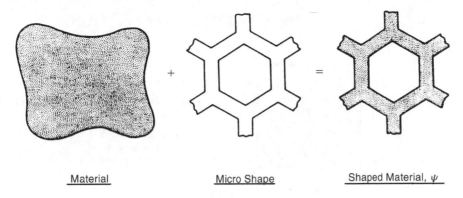

Material   Micro Shape   Shaped Material, ψ

Fig. 7.9 Mechanical efficiency can be obtained by combining material with microscopic, or internal, shape, which repeats itself to give an extensive structure. The shape is characterized by microscopic shape factors, ψ.

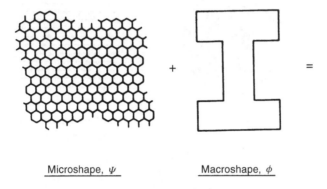

Microshape, ψ Macroshape, ϕ

Fig. 7.10 Micro-structural shape can be combined with macroscopic shape to give efficient structures. The schematic is suggested by Parkhouse (1984). The overall shape factor is the product of the microscopic and macroscopic shape factors.

structures with microscopic shape, all of which are found in nature. The first is a wood-like structure of hexagonal-prismatic cells; it has translational symmetry and is uniform, with isotropic properties in the plane of the section when the cells are regular hexagons. The second is an array of fibres separated by a foamed matrix typical of palm wood; it too is uniform in-plane and has translational symmetry. The third is an axisymmetric structure of concentric cylindrical shells separated by a foamed matrix, like the stem of some plants. And the fourth is a layered structure, a sort of multiple sandwich-panel, like the shell of the cuttle fish; it has orthotropic symmetry.

Microscopic shape factors

Consider the gain in bending stiffness when a solid cylindrical beam like that shown as a black circle in Figure 7.11 is expanded, at constant mass, to a circular beam with any one of the structures which surround it in the figure. The stiffness S_s of the original solid beam is

$$S_s = \frac{C_1 E_s I_s}{\ell^3} \tag{7.36}$$

Micro-Structured Materials

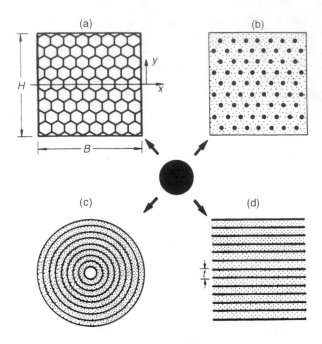

Fig. 7.11 Four extensive micro-structured materials which are mechanically efficient: (a) prismatic cells; (b) fibres embedded in a foamed matrix; (c) concentric cylindrical shells with foam between; and (d) parallel plates separated by foamed spacers.

where the subscript s means a property of the solid beam. When the beam is expanded at constant mass its density falls from ρ_s to ρ and its radius increases from r_s to

$$r = \left(\frac{\rho_s}{\rho}\right)^{1/2} r_s \tag{7.37}$$

with the result that its second moment of area increases from I_s to

$$I = \frac{\pi}{4} r^4 = \frac{\pi}{4}\left(\frac{\rho_s}{\rho}\right)^2 r_s^4 = \left(\frac{\rho_s}{\rho}\right)^2 I_s \tag{7.38}$$

If the cells, fibres, rings or plates in Figure 7.11 are extensive parallel to the axis of the beam, the modulus falls from that of the solid, E_s, to

$$E = \left(\frac{\rho}{\rho_s}\right) E_s \tag{7.39}$$

The stiffness of the expanded beam is thus

$$S = \frac{C_1 EI}{\ell^3} = \frac{C_1 E_s I_s}{\ell^3}\left(\frac{\rho_s}{\rho}\right) \tag{7.40}$$

The microscopic shape factor, ψ is defined in the same way as the macroscopic one, ϕ: it is the ratio of the stiffness of the structured beam to that of the solid one. Taking the ratio of equations (7.40) and (7.36) gives

$$\psi_B^e = \frac{S}{S_s} = \frac{\rho_s}{\rho} \tag{7.41}$$

In words: the microscopic shape factor for prismatic structures is simply the reciprocal of the relative density. Note that, in the limit of a solid (when $\rho^* = \rho_s$) ψ_B^e takes the value 1, as it obviously should. A similar analysis for failure in bending gives the shape factor

$$\psi_B^f = \left(\frac{\rho_s}{\rho}\right)^{1/2} \tag{7.42}$$

Torsion, as always, is more difficult. When the structure of Figure 7.11(c), which has circular symmetry, is twisted, its rings act like concentric tubes and for these

$$\psi_T^e = \frac{\rho_s}{\rho} \quad \text{and} \quad \psi_T^f = \left(\frac{\rho_s}{\rho}\right)^{1/2} \tag{7.43}$$

The others have lower torsion stiffness and strength (and thus lower shape factors) for the same reason that I-sections, good in bending, perform poorly in torsion.

Structuring, then, converts a solid with modulus E_s and strength $\sigma_{f,s}$ to a new solid with properties E and σ_f. If this new solid is formed to an efficient macroscopic shape (a tube, say, or an I-section) its bending stiffness, to take an example, increases by a further factor of ϕ_B^e. Then the stiffness of the beam, expressed in terms of that of the solid of which it is made, is

$$S = \psi_B^e \phi_B^e S_s \tag{7.44}$$

that is, the shape factors multiply. The same is true for strength.

This is an example of structural hierarchy and the benefits it brings. It is possible to extend it further: the individual cell walls or layers could, for instance, be structured, giving a third multiplier to the overall shape factor, and these units, too could be structured (Parkhouse, 1984). Nature does this to good effect, but for man-made structures there are difficulties. There is the obvious difficulty of manufacture, imposing economic limits on the levels of structuring. And there is the less obvious one of reliability. If the structure is optimized, then a failure of a member at one level of the structure could trigger failure of the structure as a whole. The more complex the structure, the harder it becomes to ensure the integrity at all levels.

As pointed out earlier, a micro-structured material can be thought of as a new material. It has a density, a strength, a thermal conductivity, and so on; difficulties arise only if the sample size is comparable to the cell size, when 'properties' become size dependent. This means that micro-structured materials can be plotted on the Material Selection Charts — indeed, wood appears on them already — and that all the selection criteria used for solid materials developed in Chapter 5 apply, unchanged, to the micro-structured materials.

7.7 Co-selecting material and shape

Optimizing the choice of material and shape can be done in several ways. Two are illustrated below.

Co-selection by calculation

Consider as an example the selection of a material for a stiff *shaped* beam of minimum mass. Four materials are available, listed in Table 7.4 with their properties and the shapes, characterized by ϕ_B^e, in which they are available (here, the maximum ones). We want the combination with the largest value of the index M_1 of equation (7.27) which, repeated, is

$$M_1 = \frac{(\phi_B^e E)^{1/2}}{\rho}$$

The second last column shows the simple 'fixed shape' index $E^{1/2}/\rho$: wood has the greatest value — it is more than twice as stiff as steel for the same weight. But when each material is shaped efficiently (last column) wood has the *lowest* value of M_1 — even steel is better; the aluminium alloy wins, marginally better than GFRP.

Graphical co-selection using material property charts

Shaped materials can be displayed and selected with the Material Selection Charts. The reasoning, for the case of elastic bending, goes like this. The material index for elastic bending (equation (7.27)) can be rewritten as

$$M_1 = \frac{(\phi_B^e E)^{1/2}}{\rho} = \frac{(E/\phi_B^e)^{1/2}}{\rho/\phi_B^e} \tag{7.45}$$

The equation says: a material with modulus E and density ρ, when structured, behaves like a material with modulus

$$E^* = E/\phi_B^e$$

and density

$$\rho^* = \rho/\phi_B^e$$

The $E-\rho$ chart is shown schematically in Figure 7.12. The structured material properties E^* and ρ^* can be plotted onto it. Introducing shape ($\phi_B^e = 10$, for example) moves the material M to the lower left along a line of slope 1, from the position E, ρ to the position $E/10$, $\rho/10$, as shown in the figure. The selection criteria are plotted onto the figure as before: a constant value of the index of $E^{1/2}/\rho$, for instance, plots as a straight line of slope 2; it is shown, for one value of $E^{1/2}/\rho$, as

Table 7.4 The selection of material and shape for a light, stiff, beam

Material	ρ Mg/m^3	E GPa	ϕ_{max}^*	$\dfrac{E^{1/2}}{\rho}$	$\dfrac{(\phi_{max}E)^{1/2}}{\rho}$
1020 Steel	7.85	205	65	1.8	14.7
6061-T4 Al	2.7	70	44	3.1	**20.5**
GFRP (isotropic)	1.75	28	39	2.9	19.0
Wood (oak)	0.9	13.5	5	**4.1**	9.1

$^*\phi_{max}$ means the maximum permitted value of ϕ from Table 7.3.

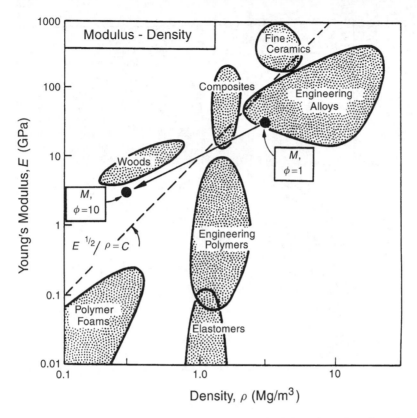

Fig. 7.12 Schematic of Materials Selection Chart 1: Young's modulus plotted against density. The best material-and-shape for a light, stiff beam is that with the greatest value of $E^{1/2}/\rho$. The structured material behaves in bending like a new material with modulus $E^* = E/\phi$ and density $\rho^* = \rho/\phi$ (where ϕ means ϕ_B^e) and can be plotted onto the charts. All the material-selection criteria still apply. A similar procedure is used for torsion.

a broken line. The introduction of shape has moved the material from a position below this line to one above; its performance has improved. Elastic twisting of shafts is treated in the same way.

Materials selection based on strength (rather than stiffness) at a minimum weight uses the chart of strength σ_f against density ρ, shown schematically in Figure 7.13. Shape is introduced in a similar way. The material index for failure in bending (equation (7.32)), can be rewritten as follows

$$M_3 = \frac{(\phi_B^f \sigma_f)^{2/3}}{\rho} = \frac{(\sigma_f/(\phi_B^f)^2)^{2/3}}{\rho/(\phi_B^f)^2} \tag{7.32}$$

The material with strength σ_f and density ρ, when shaped, behaves in bending like a material of strength

$$\sigma_f^* = \sigma_f/(\phi_B^f)^2$$

and density

$$\rho^* = \rho/(\phi_B^f)^2$$

The rest will be obvious. Introducing shape ($\phi_B^f = \sqrt{10}$, say) moves a material M along a line of slope 1, taking it, in the schematic, from a position σ_f, ρ below the material index line (the

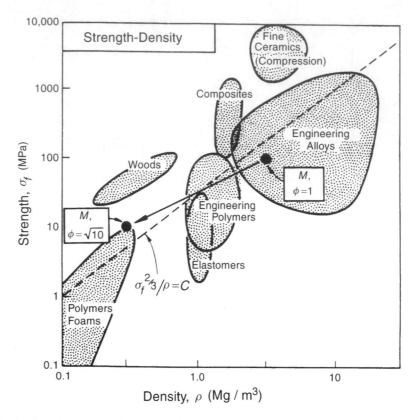

Fig. 7.13 Schematic of Materials Selection Chart 2: strength σ_f plotted against density ρ. The best material for a light, strong beam is that with the greatest value of $\sigma_f^{2/3}/\rho$. The structured material behaves in bending like a new material with strength $\sigma_f^* = \sigma_f/\phi^2$, and density ρ/ϕ^2 (where ϕ means ϕ_B^f), and can be plotted onto the chart. All the material-selection criteria still apply. A similar procedure is used for torsional strength.

broken line) to the position $\sigma_f/10$, $\rho/10$ which lies above it. The performance has again improved. Torsional failure is analysed by using ϕ_T^f in place of ϕ_B^f.

Examples of the method are given in the case studies of the next chapter.

7.8 Summary and conclusions

The designer has two groups of variables with which to optimize the performance of a load-bearing component: the material properties and the shape of the section. They are not independent. The best choice of material, in a given application, depends on the shapes in which it is available, or to which it could potentially be formed. A procedure is given for simultaneously optimizing the choice of both material and shape.

The contribution of shape is isolated by defining four shape factors. The first, ϕ_B^e, is for the elastic bending and buckling of beams; the second, ϕ_T^e, is for the elastic twisting of shafts; the third, ϕ_B^f

Table 7.5 Definitions of shape factors

Design constraint*	Bending	Torsion
Stiffness	$\phi_B^e = \dfrac{4\pi I}{A^2}$	$\phi_T^e = \dfrac{2\pi K}{A^2}$
Strength	$\phi_B^f = \dfrac{4\sqrt{\pi}Z}{A^{3/2}}$	$\phi_T^f = \dfrac{2\sqrt{\pi}Q}{A^{3/2}}$

*A = section area; I, K, Z and Q are defined in the text and tabulated in Table 7.1.

is for the plastic failure of beams loading in bending; and the last, ϕ_T^f, is for the plastic failure of twisted shafts (Table 7.5). The shape factors are dimensionless numbers which characterize the efficiency of use of the material in each mode of loading. They are defined such that all four have the value 1 for solid circular sections. With this definition, all equiaxed solid sections have shape factors of about 1, but efficient shapes which disperse the material far from the axis of bending or twisting (I-beams, hollow tubes, sandwich structures, etc.) have large values of the shape factors. They are tabulated for common shapes in Table 7.2.

The best material–shape combination for a light beam with a prescribed bending stiffness is that which maximizes the material index

$$M_1 = \frac{(E\phi_B^e)^{1/2}}{\rho}$$

A similar combination, M_2, involving ϕ_T^e, gives the lightest stiff shaft. The material–shape combination for a light beam with a prescribed strength is that which maximizes the material index

$$M_3 = \frac{(\phi_B^f \sigma_f)^{2/3}}{\rho}$$

A similar combination, M_4, involving ϕ_T^f gives the lightest strong shaft. Here, the criterion of 'performance' was that of meeting a design specification at minimum weight. Other such material–shape combinations maximize other performance criteria: minimizing cost rather than weight, for example, or maximizing energy storage. Examples are developed in Chapter 8.

The idea of micro-structural shape factors (ψ) is introduced to characterize the efficiency, in bending and torsion, of cellular, layered and other small-scale structures, common in nature. They are defined in the same way as the ϕs. The difference is that microscopic shape is repeated; structures with microscopic shape are extensive and can themselves be cut to give macroscopic shape as well. Such structures can be thought of either as a solid with properties E_s, $\sigma_{f,s}$ and ρ_s, with a microscopic shape factor of ψ; or as a new material, with a new set of properties, E_s/ψ, ρ_s/ψ, etc., with a shape-factor of 1. Wood is an example: it can be seen as solid cellulose and lignin shaped to the cells of wood, or as wood itself, with a lower density, modulus and strength than cellulose, but with greater values of indices $E^{1/2}/\rho$ and $\sigma_f^{2/3}/\rho$ which characterize structural efficiency. When micro-structured materials (ψ) are given macroscopic shape (ϕ) the total shape factor is then the product $\phi\psi$, and this can be large.

The procedure for selecting material–shape combinations is best illustrated by examples. These can be found in the next chapter.

7.9 Further reading

Books on the mechanics of materials

Gere, J.M. and Timoshenko, S.P. (1985) *Mechanics of Materials*, Wadsworth International, London.
Timoshenko, S.P. and Gere, J.M. (1961) *Theory of Elastic Stability*, McGraw-Hill Koga Kusha Ltd, London.
Young, W.C. (1989) *Roark's Formulas for Stress and Strain*, 6th edition, McGraw-Hill, New York.

Books and articles on the efficiency of structures

Ashby, M.F. (1991) Materials and shape, *Acta Metall. Mater.* **39**, 1025–1039.
Gerard, G. (1956) *Minimum Weight Analysis of Compression Structures*, New York University Press, New York.
Parkhouse, J.G. (1984) Structuring: a process of material dilution, in *3rd Int. Conf. on Space Structures*, p. 367, edited by H. Nooshin, Elsevier London.
Parkhouse, J.G. (1987) Damage accumulation in structures, *Reliability Engineering*, **17**, 97–109.
Shanley, F.R. (1960) *Weight–Strength Analysis of Aircraft Structures*, 2nd edition, Dover Publications, New York.
Weaver, P.M. and Ashby, M.F. (1996) The optimal selection of material and section shape, *Journal of Engineering Design*, **7**, 129–150.
Weaver, P.M. and Ashby, M.F. (1998) Material limits for shape efficiency, *Prog. Mater. Sci.*, **41**, 61–128.

Appendix: geometric constraints and associated shape factors

Geometric constraints

Whenever a free variable is adjusted to find an optimum, it is good practice to check that its value, when the optimum is found, is acceptable. In choosing a material and shape to meet constraints on stiffness or on strength, the scale of the section has been treated as free, choosing a value that meets the constraint. One can imagine circumstances in which this might not be acceptable — when, for instance, the outer diameter d of a tube could be chosen freely provided it was less than a critical value d; or when, to take another example, the width w of a beam was genuinely free but the height h free only so long as it was less than h^*. Dimensional constraints of this sort can change the index and the way it is used. The methods developed so far can be extended to include them.

For solid sections (cylinders, square sections) a dimensional constraint leads to a simple minimum limit for modulus or strength. Take bending stiffness as an example. The stiffness of a beam is:

$$S = \frac{C_1 EI}{\ell^3} = \frac{C_1 \pi r^4 E}{4\ell^3} \tag{A7.1}$$

(using $I = \pi r^4/4$). If there is an upper limit on r then for the stiffness constraint to be met E must exceed the value

$$E = \frac{4S\ell^3}{C_1 \pi r^4} \tag{A7.2}$$

Materials with lower moduli than this are excluded.

Limits for E for shaped sections are derived in a similar way. We take the tube as an example. Its bending stiffness is

$$S = \frac{C_1 EI}{\ell^3} = \frac{C_1 E}{\ell^3} \pi r^3 t = \frac{C_1 r^4}{\ell^3} \frac{E}{\phi} \tag{A7.3}$$

(using $\phi = r/t$). An upper limit on the radius leads to the limit

$$E = \frac{S\ell^3 \phi}{C\pi r^4} \tag{A7.4}$$

Only materials with moduli greater than this are candidates.

Constrained shapes

Constrained shapes appear when one dimension of the section is limited by the design. The idea is shown in Figure 7.14. When a 'free' shape changes scale, all the dimensions of its section scale by the same factor, as in Figure 7.1. When a constrained shape changes scale, all dimensions in one direction remain fixed, all those in the other scale by the same factor (Figure 7.14(a) and (b)). The constraint changes the material index.

When the width is constrained, we can no longer define ϕ by using a solid cylindrical section as the standard to which the other shapes are normalized. Instead — and in the same spirit as

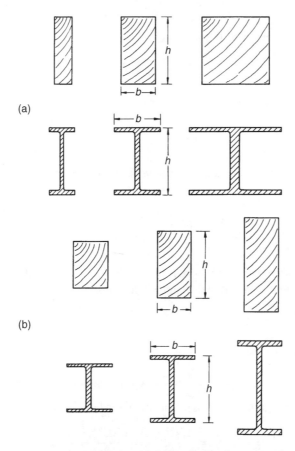

Fig. 7.14 A constrained section-shape is one in which the design fixes one dimension, but for which the other is free; all lengths in this direction change in proportion when the section changes size. It contrasts with a free shape (Figure 7.1) in which all dimensions change in proportion when the section changes in size. At (a) the height h is constrained; at (b) the width b is constrained.

before — we use the simplest solid shape that allows one dimension to be held fixed while leaving the other free: a flat plate of thickness t, width b and length ℓ. Its area A, its second moment I, and its section modulus Z are given in terms of its height t and its width b by (Table 7.1)

$$A^o = tb$$

$$I^o = \frac{bt^3}{12} \tag{A7.5}$$

$$Z^o = \frac{bt^2}{6}$$

Sections with constrained height, loaded in bending

The shape factor for elastic bending is defined, as before, as the ratio of the stiffness of the plate before (S_B^o) and after (S_B) 'structuring'. ϕ_B^e now become

$$(\phi_B^e)_b = \frac{S_B}{S_B^o} = \frac{EI}{E^o I^o} \tag{A7.6}$$

(using $I = bt^3/12$). The stiffness of the plate is

$$S = \frac{C_1 EI}{\ell^3} = \frac{C_1 \phi_B^e E bt^3}{12\ell^3} \tag{A7.7}$$

and its mass is $m = bt\ell\rho$. Eliminating t gives

$$m = \left(\frac{12b^2 S}{C_1 t^3}\right)^{1/3} \ell \left(\frac{\rho}{E^{1/3}}\right)$$

The lightest plate is that made from the material with the largest value of the index

$$\boxed{M = \frac{(\phi_B^e E)^{1/3}}{\rho}} \tag{A7.8}$$

An example will illustrate its use. Consider a plate, initially solid and of thickness t and width b which is foamed to a height h (width and length held constant). The density falls from ρ to

$$\rho^* = \rho\frac{t}{h}$$

and the modulus falls from E to

$$E^* = E\left(\frac{t}{h}\right)^2$$

(the scaling law for the modulus of foams). The stiffness falls from

$$S^o = \frac{C_1 E^o I^o}{\ell^3} = \frac{C_1 E^o bt^3}{12\ell^3}$$

to

$$S = \frac{C_1 E^* I^*}{\ell^3} = \frac{C_1 E^* bh^3}{12\ell^3} = \frac{C_1 E bt^3}{12\ell^3}\left(\frac{h}{t}\right)$$

giving, for

$$(\phi_B^e)_b = \frac{S_B}{S_B^o} = \frac{h}{t} = \frac{1}{(\rho^*/\rho)}$$

As before, we find that foaming imparts a shape factor equal to the reciprocal of the relative density. Following the same procedure for strength gives

$$(\phi_B^f)_b = \frac{M_f}{M_f^o} = \frac{Z\sigma_y}{Z^o\sigma_y^o}$$

with associated index (for minimum mass) of

$$M = \frac{(\phi_B^f \sigma_y)_b^{1/2}}{\rho}$$

Chapter 8

Shape — case studies

8.1 Introduction and synopsis

This chapter, like Chapter 6, is a collection of case studies. They illustrate the use of material indices which include shape. Remember: they are only necessary for the restricted class of problems in which section shape directly influences performance, that is, when the prime function of a component is to carry loads which cause it to bend, twist or buckle. And even then they are needed only when the shape is itself a variable, that is, when different materials come in different shapes. When all candidate-materials can be made to the *same* shapes, the indices reduce to those of Chapter 6.

Indices which include shape provide a tool for optimizing the co-selection of material-and-shape. The important ones are summarized in Table 8.1. Many were derived in Chapter 7; the others are derived here. Minimizing cost instead of weight is achieved by replacing density ρ by $C_m\rho$, where C_m is the cost per kilogram.

The selection procedure is, first, to identify candidate-materials and the section shapes in which each is available, or could be made. The relevant material properties* and shape factors for each are tabulated. The best material-and-shape combination is that with the greatest value of the appropriate index. The same information can be plotted onto Materials Selection Charts, allowing a graphical solution to the problem — one which often suggests further possibilities.

The method has other uses. It gives insight into the way in which natural materials — many of which are very efficient — have evolved. Bamboo is an example: it has both internal or microscopic shape and a tubular, macroscopic shape, giving it very attractive properties. This and other aspects are brought out in the case studies which now follow.

8.2 Spars for man-powered planes

Most engineering design is a difficult compromise: it must meet, as best it can, the conflicting demands of multiple objectives and constraints. But in designing a spar for a man-powered plane the objective is simple: the spar must be as light as possible, and still be stiff enough to maintain the aerodynamic efficiency of the wings (Table 8.2). Strength, safety, even cost, hardly matter when records are to be broken. The plane (Figure 8.1) has two main spars: the transverse spar supporting the wings, and the longitudinal spar carrying the tail assembly. Both are loaded primarily in bending (torsion cannot, in reality, be neglected, although we shall do so here).

Some 60 man-powered planes have flown successfully. Planes of the first generation were built of balsa wood and spruce. The second generation relied on aluminium tubing for the load-bearing

---

* The material properties used in this chapter are taken from the *CMS* compilation published by Granta Design, Trumpington Mews, 40B High Street, Trumpington CB2 2LS, UK.

Table 8.1 Examples of indices which include shape

(a) Stiffness and strength-limited design at minimum weight (or cost*)

Component shape, loading and constraints	Stiffness-limited design	Strength-limited design
Tie (tensile member) Load, stiffness and length specified, section-area free	$\dfrac{E}{\rho}$	$\dfrac{\sigma_f}{\rho}$
Beam (loaded in bending) Loaded externally or by self weight, stiffness, strength and length specified, section area free	$\dfrac{(\phi_B^e E)^{1/2}}{\rho}$	$\dfrac{(\phi_B^f \sigma_f)^{2/3}}{\rho}$
Torsion bar or **tube** Loaded externally , stiffness, strength and length specified, section area free	$\dfrac{(\phi_T^e E)^{1/2}}{\rho}$	$\dfrac{(\phi_T^f \sigma_f)^{2/3}}{\rho}$
Column (compression strut) Collapse load by buckling or plastic crushing and length specified, section area free	$\dfrac{(\phi_B^e E)^{1/2}}{\rho}$	$\dfrac{\sigma_f}{\rho}$

*For cost, replace ρ by $C_m\rho$ in the indices.

(a) Springs, specified energy storage at minimum volume or weight (or cost*)

Component shape, loading and constraints	Flexural springs	Torsion springs
Spring Specified energy storage, volume to be minimized	$\dfrac{(\phi_B^f \sigma_f)^2}{\phi_B^e E}$	$\dfrac{(\phi_T^f \sigma_f)^2}{\phi_T^e E}$
Spring Specified energy storage, mass to be minimized	$\dfrac{(\phi_B^f \sigma_f)^2}{\phi_B^e E \rho}$	$\dfrac{(\phi_T^f \sigma_f)^2}{\phi_T^e E \rho}$

*For cost, replace ρ by $C_m\rho$ in the indices.

Table 8.2 Design requirements for wing spars

Function	Wing spar
Objective	Minimum mass
Constraints	(a) Specified stiffness (b) Length specified

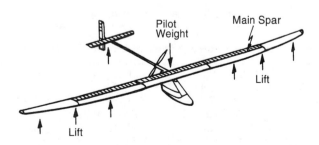

Fig. 8.1 The loading on a man-powered plane is carried by two spars, one spanning the wings and the other linking the wings to the tail. Both are designed for stiffness at minimum weight.

structure. The present, third, generation uses carbon-fibre/epoxy spars, moulded to appropriate shapes. How has this evolution come about? And how much further can it go?

The model and the selection

We seek a material-and-shape combination that minimizes weight for a given bending stiffness. The index to be maximized, read from Table 8.1, is

$$M_1 = \frac{(\phi_B^e E)^{1/2}}{\rho} \tag{8.1}$$

Data for four materials are assembled in Table 8.3. If all have the same shape, M_1 reduces to the familiar $E^{1/2}/\rho$ and the ranking is that of the second last column. Balsa and spruce are significantly better than the competition. Woods are extraordinarily efficient. That is why model aircraft builders use them now and the builders of real aircraft relied so heavily on them in the past.

The effect of shaping the section, to a rectangle for the woods, to a box-section for aluminium and CFRP, gives the results in the last column. (The shape factors listed here are typical of commercially available sections, and are well below the maximum for each material.) Aluminium is now marginally better than the woods; CFRP is best of all.

The same information is shown graphically in Figure 8.2, using the method of Chapter 7. Each shape is treated as a new material with modulus $E^* = E/\phi_B^e$ and $\rho^* = \rho/\phi_B^e$. The values of E^* and ρ^* are plotted on the chart. The superiority of both the aluminium tubing with $\phi = 20$ and the CFRP box-sections with $\phi = 10$ are clearly demonstrated.

Postscript

Why is wood so good? With no shape it does as well or better than heavily-shaped steel. It is because wood *is* shaped: its cellular structure gives it internal shape (see p. 182), increasing the performance of the material in bending; it is nature's answer to the I-beam. Bamboo, uniquely, combines microscopic and macroscoptic shape (see next section).

But the technology of drawing thin-walled aluminium tubes has improved. Aluminium itself is stiffer than balsa or spruce, but it is also nearly 10 times denser, and that makes it, as a solid, far less attractive. As a tube, though, it can be given a shape factor which cannot be reproduced in wood. An aluminium tube with a shape factor $\phi_B^e = r/t = 20$ is as good as solid balsa or spruce; one with a thinner wall is better — a fact that did not escape the designers of the second generation of man-powered planes. There is a limit, of course: tubes that are too thin will kink (a local elastic buckling); as shown in Chapter 7, this sets an upper limit to the shape factor for aluminium at about 40.

Table 8.3 Materials for wing spars

Material	Modulus E (GPa)	Density ρ (Mg/m³)	Shape factor ϕ_B^e	Index $E^{1/2}/\rho$	Index M_1^* ((GPa)$^{1/2}$/Mg/m³)
Balsa	4.2–5.2	0.17–0.24	1–2	**11**	11–15
Spruce	9.8–11.9	0.36–0.44	1–2	9	9–12
Steel	200–210	7.82–7.84	25–30	1.8	9–10
Al 7075 T6	71–73	2.8–2.82	15–25	3	12–15
CFRP	100–160	1.5–1.6	10–15	7	**23–28**

*The range of values of the indices are based on means of the material properties and corresponds to the range of values of ϕ_B^e.

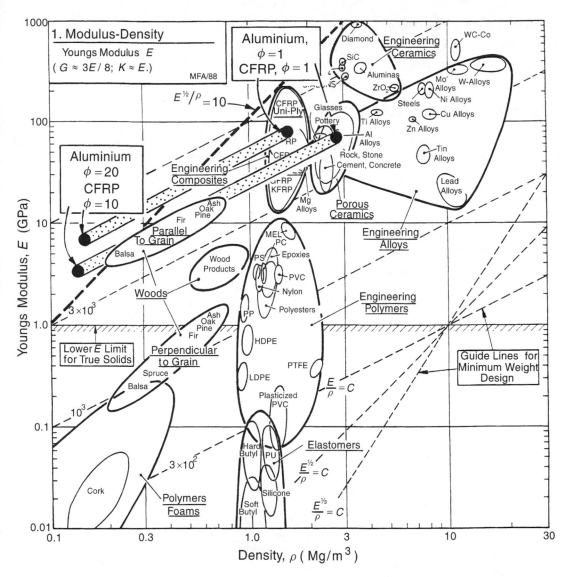

Fig. 8.2 The materials-and-shapes for wing-spars, plotted on the modulus–density chart. A spar made of CFRP with a shape factor of 10 outperforms spars made of aluminium ($\phi = 20$) and wood ($\phi \approx 1$).

The last 20 years has seen further development: carbon-fibre technology has reached the market place. As a solid beam, carbon-fibre reinforced polymer laminates are nearly as efficient as spruce. Add a bit of shape (Table 8.3) and they are better than any of the competing materials. Contemporary composite technology allows shape factors of at least 10, and that gives an increase in performance that — despite the cost — is attractive to plane builders.

Further reading: man-powered flight

Drela, M. and Langford, J.D. (1985) Man-powered flight, *Scientific American*, January issue, p. 122.

Related case studies

Case Study 8.3: Forks for a racing bicycle
Case Study 8.4: Floor joists

8.3 Forks for a racing bicycle

The first consideration in bicycle design (Figure 8.3) is strength. Stiffness matters, of course, but the initial design criterion is that the frame and forks should not yield or fracture in normal use. The loading on the forks is predominantly *bending*. If the bicycle is for racing, then the mass is a primary consideration: the forks should be as light as possible. What is the best choice of material and shape? Table 8.4 lists the design requirements.

The model and the selection

We model the forks as beams of length ℓ which must carry a maximum load P (both fixed by the design) without plastic collapse or fracture. The forks are tubular, of radius r and fixed wall-thickness t. The mass is to be minimized. The fork is a light, strong beam. Further details of load and geometry are unnecessary: the best material and shape, read from Table 8.1, is that with the

Fig. 8.3 The bicycle. The forks are loaded in bending. The lightest forks which will not collapse plastically under a specified design load are those made of the material and shape with the greatest value of $(\phi_B^f \sigma_f)^{2/3}/\rho$.

Table 8.4 Design requirements for bicycle forks

Function	Bicycle forks
Objective	Minimize mass
Constraints	(a) Must not fail under design loads — a strength constraint
	(b) Length specified

Table 8.5 Material for bicycle forks

Material	Strength σ_f (MPa)	Density ρ (Mg/m³)	Shape factor ϕ_B^f	Index $\sigma_f^{2/3}/\rho$	Index M_3^* ((MPa)²/³/Mg/m³)
Spruce (Norwegian)	70–80	0.46–0.56	1–1.5	**36**	36–50
Bamboo	80–160	0.6–0.8	2.4–2.8	(33)	59–65
Steel (Reynolds 531)	770–990	7.82–7.83	7–8	12	44–48
Alu (6061–T6)	240–260	2.69–2.71	5.5–6.3	15	47–51
Titanium 6-4	930–980	4.42–4.43	5.5–6.3	22	69–75
Magnesium AZ 91	160–170	1.80–1.81	4–4.5	17	42–46
CFRP	300–450	1.5–1.6	4–4.5	33	**83–90**

*The range of values of the indices are based on means of the material properties and corresponds to the range of values of ϕ_B^e.

greatest value of

$$M_3 = \frac{(\phi_B^f \sigma_f)^{2/3}}{\rho} \tag{8.2}$$

Table 8.5 lists seven candidate materials. Solid spruce or bamboo are remarkably efficient; without shape (second last column) they are better than any of the others. Bamboo is special because it grows as a hollow tube with a macroscopic shape factor ϕ_B^f between 3 and 5, giving it a bending strength which is much higher than solid spruce (last column). When shape is added to the other materials, however, the ranking changes. The shape factors listed in the table are achievable using normal production methods. Steel is good; CFRP is better; Titanium 6-4 is better still. In strength-limited applications magnesium is poor despite its low density.

Postscript

Bicycles have been made of all seven of the materials listed in the table — you can still buy bicycles made of six of them (the magnesium bicycle was discontinued in 1997). Early bicycles were made of wood; present-day racing bicycles of steel, aluminium or CFRP, sometimes interleaving the carbon fibres with layers of glass or Kevlar to improve the fracture-resistance. Mountain bicycles, for which strength and impact resistance are particularly important, have steel or titanium forks.

The reader may be perturbed by the cavalier manner in which theory for a straight beam with an end load acting normal to it is applied to a curved beam loaded at an acute angle. No alarm is necessary. When (as explained in Chapter 5) the variables describing the functional requirements (F), the geometry (G) and the materials (M) in the performance equation are separable, the details of loading and geometry affect the terms F and G but not M. This is an example: beam curvature and angle of application of load do not change the material index, which depends only on the design requirement of strength in bending at minimum weight.

Further reading: bicycle design

Sharp, A. (1993) *Bicycles and Tricycles, an Elementary Treatise on their Design and Construction*, The MIT Press, Cambridge, MA.
Watson, R. and Gray, M. (1978) *The Penguin Book of the Bicycle*, Penguin Books, Harmondsworth.
Whitt, F.R. and Wilson, D.G. (1985) *Bicycling Science*, 2nd edition, The MIT Press, Cambridge, MA.
Wilson, D.G. (1986) A short history of human powered vehicles, *The American Scientist*, **74**, 350.

Related case studies

Case Study 8.2: Wing spars for man powered planes
Case Study 8.4: Floor joists: wood or steel?

8.4 Floor joists: wood or steel?

Floors are supported on *joists:* beams which span the space between the walls. Let us suppose that a joist is required to support a specified bending load (the 'floor loading') without sagging excessively or failing; and it must be cheap. Traditionally, joists are made of wood with a rectangular section of aspect ratio 2:1, giving an elastic shape factor (Table 7.2) of $\phi_B^e = 2.1$. But steel, shaped to an I-section, could be used instead (Figure 8.5). Standard steel I-section joists have shape factors in the range $15 \leq \phi_B^e \leq 25$ (special I- sections can have much larger values). Are steel I-joists a better choice than wooden ones? Table 8.6 summarizes the design requirements.

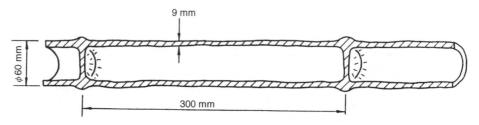

Fig. 8.4 The cross-section of a typical bamboo cane. The tubular shape shown here gives 'natural' shape factors of $\phi_B^e = 3.3$ and $\phi_B^f = 2.6$. Because of this (and good torsional shape factors also) it is widely used for oars, masts, scaffolding and construction. Several bamboo bicycles have been marketed.

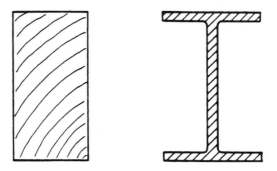

Fig. 8.5 The cross-sections of a wooden beam ($\phi_B^e = 2$) and a steel I-beam ($\phi_B^e = 10$). The values of ϕ are calculated from the ratios of dimensions of each beam, using the formulae of Table 7.2.

Table 8.6 Design requirements for floor joists

Function	Floor joist
Objective	Minimum material cost
Constraints	(a) Length specified
	(b) Minimum stiffness specified
	(c) Minimum strength specified

The model and the selection

Consider stiffness first. The cheapest beam, for a given stiffness, is that with the largest value of the index (read from Table 8.1 with ρ replaced by $C_m\rho$ to minimize cost):

$$M_1 = \frac{(\phi_B^e E)^{1/2}}{C_m\rho} \tag{8.3}$$

Data for the modulus E, the density ρ, the material cost C_m and the shape factor ϕ_B^e are listed in Table 8.7, together with the values of the index M_1 with and without shape. The steel beam with $\phi_B^e = 25$ has a slightly larger value M_1 than wood, meaning that it is a little cheaper for the same stiffness.

But what about strength? The best choice for a light beam of specified strength is that which maximizes the material index:

$$M_2 = \frac{(\phi_B^f \sigma_f)^{2/3}}{C_m\rho} \tag{8.4}$$

The quantities of failure strength σ_f, shape factor ϕ_B^f and index M_3 are also given in the table. Wood performs better than even the most efficient steel I-beam.

As explained in Chapter 7, a material with a modulus E and cost per unit volume $C_m\rho$, when shaped, behaves in bending like a material with modulus $E^* = E/\phi_B^e$ and cost $(C_m\rho)^* = C_m\rho/\phi_B^e$. Figure 8.6 shows the E–$C_m\rho$ chart with data for the wooden joists and the steel I-beams plotted onto it. The heavy broken line shows the material index $M_1 = (\phi_B^e E)^{1/2}/C_m\rho$, positioned to leave a small subset of materials above it. Woods with a solid circular section ($\phi_B^e = 1$) lie comfortably above the line; solid steel lies far below it. Introducing the shape factors moves the wood slightly (the shift is not shown) but moves the steel a lot, putting it in a position where it performs as well as wood.

Strength is compared in a similar way in Figure 8.7. It shows the σ_f–$C_m\rho$ chart. The heavy broken line, this time, is the index $M_3 = (\phi_B^f \sigma_f)^{2/3}/C_m\rho$, again positioned just below wood. Introducing shape shifts the steel as shown, and this time it does not do so well: even with the largest shape factor ($\phi_B^f = 10$) steel performs less well than wood. Both conclusions are exactly the same as those of Table 8.7.

Table 8.7 Materials for floor joists

Property	Wood (pine)	Steel (standard)
Density (Mg/m^3)	0.52–0.64	7.9–7.91
Flexural modulus (GPa)	9.8–11.9	208–212
Failure strength — MOR (MPa)	56–70	350–360
Material cost ($/kg)	0.8–1.0	0.6–0.7
ϕ_B^e	2.0–2.2	15–25
ϕ_B^f	1.6–1.8	5.5–7.1
$E^{1/2}/C_m\rho$ (GPa)$^{1/2}$/(k$/m^3)*	6.3	2.8
$\sigma_f^{2/3}/C_m\rho$ (MPa)$^{2/3}$/(k$/m^3)*	30	9.7
M_1 (GPa)$^{1/2}$/(k$/m^3)*	8.9–9.3	10.8–14.0
M_2 (MPa)$^{2/3}$/(k$/m^3)*	41–44	30–36

*The range of values of the indices are based on means of the material properties and corresponds to the range of values of ϕ_B^e.

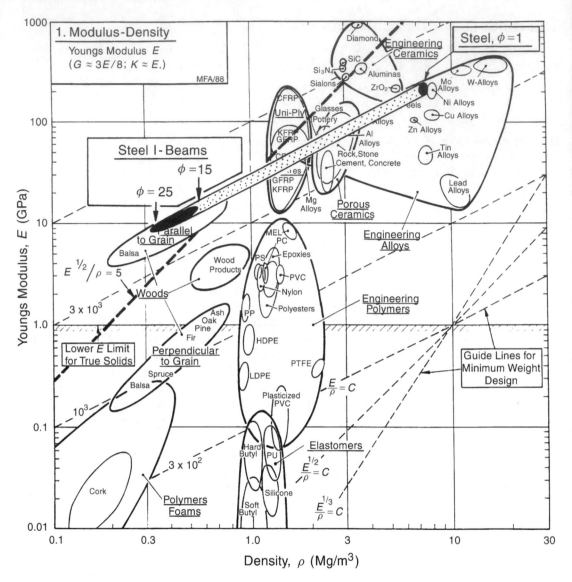

Fig. 8.6 A comparison of light, stiff beams. The heavy broken line shows the material index $M_1 = 5\,(\text{GPa})^{1/2}/(\text{Mg/m}^3)$. Steel I-beams are slightly more efficient than wooden joists.

Postscript

So the conclusion: as far as performance per unit material-cost is concerned, there is not much to choose between the standard wood and the standard steel sections used for joists. As a general statement, this is no surprise — if one were much better than the other, the other would no longer exist. But — looking a little deeper — wood dominates certain market sectors, steel dominates others. Why?

Wood is indigenous to some countries, and grows locally; steel has to come further, with associated transport costs. Assembling wood structures is easier than those of steel; it is more forgiving

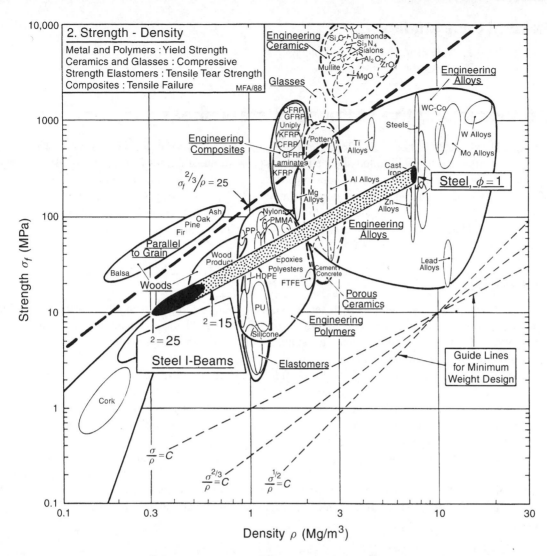

Fig. 8.7 A comparison of light, strong beams. The heavy broken line shows the material index $M_2 = 25 \, (MPa)^{2/3}/(Mg/m^3)$. Steel I-beams are less efficient than wooden joists.

of mismatches of dimensions, it can be trimmed on site, you can hammer nails into it anywhere. It is a user-friendly material.

But wood is a variable material, and, like us, is vulnerable to the ravishes of time, prey to savage fungi, insects and small mammals. The problems so created in a small building — family home, say — are easily overcome, but in a large commercial building — an office block, for instance — they create greater risks, and are harder to fix. Here, steel wins.

Further reading

Cowan, H.J. and Smith, P.R. (1988) *The Science and Technology of Building Materials*, Van Nostrand Reinhold, New York.

Related case studies

Case Study 8.2: Spars for man-powered planes
Case Study 8.3: Forks for a racing bicycle

8.5 Increasing the stiffness of steel sheet

How could you make steel sheet stiffer? There are many reasons you might wish to do so. The most obvious: to enable stiffness-limited sheet structures to be lighter than they are now; to allow panels to carry larger compressive loads without buckling; and to raise the natural vibration frequencies of sheet structures. Bending stiffness is proportional to EI (E is Young's modulus, I is the second moment of area of the sheet, equal to $t^3/12$ per unit width). There is nothing much you can do to change the modulus of steel, which is always close to 210 GPa. But you can add a bit of shape. So consider the design brief of Table 8.8.

The model

The age-old way to make sheet steel stiffer is to corrugate it, giving it a roughly sinusoidal profile. The corrugations increase the second moment of area of the sheet about an axis normal to the corrugations themselves. The resistance to bending in one direction is thereby increased, but in the cross-direction it is not changed at all.

Corrugations are the clue, but — to be useful — they must stiffen the sheet in all directions, not just one. A hexagonal grid of dimple (Figure 8.8) achieves this. There is now no direction of bending that is not dimpled. The dimples need not be hexagons; any pattern arranged in such a way that you cannot draw a straight line across it without intersecting dimples will do. But hexagons are probably about the best.

Dimples improve all the section-properties of a sheet, in a way that can be estimated as follows. Consider an idealized cross-section as in the lower part of Figure 8.8, which shows the section A–A, enlarged. As before, we define the shape factor as the ratio of the stiffness of the dimpled sheet to that of the flat sheet from which it originated. The second moment of area of the flat sheet is

$$I_o = \frac{t^3}{12}\lambda \tag{8.5}$$

That of the dimpled sheet with amplitude a is

$$I \approx \frac{1}{12}(2a + t)^2 \lambda t \tag{8.6}$$

Table 8.8 Design requirements for stiffened steel sheet

Function	Steel sheet for stiffness-limited structures
Objective	Maximize bending stiffness of sheet
Constraints	(a) Profile limited to a maximum deviation ± 5 times the sheet thickness from flatness
	(b) Cheap to manufacture

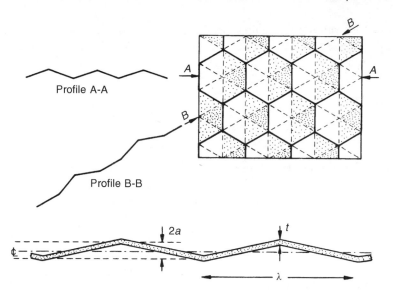

Profile A-A

Profile B-B

Fig. 8.8 A sheet with a profile of adjacent hexagonal dimples which increases its bending stiffness and strength. Shape factors for the section A–A are calculated in the text. Those along other trajectories are lower but still significantly greater than 1.

giving a shape factor, defined as before as the ratio of the stiffness of the sheet before and after corrugating (see the Appendix of Chapter 7):

$$\phi_B^e = \frac{I}{I_o} = \frac{(2a + t)^2}{t^2} \tag{8.7}$$

Note that the shape factor has the value unity when the amplitude is zero, but increases as the amplitude increases. The equivalent shape factor for failure in bending is

$$\phi_B^f = \frac{Z}{Z_o} = \frac{(2a + t)}{t} \tag{8.8}$$

These equations predict large gains in stiffness and strength. The reality is a little less rosy. This is because, while all cross-sections of the sheet are dimpled, only those which cut through the peaks of the dimples have an amplitude equal to the peak height (all others have less) and, even among these, only some have adjacent dimples; the section B–B, for example does not. Despite this, and limits set by the onset of local buckling, the gain is real.

Postscript

Dimpling can be applied to most rolled-sheet products. It is done by making the final roll-pass through mating rolls with meshing dimples, adding little to the cost. It is most commonly applied to sheet steel. Here it finds applications in the automobile industry including bumper armatures, seat frames, side impact bars: the material offers weight saving without loss of mechanical performance. Stiffening sheet also raises its natural vibration frequencies, making them harder to excite, thus helping to suppress vibration in panels.

But a final word of warning: stiffening the sheet may change its failure mechanism. Flat sheet yields when bent; dimpled sheet, if thin, could fail by a local buckling mode. It is this which ultimately limits the useful extent of dimpling.

Further reading

Fletcher, M. (1998) Cold-rolled dimples improve gauge strength, *Eureka*, May, p. 28.

8.6 Ultra-efficient springs

Springs, we deduced in Case Study 6.7, store energy. They are best made of a material with a high value of σ_f^2/E, or, if mass is more important than volume, then of $\sigma_f^2/\rho E$. Springs can be made more efficient still by shaping their section. Just how much more is revealed below.

We take as a measure of performance the energy stored per unit volume of solid of which the spring is made; we wish to maximize this energy. Energy per unit weight and per unit cost are maximized by similar procedures (Table 8.9).

The model

Consider a leaf spring first (Figure 8.9(a)). A leaf spring is an elastically bent beam. The energy stored in a bent beam, loaded by a force F, is

$$U = \frac{1}{2}\frac{F^2}{S_B} \tag{8.9}$$

where S_B, the bending stiffness of the spring, is given by equation (7.1), or, after replacing I by ϕ_B^e, by equation (7.25), which, repeated, is

$$S_B = \frac{C_1}{4\pi}\phi_B^e\frac{A^2}{\ell^3}E \tag{8.10}$$

Table 8.9 Design requirements for ultra-efficient springs

Function	Material-efficient spring
Objective	Maximum stored energy per unit volume (or mass, or cost)
Constraint	Must remain elastic under design loads

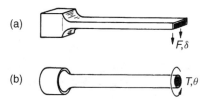

(a)

F,δ

(b)

T,θ

Fig. 8.9 Hollow springs use material more efficiently than solid springs. Best in bending is the hollow elliptical section; best in torsion is the tube.

The force F in equation (8.9) is limited by the onset of yield; its maximum value is

$$F_f = C_2 Z \frac{\sigma_f}{\ell} = \frac{C_2}{4\sqrt{\pi}\ell} \sigma_f \phi_B^f A^{3/2} \tag{8.11}$$

(The constants C_1 and C_2 are tabulated in Appendix A Section A3 and A4). Assembling these gives the maximum energy the spring can store:

$$\frac{U_{max}}{V} = \frac{C_2^2}{8C_1} \left(\frac{(\phi_B^f \sigma_f)^2}{\phi_B^e E} \right) \tag{8.12}$$

where $V = A\ell$ is the volume of solid in the spring. The best material and shape for the spring — the one that uses the least material — is that with the greatest value of the quantity

$$M_5 = \frac{(\phi_B^f \sigma_f)^2}{\phi_B^e E} \tag{8.13}$$

For a fixed section shape, the ratio involving the two ϕs is a constant: then the best choice of material is that with the greatest value of σ_f^2/E — the same result as before. When shape is a variable, the most efficient shapes are those with large $(\phi_B^f)^2/\phi_B^e$. Values for these ratios are tabulated for common section shapes in Table 8.10; hollow elliptical sections are up to three times more efficient than solid shapes.

Torsion bars and helical springs are loaded in torsion (Figure 8.9(b)). The same calculation, but using equations (7.28) and (7.33), in the way that equations (8.10) and (8.11) were used, gives

$$\frac{U_{max}}{V} = \frac{1}{16} \frac{(\phi_T^f \sigma_f)^2}{\phi_T^e G} \tag{8.14}$$

The most efficient material and shape for a torsional spring is that with the largest value of

$$M_6 = \frac{(\phi_T^f \sigma_f)^2}{\phi_T^e E} \tag{8.15}$$

(where G has been replaced by $3E/8$). The criteria are the same: when shape is not a variable, the best torsion-bar materials are those with high values of σ_f^2/E. Table 8.10 shows that the best shapes are hollow tubes, which have a ratio of $(\phi_T^f)^2/\phi_T^e$ which is twice that of a solid cylinder; all other shapes are less efficient. Springs which store the maximum energy per unit weight (instead of unit volume) are selected with indices given by replacing E by $E\rho$ in equations (8.13) and (8.15). For maximum energy per unit cost, replace $E\rho$ by $EC_m\rho$ where C_m is the cost per kg.

Postscript

Hollow springs are common in vibrating and oscillating devices and for instruments in which inertial forces must be minimized. The hollow elliptical section is widely used for springs loaded in bending; the hollow tube for those loaded in torsion. More about this problem can be found in the classic paper by Boiten.

Table 8.10 Shape factors for the efficiency of springs

Section shape	$(\phi_B^f)^2/\phi_B^e$	$(\phi_T^f)^2/\phi_T^e$
	1	1
	1.33	0.63
	1	$\dfrac{(1 + a^2/b^2)}{2}$ $(a < b)$
	1.33	$\dfrac{2}{3}\left(1 - 0.6\left(\dfrac{b}{h} - \dfrac{b^2}{h^2}\right)\right)$ $(h > b)$
	0.5	0.53
	2 $(t \ll r)$	2 $(t \ll r)$
	2.67 $(t \ll b)$	2
	$\dfrac{1 + 3b/a}{1 + b/a}$ $(a < b)$	$\dfrac{2b^3(1 + a^2/b^2)}{(ab)^{3/2}(1 + a/b)}$ $(a < b)$
	4	—
	$\dfrac{4}{3}\dfrac{(1 + 3b/h)}{(1 + b/h)}$	$\dfrac{2}{3}\dfrac{(1 + 4h/b)}{(1 + h/b)}$
	$\dfrac{3}{4}\dfrac{(1 + 4bt^2/h^3)}{(1 + b/h)}$	$\dfrac{2}{3}\dfrac{(1 + 4b/h)}{(1 + b/h)}$
	2	—

Further reading: design of efficient springs

Boiten, R.G. (1963) Mechanics of instrumentation, *Proc. I. Mech. E.*, **177**, p. 269.

Related case studies

Case Study 6.9: Materials for springs

8.7 Summary and conclusions

In designing components which are loaded such that they bend, twist or buckle, the designer has two groups of variables with which to optimize performance: the *material properties* and the *shape of the section*. The best choice of material depends on the shapes in which it is available, or to which it could potentially be formed. The procedure of Chapter 7 gives a method for optimizing the choice of material and shape.

Its use is illustrated in this chapter. Often the designer has available certain stock materials in certain shapes. Then that with the greatest value of the appropriate material index (of which a number were listed in Table 8.1) maximizes performance. Sometimes sections can be specially designed; then material properties and design loads determine a maximum practical value for the shape factor above which local buckling leads to failure; again, the procedure gives an optimal choice of material and shape. Further gains in efficiency are possible by combining microscopic with macroscopic shape.

Chapter 9

Multiple constraints and compound objectives

9.1 Introduction and synopsis

Most decisions you make in life involve trade-offs. Sometimes the trade-off is to cope with conflicting constraints: I must pay this bill but I must also pay that one — you pay the one which is most pressing. At other times the trade-off is to balance divergent objectives: I want to be rich but I also want to be happy — and resolving this is harder since you must balance the two, and wealth is not measured in the same units as happiness.

So it is with selecting materials. Commonly, the selection must satisfy several, often conflicting, constraints. In the design of an aircraft wing-spar, weight must be minimized, with constraints on stiffness, fatigue strength, toughness and geometry. In the design of a disposable hot-drink cup, cost is what matters; it must be minimized subject to constraints on stiffness, strength and thermal conductivity, though painful experience suggests that designers sometimes neglect the last. In this class of problem there is one design objective (minimization of weight or of cost) with many constraints. Nature being what it is, the choice of material which best satisfies one constraint will not usually be that which best meets the others.

A second class of problem involves divergent objectives, and here the conflict is more severe. The designer charged with selecting a material for a wing-spar that must be both as light *and* as cheap as possible faces an obvious difficulty: the lightest material will certainly not be the cheapest, and vice versa. To make any progress, the designer needs a way of trading off weight against cost. Strategies for dealing with both classes of problem are summarized in Figure 9.1 on which we now expand.

There are a number of quick although subjective ways of dealing with conflicting constraints and objectives: the *sequential index* method, the *method of weight-factors*, and methods employing *fuzzy logic*. They are a good way of getting into the problem, so to speak, but their limitations must be recognized. Subjectivity is eliminated by employing the *active constraint method* to resolve conflicting constraints, and by combining objectives, using *exchange constants*, into a single *value function*.

We use the beam as an example, since it is now familiar. For simplicity we omit shape (or set all shape factors equal to 1); reintroducing it is straightforward.

9.2 Selection by successive application of property limits and indices

Suppose you want a material for a light beam (the objective) which is both stiff (constraint 1) and strong (constraint 2), as in Figure 9.2. You could choose materials with high modulus E for

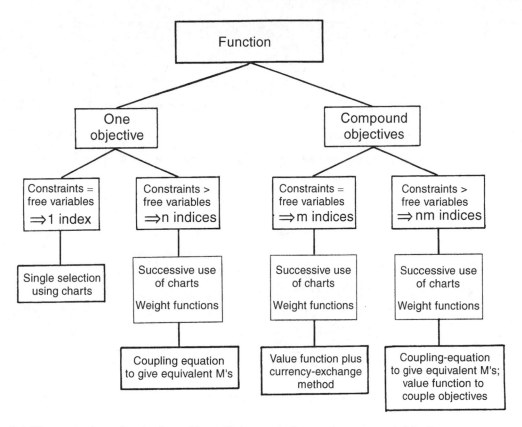

Fig. 9.1 The procedures for dealing with multiple constraints and compound objectives.

stiffness, and then the subset of these which have high elastic limits σ_y for strength, and the subset of those which have low density ρ for light weight. Some selection systems work that way, but it is not a good idea because there is no guidance in deciding the relative importance of the limits on E, σ_y and ρ.

A better idea: first select the subset of materials which is light and stiff (index $E^{1/2}/\rho$), then the subset which is light and strong (index $\sigma_y^{2/3}/\rho$), and then seek the common members of the two subsets. Then you have combined some of the properties in the right way.

Put more formally: an objective function is identified; each constraint is used in turn to eliminate the free variable, temporarily ignoring the others, giving a set of material-indices (which we shall call M_i) which are ranked according to the importance, in your judgement, of the constraints from which they arise. Then a subset of materials is identified which has large values of the first index, M_1, either by direct calculation or by using the appropriate selection chart. The subset is left large enough to allow the remaining constraints to be applied to it.

The second index M_2 is now applied, identifying a second subset of materials. Common members of the two subsets are identified and ranked according to their success in maximizing the two indices. It will be necessary to iterate, narrowing the subset controlled by the hard constraints, broadening that of the softer ones. The procedure can be repeated, using further constraints, as often as needed provided the initial subsets are not made too small. The same method can be applied to multiple objectives.

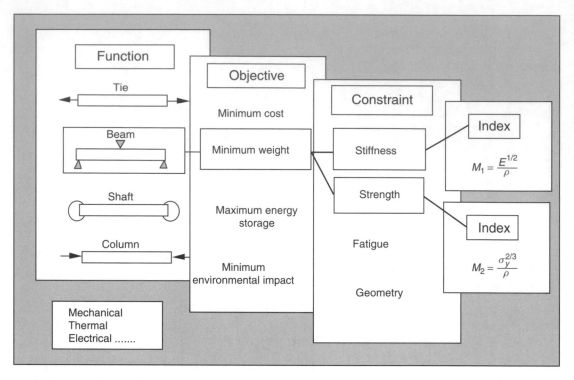

Fig. 9.2 One objective (here, minimizing mass) and two constraints (stiffness and strength) lead to two indices.

This approach is quick (particularly if it is carried out using computer-based methods*), and it is a good way of getting a feel for the way a selection exercise is likely to evolve. But it is far from perfect, because it involves judgement in placing the boundaries of the subsets. Making judgements is a part of materials selection — the context of any real design is sufficiently complex that expert judgmental skills is always needed. But there are problems with the judgements involved in the successive use of indices. The greatest is that of avoiding subjectivity. Two informed people applying the same method can get radically different results because of the sensitivity of the outcome to the way the judgements are applied.

9.3 The method of weight-factors

Weight-factors express judgements in a more formal way. They provide a way of dealing with quantifiable properties (like E, or ρ, or $E^{1/2}/\rho$) and also with properties which are difficult to quantify, like corrosion and wear.

The method, applied to material selection, works like this. The key properties or indices are identified and their values M_i are tabulated for promising candidates. Since their absolute values can differ widely and depend on the units in which they are measured, each is first scaled by dividing it by the largest index of its group, $(M_i)_{max}$, so that the largest, after scaling, has the value 1. Each is

---

* See, for example, the CMS selection software marketed by Granta Design (1995).

then multiplied by a weight-factor, w_i, which expresses its relative importance for the performance of the component, to give a weighted index W_i:

$$W_i = w_i \frac{M_i}{(M_i)_{\text{max}}} \tag{9.1}$$

For properties that are not readily expressed as numerical values, such as weldability or wear resistance, rankings such as A to E are expressed instead by a numeric rating, A = 5 (very good) to E = 1 (very bad) and then, as before, dividing by the highest rating value. For properties that are to be minimized, like corrosion rate, the scaling uses the minimum value $(M_i)_{\text{min}}$, expressed in the form

$$W_i = w_i \frac{(M_i)_{\text{min}}}{M_i}$$

The weight-factors w_i are chosen such that they add up to 1, that is: $w_i < 1$ and $\Sigma w_i = 1$. There are numerous schemes for assigning their values (see Further Reading: Weight factors). All require, in varying degrees, the use of judgement. The most important property or index is given the largest w, the second most important, the second largest and so on. The W_i are calculated from equation (9.1) and summed. The best selection is the material with the largest value of the sum

$$W = \sum_i W_i = \sum_i w_i \frac{M_i}{(M_i)_{\text{max}}} \tag{9.2}$$

But there are problems with the method, some obvious (like that of assigning values for the weight factors), some more subtle*. Here is an example: the selection of a material for a light beam which must meet constraints on both stiffness (index $M_1 = E^{1/2}/\rho$) and strength (index $M_2 = \sigma_y^{2/3}/\rho$). The values of these indices are tabulated for four materials in Table 9.1. Stiffness, in our judgement, is more important than strength, so we assign it the weight factor

$$w_1 = 0.7$$

That for strength is then

$$w_2 = 0.3$$

Normalize the index values (as in equation (9.1)) and sum them (equation (9.2)) to give W. The second last column of Table 9.1 shows the result: beryllium wins easily; Ti-6-4 comes second, 6061 aluminium third. But observe what happens if beryllium (which can be toxic) is omitted from the selection, leaving only the first three materials. The same procedure now leads to the values of W in the last column: 6061 aluminium wins, Ti-6-4 is second. Removing one, non-viable, material

Table 9.1 Example of use of weight factors

Material	ρ Mg/m^3	E GPa	σ_y MPa	$\dfrac{E^{1/2}}{\rho}$	$\dfrac{\sigma_y^{2/3}}{\rho}$	W (inc. Be)	W (excl. Be)
1020 Steel	7.85	205	320	1.82	6.0	0.24	0.52
6061 Al (T4)	2.7	70	120	3.1	9.0	0.39	**0.86**
Ti-6-4	4.4	115	950	2.4	**17.1**	0.48	0.84
Beryllium	1.86	300	170	**9.3**	16.5	**0.98**	—

*For a fuller discussion see de Neufville and Stafford (1971) or Field and de Neufville (1988).

from the selection has reversed the ranking of those which remain. Even if the weight factors could be chosen with accuracy, this dependence of the outcome on the population from which the choice is made is disturbing. The method is inherently unstable, sensitive to irrelevant alternatives.

The most important factor, of course, is the set of values chosen for the weight-factors. The schemes for selecting them are structured to minimize subjectivity, but an element of personal judgement inevitably remains. The method gives pointers, but is not a rigorous tool.

9.4 Methods employing fuzzy logic

Fuzzy logic takes weight-factors one step further. Figure 9.3 at the upper left, shows the probability $P(R)$ of a material having a property or index-value in a given range of R. Here the property has a well-defined range for each of the four materials A, B, C and D (the values are *crisp* in the terminology of the field). The selection criterion, shown at the top right, identifies the range of R which is sought for the properties, and it is *fuzzy*, that is to say, it has a well-defined *core* defining the ideal range sought for the property, with a wider *base*, extending the range to include boundary regions in which the value of the property or index is allowable, but with decreasing acceptability as the edges of the base are approached.

The superposition of the two figures, shown at the centre of Figure 9.3, illustrates a single selection stage. Desirability is measured by the product $P(R)S(R)$. Here material B is fully acceptable — it acquires a weight of 1. Material A is acceptable but with a lower weight, here 0.5; C is acceptable with a weight of roughly 0.25, and D is unacceptable — it has a weight of 0. At the end

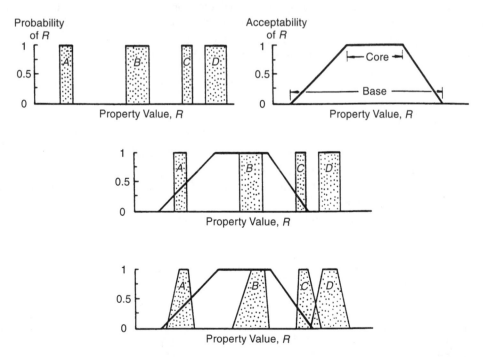

Fig. 9.3 Fuzzy selection methods. Sharply-defined properties and a fuzzy selection criterion, shown at (a), are combined to give weight-factors for each material at (b). The properties themselves can be given fuzzy ranges, as shown at (c).

of the first selection stage, each material in the database has one weight-factor associated with it. The procedure is repeated for successive stages, which could include indices derived from other constraints or objectives. The weights for each material are aggregated — by multiplying them together, for instance — to give each a super-weight with a value between 0 (totally unacceptable) to 1 (fully acceptable by all criteria). The method can be refined further by giving fuzzy boundaries to the material properties or indices as well as to the selection criteria, as illustrated in the lower part of Figure 9.3. Techniques exist to choose the positions of the cores and the bases, but despite the sophistication the basic problem remains: the selection of the ranges $S(R)$ is a matter of judgement.

Successive selection, weight factors and fuzzy methods all have merit when more rigorous analysis, of the sort described next, is impractical. And they can be fast. They are a good first step. But if you really want to identify the best material for a complex design, you need to go further. Ways of doing that come next.

9.5 Systematic methods for multiple constraints

Commonly, the specification of a component results in a design with multiple constraints, as in the second column of Figure 9.1. Here the *active constraint method* is the best way forward. It is systematic — it removes the dependence on judgement. The idea is simple enough. Identify the most restrictive constraint. Base the design on that. Since it is the most restrictive, all other constraints will automatically be satisfied.

The method is best illustrated through an example. We stay with that of the light, stiff, strong beam. For simplicity, we leave out shape (including it involves no new ideas). The objective function is

$$m = A\ell\rho \tag{9.3}$$

where $A = t^2$ is the area of the cross-section. The first constraint is that on stiffness, S

$$S = \frac{C_1 EI}{\ell^3} \tag{9.4}$$

with $I = t^4/12$ and $C_1 = 48$ for the mode of loading shown in Figure 9.4; the other variables have the same definitions as in Chapter 5. Using this to eliminate A in equation (9.3) gives the mass of the beam which will just provide this stiffness S (equation (5.10), repeated here):

$$m_1 = \left(\frac{12}{C_1} \frac{S}{\ell} \right)^{1/2} \ell^3 \left[\frac{\rho}{E^{1/2}} \right] \tag{9.5}$$

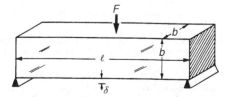

Fig. 9.4 A square-section beam loaded in bending. It has a second moment of area $I = t^4/12$. It must have a prescribed stiffness S and strength F_f, and be as light as possible.

The second constraint is that on strength. The collapse load of a beam is

$$F_f = C_2 \frac{I\sigma_y}{y_m \ell} \tag{9.6}$$

where $C_2 = 4$ and $y_m = t/2$ for the configuration shown in the figure. Using this instead of equation (9.4) to eliminate A in equation (9.3) gives the mass of the beam which will just support the load F_f:

$$m_2 = \left(\frac{6}{C_2} \frac{F_f}{\ell^2} \right)^{2/3} \ell^3 \left[\frac{\rho}{\sigma_y^{2/3}} \right] \tag{9.7}$$

More constraints simply lead to more such equations for m.

If the beam is to meet both constraints, its weight is determined by the larger of m_1 and m_2; if there are i constraints, then it is determined by the largest of all the m_i. Define $\tilde{m}$ as

$$\tilde{m} = \max(m_1, m_2, m_3, \ldots) \tag{9.8}$$

The best choice is that of the material with the smallest value of $\tilde{m}$. It is the lightest one that meets or exceeds all the constraints.

That is it. Now the ways to use it.

The analytical method

Table 9.2 illustrates the use of the method to select a material for a light, stiff, strong beam of length ℓ, stiffness S and collapse load F_f with the values

$$\ell = 1\,\text{m} \qquad S = 10^6\,\text{N/m} \qquad F_f = 2 \times 10^4\,\text{N}$$

Substituting these values and the material properties shown in the table into equations (9.5) and (9.7) gives the values for m_1 and m_2 shown in the table. The last column shows $\tilde{m}$ calculated from equation (9.8). For these design requirements Ti-6-4 is emphatically the best choice: it allows the lightest beam which satisfies both constraints.

The best choice depends on the details of the design requirements; a change in the prescribed values of S and F_f alters the selection. This is an example of the power of using a systematic method: it leads to a selection which does not rely on judgement; two people using it independently will reach exactly the same conclusion. And the method is robust: the outcome is not influenced by irrelevant alternatives. It can be generalized and presented on selection charts (allowing a clear graphical display even when the number of materials is large) as described next.

Table 9.2 Selection of a material for a light, stiff, strong beam

Material	ρ kg/m^3	E GPa	σ_y MPa	m_1 kg	m_2 kg	$\tilde{m}$ kg
1020 Steel	7850	205	320	8.7	**16.2**	16.2
6061 Al	2700	70	120	5.1	**10.7**	10.7
Ti-6-4	4400	115	950	**6.5**	4.4	**6.5**

The graphical method

Stated more formally, the steps of the example in the last section were these.

(a) Express the objective as an equation, here equation (9.3).
(b) Eliminate the free variable using each constraint in turn, giving sets of performance equations (objective functions) with the form.

$$P_1 = f_1(F)g_1(G)M_1 \qquad \text{9.9(a)}$$

$$P_2 = f_2(F)g_2(G)M_2 \qquad \text{9.9(b)}$$

$$P_3 = f_3(F)\ldots \text{ etc.}$$

where f and g are expressions containing the functional requirements F and geometry G, and M_1 and M_2 are material indices. In the example, these are equations (9.5) and (9.7).

(c) If the first constraint is the most restrictive (that is, it is the *active* constraint), the performance is given by equation (9.9a), and this is maximized by seeking materials with the best values of M_1 ($E^{1/2}/\rho$ in the example). When the second constraint is the active one, the performance equation is given by equation (9.9b) and the highest values of M_2 (here, $\sigma_y^{2/3}/\rho$) must be sought. And so on.

In the example above, performance was measured by the mass m. The selection was made by evaluating m_1 and m_2 and comparing them to identify the active constraint, which, as Table 9.2 shows, depends on the material itself. The same thing can be achieved graphically for two constraints (and more if repeated), with the additional benefit that it displays, in a single picture, the active constraint and the best material choice even when the number of materials is large. It works like this.

Imagine a chart with axes of M_1 and M_2, as in Figure 9.5. It can be divided into two domains in each of which one constraint is active, the other inactive. The switch of active constraint lies at the boundary between the two regimes; it is the line along which the equations (9.9a) and (9.9b)

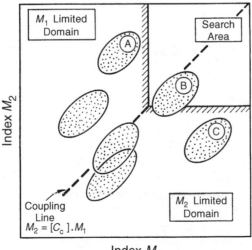

Fig. 9.5 A chart with two indices as axes, showing a box-shaped contour of constant performance. The corner of the box lies on the coupling line. The best choices are the materials which lie in the box which lies highest up the coupling line.

are equal. Equating them and rearranging gives:

$$M_2 = \left[\frac{f_1(F)g_1(G)}{f_2(F)g_2(G)}\right] M_1 \tag{9.10}$$

or

$$M_2 = [C_c]M_1 \tag{9.11}$$

This equation couples the two indices M_1 and M_2; we shall call it the *coupling equation*. The quantity in square brackets — the *coupling constant*, C_c — is fixed by the specification of the design. Materials with M_2/M_1 larger than this value lie in the M_1-limited domain. For these, the first constraint is active and performance limited by equation (9.9a) and thus by M_1. Those with M_2/M_1 smaller than C_c lie in the M_2-limited domain; the second constraint is active and performance limited by equation (9.9b) and thus by M_2. It is these conditions which identify the box-shaped search region shown in Figure 9.5. The corner of the box lies on the coupling line (equation (9.11)); moving the box up the coupling line narrows the selection, identifying the subset of materials which maximize the performance while simultaneously meeting both constraints. Change in the value of the functional requirements F or the geometry G changes the coupling constant, shifts the line, moves the box and changes the selection.

Taking the example earlier in this section and equating m_1 to m_2 gives:

$$M_2 = \left[\left(\frac{6F_f}{C_2\ell^2}\right)^{2/3}\left(\frac{C_1}{12}\frac{\ell}{S}\right)^{1/2}\right] M_1 \tag{9.12}$$

with $M_1 = E^{1/2}/\rho$ and $M_2 = \sigma_y^{2/3}/\rho$. The quantity in square brackets is the coupling constant. It depends on the values of stiffness S and collapse load F_f, or more specifically, on the two structural loading coefficients* S/ℓ and F_f/ℓ^2. They define the position of the coupling line, and thus the selection.

Worked examples are given in Chapter 10.

9.6 Compound objectives, exchange constants and value-functions

Cost, price and utility

Almost always, a design requires the coupled optimization of two or more measures of performance; it has *compound objectives* (Figure 9.1, third column and Figure 9.6). The designer's objective for a performance bicycle might be to make it as light as possible; his marketing manager might insist that it be as cheap as possible. The owner's objective in insulating his house might be to minimize heat loss, but legislation might require that the environmental impact of the blowing agent contained in the insulation be minimized instead. These examples reveal the difficulties: the individual objectives conflict, requiring that a compromise be sought; and in seeking it, how is weight to be compared with cost, or heat flow with environmental impact? Unlike the Ps of the last section, each is measured in different units; they are incommensurate. As mentioned earlier, the judgement-based methods described earlier in this chapter can be used. The 'successive selection' procedure using the charts

---

* See Section 5.5 for discussion of structural loading coefficients

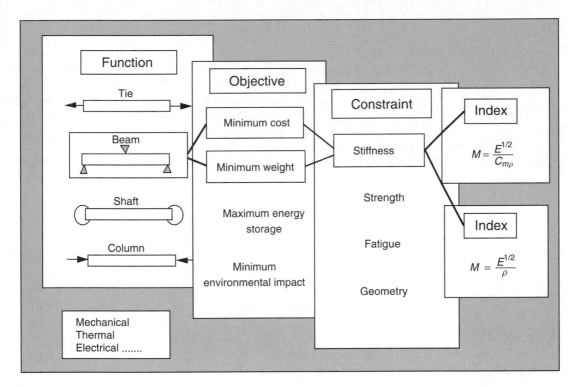

Fig. 9.6 Two objectives (here, minimizing mass and cost) and one constraint (stiffness) lead to two indices.

('first choose the subset of materials which minimizes mass then the subset which minimizes cost, then seek the common members of the two subsets'), and the refinements of it by applying weight-factors or fuzzy logic lead to a selection, but because dissimilar quantities are being compared, the reliance on judgement and the attendant uncertainty is greater than before.

The problem could be overcome if we had a way of relating mass to cost, or energy to environ-mental impact. With this information a 'compound-objective' or *value function* can be formulated in which the two objectives are properly coupled. A method based on this idea is developed next. To do so, we require *exchange constants* between the objectives which, like exchange-rates between currencies, allows them to be expressed in the same units — in a common currency, so to speak. Any one of those just listed — mass, cost, energy or environmental impact — could be used as the common measure, but the obvious one is cost. Then the exchange constant is given the symbol $E^\$$.

First, some definitions. A product has a *cost*, C; it is the sum of the costs to the manufacturer of materials, manufacture and distribution. To the consumer, the product has a *utility U*, a measure, in his or her mind, of the worth of the product. The consumer will be happy to purchase the product if the *price*, P, is less than U; and provided P is greater than C, the manufacturer will be happy too. This desirable state of affairs is summed up by

$$C < P < U \tag{9.13}$$

Exchange any two terms in this equation, and someone is unhappy. The point is that utility is not the same as cost. In some situations a given product can have a high utility, in others it is worthless, even though the cost has not changed. More specific examples in a moment.

Value functions and exchange constants

First, a formal definition; then examples.

A design requires that several (i) objectives must be met. Each objective relates to a performance characteristic P_i with the general form of equations (9.9). The first of these might (for example) describe the mass of the component; the second, the energy consumed in making it; the third, its cost. Ideally we would like to minimize all three, but the cheapest is not the lightest or the most energy efficient; minimizing one does not minimize the others.

To overcome this, define a value function, V, such that

$$V = E_1^\$ P_1 + E_2^\$ P_2 + E_3^\$ P_3 \cdots \tag{9.14}$$

The quantities $E_1^\$, E_2^\$$, etc. are the exchange constants. They convert performances P_i (here with units of kg, MJ and \$) into value (measured in \$, say). Differentiating equation (9.14) gives

$$E_1^\$ = \left(\frac{\partial V}{\partial P_1} \right)_{P_2, P_3 \ldots} \tag{9.14a}$$

$$E_2^\$ = \left(\frac{\partial V}{\partial P_2} \right)_{P_1, P_3 \ldots} \tag{9.14b}$$

and so on. If P_1 is the mass of the component, then $E_1^\$$ is the change in value associated with unit change in mass. If P_2 is the energy content, then $E_2^\$$ is the change in value associated with unit change in energy content. And if P_3 is the cost of the component, then $E_3^\$ = -1$ because unit increase in cost give unit decrease in value.

The value of the exchange constant depends on the application. Its value is influenced by many factors, some of them based on sound engineering reasoning, some on market forces, and still others on perceived value. Approximate values for the exchange constant relating mass and cost are listed in Table 9.3. Their values are negative because an increase in mass leads, in these applications, to a decrease in value. In a space-vehicle the value of a mass reduction is high; that of the same mass reduction in a family car is much lower. The ranges given in the top part of the table are related to their applications. They can be estimated approximately in various ways. The cost of launching a payload into space lies in the range \$3000 to \$10 000/kg; a reduction of 1 kg in the weight of the launch structure would allow a corresponding increase in payload, giving the value-range in the table. Similar arguments based on increased payload or decreased fuel consumption give the values shown for civil aircraft, commercial trucks and automobiles. The values change slowly with time, reflecting changes in fuel costs, legislation to increase fuel economy and such like. Special

Table 9.3 Exchange constant for mass saving in transport systems

Transport system	Exchange constant $E^\$$ (US\$/kg)
	(note negative sign)
Family car (based on fuel saving)	−0.5 to −1.5
Truck (based on payload)	−5 to −10
Civil aircraft (based on payload)	−100 to −500
Military aircraft (performance, payload)	−500 to −2000
Space vehicle (based on payload)	−3000 to −10 000
Bicycle (based largely on perceived value)	−80 to −2000

circumstances can change them dramatically — an aero-engine builder who has guaranteed a certain power/weight ratio for his engine, may be willing to pay more than $1000 to save a kilogram if it is the only way in which the guarantee can be met — but we shall ignore these and stay with the more usual situations.

The change in value associated with a unit change in energy consumption has an even larger range (Table 9.4). Here the exchange constant is the change in value associated with unit increase in the energy consumed, and that is simply the cost of energy (with the sign reversed, since an increase in energy consumption, all other factors held constant, gives a decrease value). The exchange constant depends on the form in which the energy is provided, the country in which it is purchased, and — in the case of electricity — the time of day. The example of electricity illustrates just how great can be the variations in exchange constant: grid power for industrial use costs about $0.02/MJ (1 MJ is 3.6 kWh); energy in the form of an AA battery for your walkman costs more than 1000 times more; the energy source in your watch cost you, per MJ, 100 times more than that. But you pay it, because, on your scale of values, it is worth it.

These values for the exchange constant are based on engineering criteria. More difficult to assess are those based on perceived value. That for the weight/cost trade-off for a bicycle is an example. To the enthusiast, a lighter bike is a better bike. Figure 9.7 shows just how much the cyclist values reduction in weight. The tangents give the exchange constant: it ranges from $80/kg to $2000/kg, depending on the mass: the exchange constant depends on the value of the performance characteristic, here mass. Over any small part of the curve it can be linearized giving the tangents, and this is usually acceptable; but if large changes of mass become possible, this dependence must be included (by expressing $E^\$$ as $E^\$(m)$). Does it make sense for the ordinary cyclist to pay $2000 to reduce the mass of the bike by 1 kg when, by dieting, he could reduce the mass of the system (himself plus the bike) by more without spending a penny? Possibly. But mostly it is perceived value. Advertising aims to increase the perceived value of a product, increasing its value without increasing its cost. It influences the exchange constants for family cars and it is the driver for the development of titanium watches, carbon fibre spectacle frames and much more. Perceived, rather than rational, values are frequently associated with the choice of material for sports equipment. And they are harder to pin down.

Determining exchange constants

Exchange constants based on engineering criteria are determined by analysing the economics of the way in which each performance characteristic changes the life-cost of the product. Simple

Table 9.4 Approximate exchange constant of energy

Energy source	Exchange constant $E^\$$ (US$/MJ)
Coal	−0.003 to −0.006
Oil	−0.007 to −0.012
Gas	−0.003 to −0.005
Gasoline (US)	−0.012 to −0.015
Gasoline (Europe)	−0.03 to −0.04
Electricity (national grid, US)	−0.02 to −0.03
Electricity (national grid, Europe)	−0.03 to −0.04
Electricity (lead-acid battery, 1000 recharges)	−0.1 to −0.3
Electricity (alkaline AA battery)	−35 to −150
Electricity (silver oxide battery)	−1000 to −3500

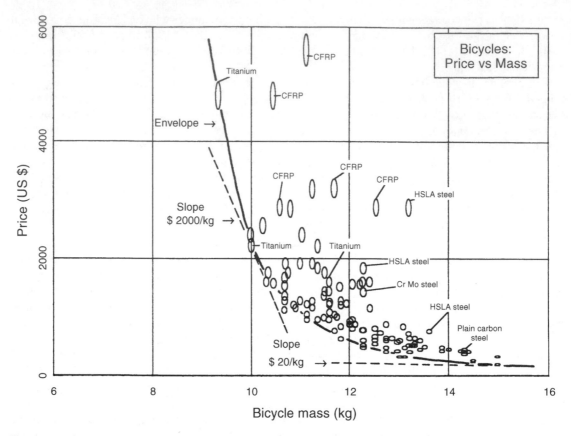

Fig. 9.7 A plot of cost against mass for bicycles. The lower envelope of price is here treated as a contour of constant value. The exchange constant is the slope of a tangent to this curve. It varies with mass.

examples, given earlier, led to the values in Tables 9.3 and 9.4. When — as with bicycles — a range of products has existed long enough for the prices to have stabilized, the exchange constant can be estimated from the appropriate plots, as in Figure 9.7. When this is not so, it may be possible to build up a plot like Figure 9.7 by using interviewing techniques which elicit the change in value that a potential purchaser might associate with a given change in performance*.

Sometimes, however, establishing the exchange constant can be very difficult. An example is that for environmental impact — the damage to the environment caused by manufacture, or use, or disposal of a given product. Minimizing environmental damage could be made an objective, like minimizing cost. Ingenious design can reduce the first without driving the second up too much, but until the exchange constant is defined — by legislation, perhaps, or by necessity — it is difficult for the designer to respond.

An example: value function for mass and cost

We will use the same example as before: the now-very-boring beam. Consider, then, an application for which is sought a material for a *light*, *cheap*, stiff beam (objectives italicized) of prescribed

---

* For a fuller discussion see de Neufville and Stafford (1971) and Field and de Neufville (1988).

length ℓ and stiffness S. Ignoring shape, its mass, m, is given by equation (9.5) which, repeated, is

$$m = \left(\frac{12}{C_1}\frac{S}{\ell}\right)^{1/2} \ell^3 \left[\frac{\rho}{E^{1/2}}\right] \tag{9.15a}$$

The first objective is to minimize m. The cost of the beam is

$$C = C_m m \tag{9.15b}$$

where C_m is the material cost (in shaped form if necessary) and m is defined by equation (9.15a). The second objective is to minimize C.

To proceed further we need the exchange constant, $E^\$$, and this, we know, depends on the application. Given this, we construct a value function, V, following equation (9.14):

$$\boxed{V = E^\$ m - C} \tag{9.16}$$

Think of it this way: the term $E^\$ m$ measures the value V to you of a beam of mass m and stiffness S; the term C measures its cost — its exchange constant is simply -1, giving the negative sign. The best choice of material is that with the largest (least negative) value of this function.

Analytical evaluation

Table 9.5 illustrates the use of the method to select a beam with

$$\ell = 1\,\text{m} \qquad S = 10^6\,\text{N/m} \qquad C_1 = 48$$

Substituting these and the material properties shown in the table into equations (9.15a) and (9.15b) gives the values of m and C shown in the table. Forming the value function V of equation (9.16) with $E^\$ = -1$ \$/kg gives the values shown in the second last column of the table; for this exchange constant, steel wins. The last column shows what happens if $E^\$ = -100$ \$/kg: the 6061 aluminium alloy maximizes the value function.

Although we have used the values of the design requirements ℓ, S and C_1 to evaluate V, they were not, in fact, necessary to make the selection, which remains unchanged for all values of these variables. This remarkable and useful fact can be understood in the following way. Substituting equations (9.15a) and (9.15b) into (9.16) gives

$$V = (E^\$ - C_m)m = (E^\$ - C_m)\left(\frac{12S}{C_1\ell}\right)^{1/2} \ell^3 \left[\frac{\rho}{E^{1/2}}\right] \tag{9.17}$$

Table 9.5 Value functions, V, for two values of exchange constant, $E^\$$

Material	ρ kg/m³	E GPa	C_m^* \$/kg	$m,$ kg	t mm	C \$	$V, E^\$ =$ -1 \$/kg	$V, E^\$ =$ -100 \$/kg
1020 Steel	7850	205	0.5	8.7	33	**4.35**	−13	−8700
6061 Al	2700	70	1.9	**5.1**	43	9.7	−15	**−5100**
Ti-6-4	4400	115	22	6.5	38	143	−150	−6640

*Cost of material in shape of beam.

This is the quantity we wish to maximize. Rearranging gives

$$\tilde{V} = \frac{V}{\ell^3(12S/C_1\ell)^{1/2}} = E^\$ \left[\frac{\rho}{E^{1/2}}\right] - \left[\frac{C_m\rho}{E^{1/2}}\right]$$

which we write as

$$\tilde{V} = E^\$ M_1^* - M_2^* \tag{9.18}$$

Ranking materials by V gives the same order as ranking them by $\tilde{V}$, and this is independent of the values of ℓ, S and C_1. It depends only on $E^\$$ and on two material indices $M_1^* = \rho/E^{1/2}$ and $M_2^* = C_m\rho/E^{1/2}$. (These are the reciprocals of indices used earlier; the asterisk on the Ms are a reminder of this.)

Graphical analysis

The graphical method involves a selection chart with axes M_1^* and M_2^*. Consider first the use of the value function to seek a *substitute* for an existing material. The incumbent is material A. On a plot of M_1^* against M_2^* (Figure 9.8) materials which lie below A have a lower value of M_1^*; those which lie to the left have a lower value of M_2^*. It is clear that the materials which lie in the lower left quadrant have lower values of $\tilde{V}$, regardless of the value of $E^\$$, and thus are superior to A in performance.

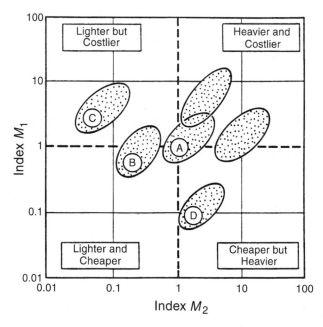

Fig. 9.8 A schematic chart showing two indices, M_1^* measuring cost and M_2^* measuring weight. If the currently used material is A, then B is both cheaper and lighter. The material C is lighter but costs more; D is cheaper but heavier. This selection ignores the trade-off between weight and cost.

That argument is correct, but incomplete. For a given value of the exchange constant, materials in the two neighbouring quadrants can also be viable substitutes. A line of constant $\tilde{V}$, through the material A links materials which have the *same* value of $\tilde{V}$. This line has a slope, found by differentiating equation (9.18), of

$$\left(\frac{\partial M_2^*}{\partial M_1^*}\right)_{\tilde{V}} = E^{\$} \tag{9.19}$$

and this, on linear scales, is a straight line with a slope of $E^{\$}$. Materials below this line perform better than those on or above it — and this now includes materials in the neighbouring quadrants.

In practice, the ranges of M_1^* and M_2^* are large, and it becomes more attractive to use logarithmic scales. Then the line becomes curved. Figure 9.9 shows lines of constant $\tilde{V}$ for three values of $E^{\$}$; it is now apparent that the straight, horizontal and vertical lines on the figure are the extremes, corresponding to $E^{\$} = 0$ and $E^{\$} = \pm\infty$. As Figure 9.9 shows, a low value of $E^{\$}$ makes material D a better choice than A, and a high one makes material C a better choice, although neither lie in the bottom left quadrant.

For *selection* we require contours for different values of $\tilde{V}$ (equation (9.18)) not just for the one on which material A sits. Figure 9.10 shows, for two values of $E^{\$}$, the way the contours look. They all have the same shape; as $\tilde{V}$ decreases, they move downwards along the line

$$\tilde{M}_2 = E^{\$}\tilde{M}_1 - \tilde{V}$$

The best choice of material, for a given $E^{\$}$, is that with the lowest $\tilde{V}$. Note that the absolute value of $\tilde{V}$ is unimportant, either for selection or for substitution; it is only the relative value that is

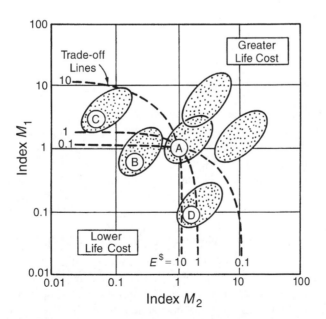

Fig. 9.9 The proper comparison of A with competing materials is made by constructing a value-function, $\tilde{V}$, which combines the cost of material with that associated with weight, using the weight–cost exchange constant $E^{\$}$. When $E^{\$} = 10$ \$/kg, C is a better choice than A; when $E^{\$} = 0.1$ \$/kg, D is better than A.

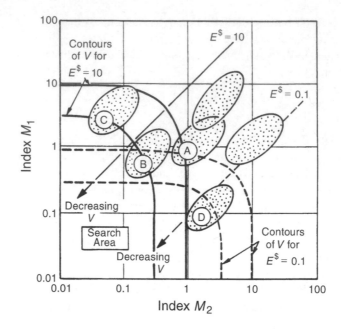

Fig. 9.10 Selection to minimize total life-cost, using the exchange constant method. The contours show the value-function, $\tilde{V}$, for two values of $E^\$$.

required. Thus the constant $\ell^3(12S/C_1\ell)^{1/2}$ in equation (9.15) can be dropped; and the only matter of importance is to ensure that M_1^*, M_2^* and $E^\$$ are expressed in consistent units.

Described in the abstract, as here, these methods sound a little complicated. The case studies of Chapter 10 will illustrate how they work.

9.7 Summary and conclusions

The method of material indices allows a simple, transparent procedure for selecting materials to meet an objective (like minimizing component weight) while meeting a constraint (safely carrying a design load, for instance). But most designs have several constraints, and it is usual that the selection is driven by divergent objectives.

Judgement can be used to rank the importance of the competing constraints and objectives, and this is often the simplest route. To do this, an index is derived for the most important of these, which is then used to select the first, broad, subset of materials. The members of the subset are now examined for their ability to satisfy the remaining constraints and objectives. Weight-factors or fuzzy logic put the judgement on a more formal footing, but can also obscure its consequences.

Judgement can, sometimes, be replaced by analysis. For multiple constraints, this is done by identifying the active constraint and basing the design on this. The procedure can be made graphical by deriving coupling equations which link the material indices; then simple constructions on material selection charts with the indices as axes identify unambiguously the subset of materials which maximize performance while meeting all the constraints. Compound objectives require consideration of the exchange constant; it allows both objectives to be expressed in the same units (usually cost) and the formulation of a value-function V which guides material choice. Here, too, simple constructions on charts with material indices as axes allow the optimum subset of materials to be selected.

When multiple constraints operate, or a compound objective is involved, the best choice of material is far from obvious, and can be counter-intuitive. It is here that the methods developed have real power.

9.8 Further reading

Multiple constraints and compound objectives

Ashby, M.F. and Cebon, D. (1996) *Case Studies in Materials Selection*, Granta Design, Trumpington Mews, 40B High Street, Trumpington, Cambridge CB2 2LS, UK.

Ashby, M.F. (1997) Materials Selection: Multiple Constraints and Compound Objectives, ASTM STP 1311 on Computerisation/Networking of Material Databases, ASTM Publications, USA.

Weight-factors and fuzzy logic

Bassetti, D. (1995). Fuzzymat User Guide, Laboratoire de Thermodynamique et Physico-Chimie Métallurgiques, ENSEEG - 1130 rue de la Piscine, Domaine Universitaire, BP 75 - 38402 Saint Martin d'Hères, Cedex, France.

Dieter, G.E. (1991) *Engineering Design, A Materials and Processing Approach*, 2nd edition, McGraw-Hill, New York, pp. 150–153 and 255–257.

Dubois, D. and Prade, H. (1988) *Possibility Theory: An Approach to Computerised Processing of Uncertainty*, Plenum Press, New York.

Pechambert, P. and Brechet, Y. (1995) Etude d'une Methodologie de Choix des Materiaux Composites and Conception d'un Logical d'Aide à la Formulation des Verres. Memoire de DEA, Laboratoire de Thermodynamique et Physico-Chimie Métallurgiques, ENSEEG - 1130 rue de la Piscine, Domaine Universitaire, BP 75 - 38402 Saint Martin d'Hères, Cedex, France.

Sargent, P.M. (1991) *Materials Information for CAD/CAM*, Butterworth-Heinemann, Oxford.

Exchange constants and value functions

Bader, M.G. (1977) Proc of ICCM-11, Gold Coast, Australia, Vol. 1: Composites Applications and Design, ICCM, London.

Clark, J.P., Roth, R. and Field, F.R. (1997) Techno-economic Issues in Materials Science, AMS Handbook Vol. 20, *Materials Selection and Design*, ASM International, Materials Park, Ohio 44073-0002, USA

Field, F.R. and de Neufville, R. (1988) Materials selection — maximizing overall utility, *Metals and Materials*, June, pp. 378–382.

Keeney, R. and Raiffa, H. (1976) *Decisions with Multiple Objectives: Preference and Trade-Offs*, Wiley, New York.

de Neufville, R. and Stafford, J.H. (1971) *Systems Analysis for Engineers and Managers*, McGraw-Hill, New York.

Software

CMS Software (1995) Granta Design, Trumpington Mews, 40B High Street, Trumpington, Cambridge CB2 2LS, UK.

_____

Chapter 10

Case studies: multiple constraints and compound objectives

_____

10.1 Introduction and synopsis

These case studies illustrate how the techniques described in the previous chapter really work. Two were sketched out there: the light, *stiff*, *strong* beam, and the *light*, *cheap*, stiff beam. Here we develop four more. The first pair illustrate multiple constraints; here the *active constraint* method is used. The second pair illustrate compound objectives; here a *value function* containing an exchange constant, $E^\$$, is formulated. The examples are deliberately simplified to avoid clouding the illustration with unnecessary detail. The simplification is not nearly as critical as it may at first appear: the choice of material is determined primarily by the physical principles of the problem, not by details of geometry. The principles remain the same when much of the detail is removed so that the selection is largely independent of these.

Further case studies can be found in the sources listed under Further reading.

10.2 Multiple constraints — con-rods for high-performance engines

A connecting rod in a high performance engine, compressor or pump is a critical component: if it fails, catastrophe follows. Yet — to minimize inertial forces and bearing loads — it must weigh as little as possible, implying the use of light, strong materials, stressed near their limits. When cost, not performance, is the design goal, con-rods are frequently made of cast iron, because it is so cheap. But what are the best materials for con-rods when performance is the objective?

The model

Table 10.1 summarizes the design requirements for a connecting rod of minimum weight with two constraints: that it must carry a peak load F without failing either by fatigue or by buckling elastically. For simplicity, we assume that the shaft has a rectangular section $A = bw$ (Figure 10.1).

The objective function is an equation for the mass which we approximate as

$$m = \beta AL\rho \tag{10.1}$$

where L is the length of the con-rod and ρ the density of the material of which it is made, A the cross-section of the shaft and β a constant multiplier to allow for the mass of the bearing housings.

Table 10.1 The design requirements: connecting rods

Function	Connecting rod for reciprocating engine or pump
Objective	Minimize mass
Constraints	(a) Must not fail by high-cycle fatigue, or
	(b) Must not fail by elastic buckling
	(c) Stroke, and thus con-rod length L, specified

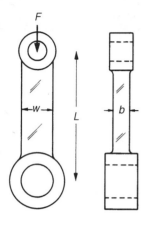

Fig. 10.1 A connecting rod. The rod must not buckle, fail by fatigue or by fast fracture (an example of multiple constraints). The objective is to minimize mass.

The fatigue constraint requires that

$$\frac{F}{A} \leq \sigma_e \tag{10.2}$$

where σ_e is the endurance limit of the material of which the con-rod is made. (Here, and elsewhere, we omit the safety factor which would normally enter an equation of this sort, since it does not influence the selection.) Using equation (10.2) to eliminate A in equation (10.1) gives the mass of a con-rod which will just meet the fatigue constraint:

$$m_1 = \beta F L \left(\frac{\rho}{\sigma_e} \right) \tag{10.3}$$

containing the material index

$$\boxed{M_1 = \frac{\sigma_e}{\rho}} \tag{10.4}$$

The buckling constraint requires that the peak compressive load F does not exceed the Euler buckling load:

$$F \leq \frac{\pi^2 EI}{L^2} \tag{10.5}$$

with $I = b^3 w / 12$. Writing $b = \alpha w$, where α is a dimensionless 'shape-constant' characterizing the proportions of the cross-section, and eliminating A from equation (10.1) gives a second equation for the mass

$$m_2 = \beta \left(\frac{12F}{\alpha \pi^2} \right)^{1/2} L^2 \left(\frac{\rho}{E^{1/2}} \right) \tag{10.6}$$

containing the material index (the quantity we wish to maximize to avoid buckling):

$$M_2 = \frac{E^{1/2}}{\rho} \qquad (10.7)$$

The con-rod, to be safe, must meet both constraints. For a given stroke, and thus length, L, the active constraint is the one leading to the largest value of the mass, m. Figure 10.2 shows the way in which m varies with L (a sketch of equations (10.3) and (10.6)), for a single material: short con-rods are liable to fatigue failure, long ones are prone to buckle.

The selection: analytical method

Consider first the selection of a material for the con-rod from among those listed in Table 10.2. The specifications are

$$L = 150\,\text{mm} \qquad F = 50\,\text{kN} \qquad \alpha = 0.5 \qquad \beta = 1$$

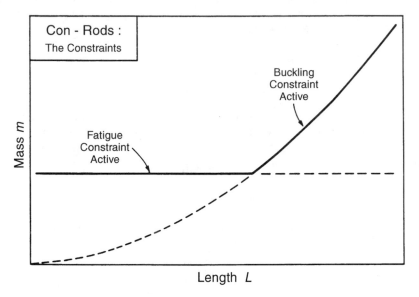

Fig. 10.2 The equations for the mass m of the con-rod are shown schematically as a function of L.

Table 10.2 Selection of a material for the con-rod

Material	ρ kg/m^3	E GPa	σ_e MPa	m_1 kg	m_2 kg	$\tilde{m}$ kg
Nodular cast iron	7150	178	250	**0.21**	0.13	0.21
HSLA steel 4140 (o.q. T-315)	7850	210	590	0.1	**0.13**	0.13
Al 539.0 casting alloy	2700	70	75	**0.27**	0.08	0.27
Duralcan Al-SiC(p) composite	2880	110	230	**0.09**	0.07	**0.09**
Ti-6-4	4400	115	530	0.06	**0.1**	0.1

The table lists the mass m_1 of a rod which will just meet the fatigue constraint, and the mass m_2 which will just meet that on buckling (equations (10.3) and (10.6)). For three of the materials the active constraint is that of fatigue; for two it is that of buckling. The quantity $\tilde{m}$ in the last column of the table is the larger of m_1 and m_2 for each material; it is the lowest mass which meets both constraints. The material offering the lightest rod is that with the smallest value of $\tilde{m}$. Here it is the metal-matrix composite Duralcan 6061–20% SiC(p). The titanium alloy is a close second. Both weigh about half as much as a cast-iron rod.

The selection: graphical method

The mass of the rod which will survive both fatigue and buckling is the larger of the two masses m_1 and m_2 (equations (10.3) and (10.6)). Setting them equal gives the equation of the coupling line:

$$M_2 = \left[\left(\frac{12L^2}{\pi^2 \alpha F} \right)^{1/2} \right] M_1 \tag{10.8}$$

The quantity in square brackets is the coupling constant: it contains the quantity F/L^2 — the 'structural loading coefficient' of Chapter 5.

Materials with the optimum combination of M_1 and M_2 are identified by creating a chart with these indices as axes. Figure 10.3 illustrates this, using a database of light alloys. Coupling lines for two values of F/L^2 are plotted on it, taking $\alpha = 0.5$. Two extreme selections are shown, one isolating the best subset when the structural loading coefficient F/L^2 is high, the other when it is low. For the high value ($F/L^2 = 0.5$ MPa), the best materials are high-strength Mg-alloys, followed by high-strength Ti-alloys. For the low value ($F/L^2 = 0.05$ MPa), beryllium alloys are the optimum choice. Table 10.3 lists the conclusions.

Postscript

Con-rods have been made from all the materials in the table: aluminium and magnesium in family cars, titanium and (rarely) beryllium in racing engines. Had we included CFRP in the selection, we would have found that it, too, performs well by the criteria we have used. This conclusion has been reached by others, who have tried to do something about it: at least three designs of CFRP con-rods have been prototyped. It is not easy to design a CFRP con-rod. It is essential to use continuous fibres, which must be wound in such a way as to create both the shaft and the bearing housings; and the shaft must have a high proportion of fibres which lie parallel to the direction in which F acts. You might, as a challenge, devise how you would do it.

Table 10.3 Materials for high-performance con-rods

Material	Comment
Magnesium alloys	ZK 60 and related alloys offer good all-round performance.
Titanium alloys	Ti-6-4 is the best choice for high F/L^2.
Beryllium alloys	The ultimate choice when F/L^2 is small. Difficult to process.
Aluminium alloys	Cheaper than titanium or magnesium, but lower performance.

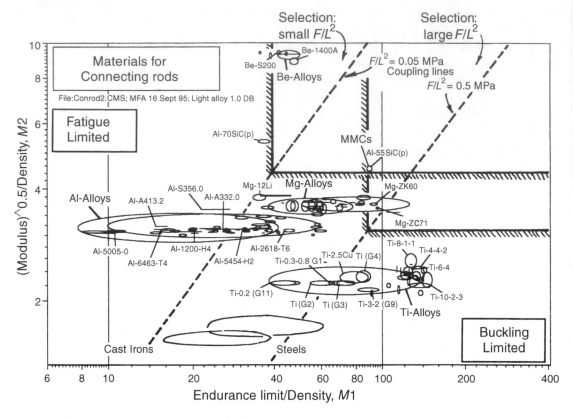

Fig. 10.3 Over-constrained design leads to two or more performance indices linked by coupling equations. The diagonal broken lines show the coupling equations for two values of the coupling constant, determined by the 'structural loading coefficient' F/L^2. The two selection lines must intersect on the appropriate coupling line giving the box-shaped search areas. (Figure created using *CMS* (1995) software.)

Related case studies

Case Study 10.3: Multiple constraints — windings for high field magnets

10.3 Multiple constraints — windings for high field magnets

Physicists, for reasons of their own, like to see what happens to things in high magnetic fields. 'High' means 50 tesla or more. The only way to get such fields is the old-fashioned one: dump a huge current through a wire-wound coil; neither permanent magnets (practical limit: 1.5 T), nor super-conducting coils (present limit: 25 T) can achieve such high fields. The current generates a field-pulse which lasts as long as the current flows. The upper limits on the field and its duration are set by the material of the coil itself: if the field is too high, the coil blows itself apart; if too long, it melts. So choosing the right material for the coil is critical. What should it be? The answer depends on the pulse length.

Table 10.4 Duration and strengths of pulsed fields

Classification	Duration	Field strength
Continuous	$1\,\text{s}-\infty$	$<30\,\text{T}$
Long	$100\,\text{ms}-1\,\text{s}$	$30-60\,\text{T}$
Standard	$10-100\,\text{ms}$	$40-70\,\text{T}$
Short	$10-1000\,\mu\text{s}$	$70-80\,\text{T}$
Ultra-short	$0.1-10\,\mu\text{s}$	$>100\,\text{T}$

Pulsed fields are classified according to their duration and strength as in Table 10.4.

The model

The magnet is shown, very schematically, in Figure 10.4. The coils are designed to survive the pulse, although not all do. The requirements for survival are summarized in Table 10.5. There is one objective — to maximize the field — with two constraints which derive from the requirement of survivability for a given pulse length.

Consider first destruction by magnetic loading. The field, B (units: weber/m²), in a long solenoid like that of Figure 10.4 is:

$$B = \frac{\mu_o Ni}{\ell}\lambda_f F(\alpha, \beta) \tag{10.9}$$

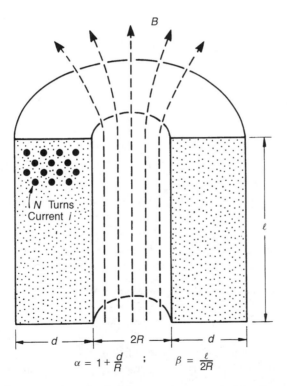

$$\alpha = 1 + \frac{d}{R} \quad ; \quad \beta = \frac{\ell}{2R}$$

Fig. 10.4 Windings for high-powered magnets. There are two constraints: the magnet must not overheat; and it must not fail under the radial magnetic forces.

Table 10.5 The design requirements: high field magnet

Function	Magnet windings
Objective	Maximize magnetic field
Constraints	(a) No mechanical failure
	(b) Temperature rise $<150°C$
	(c) Radius R and length ℓ of coil specified

where μ_o is the permeability of air ($4\pi \times 10^{-7}$ Wb/Am), N is the number of turns, i is the current, ℓ is the length of the coil, λ_f is the filling-factor which accounts for the thickness of insulation (λ_f = cross-section of conductor/cross section of coil), and $F(\alpha, \beta)$ is a geometric constant (the 'shape factor') which depends on the proportions of the magnet (defined on Figure 10.4), the value of which need not concern us. The field creates a force on the current-carrying coil. It acts radially outwards, rather like the pressure in a pressure vessel, with a magnitude

$$p = \frac{B^2}{2\mu_o F(\alpha, \beta)} \qquad (10.10)$$

though it is actually a body force, not a surface force. The pressure generates a stress σ in the windings and their casing

$$\sigma = \frac{pR}{d} = \frac{B^2}{2\mu_o F(\alpha, \beta)} \frac{R}{d} \qquad (10.11)$$

This must not exceed the yield strength σ_y of the windings, giving the first limit on B:

$$B_1 \leq \left(\frac{2\mu_o d\sigma_y F(\alpha, \beta)}{R} \right)^{1/2} \qquad (10.12)$$

The field is maximized by maximizing

$$\boxed{M_1 = \sigma_y} \qquad (10.13)$$

One could have guessed this: the best material to carry a stress σ is that with the largest yield strength σ_y.

Now consider destruction by overheating. High-powered magnets are initially cooled in liquid nitrogen to $-196°C$ in order to reduce the resistance of the windings; if the windings warm above room temperature, the resistance, R_e, in general, becomes too large. The entire energy of the pulse, $\int i^2 R_e \, dt \approx i^2 \overline{R}_e t_p$ is converted into heat (here $\overline{R}_e$ is the average of the resistance over the heating cycle and t_p is the length of the pulse); and since there is insufficient time for the heat to be conducted away, this energy causes the temperature of the coil to rise by ΔT, where

$$\Delta T = \frac{B^2}{\mu_o^2} \frac{\rho_e t_p}{d^2 C_p \rho} \qquad (10.14)$$

Here ρ_e is the resistivity of the material, C_p its specific heat (J/kg K) and ρ its density. The resistance of the coil, R_e, is related to the resistivity of the material of the windings by

$$R_e = \frac{4\ell \rho_e}{\pi d^2}$$

where d is the diameter of the conducting wire. If the upper limit for the temperature is 200 K, $\Delta T_{max} \leq 100$ K, giving the second limit on B:

$$B_2 \leq \left(\frac{\mu_o^2 d^2 C_p \rho \lambda_f \Delta T_{max}}{t_p \rho_e} \right)^{1/2} F(\alpha, \beta) \qquad (10.15)$$

The field is maximized by maximizing

$$M_2 = \frac{C_p \rho}{\rho_e} \qquad (10.16)$$

The two equations for B are sketched, as a function of pulse-time, t_p, in Figure 10.5. For short pulses, the strength constraint is active; for long ones, the heating constraint is dominant.

The selection: analytical method

Table 10.6 lists material properties for three alternative windings. The sixth column gives the strength-limited field strength, B_1; the seventh column, the heat-limited field B_2 evaluated for the following values of the design requirements:

$$t_p = 10\,\text{ms} \qquad \lambda_f = 0.5 \qquad \Delta T_{max} = 100\,\text{K}$$
$$F(\alpha, \beta) = 1 \qquad R = 0.05\,\text{m} \qquad d = 0.1\,\text{m}$$

Strength is the active constraint for the copper-based alloys; heating for the steels. The last column lists the limiting field $\tilde{B}$ for the active constraint. The Cu–Nb composites offer the largest $\tilde{B}$.

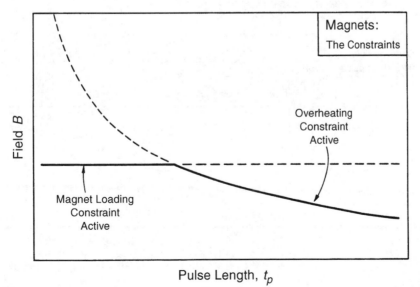

Fig. 10.5 The two equations for B are sketched, indicating the active constraint.

Table 10.6 Selection of a material for a high field magnet, pulse length 10 ms

Material	ρ Mg/m³	σ_y MPa	C_p J/kg K	ρ_e $10^{-8}\,\Omega\,m$	B_1 Wb/m²	B_2 Wb/m²	$\tilde{B}$ Wb/m²
High-conductivity copper	8.94	250	385	1.7	35	**113**	35
Cu−15% Nb composite	8.90	780	368	2.4	62	92	**62**
HSLA steel	7.85	1600	450	25	**89**	30	30

The selection: graphical method

The cross-over lies along the line where equations (10.12) and (10.15) are equal, giving the coupling the line

$$M_1 = \left[\frac{\mu_o Rd\lambda_f F(\alpha,\,\beta)\Delta T_{\max}}{2t_p} \right] M_2 \tag{10.17}$$

The quantity in square brackets is the coupling constant; it depends on the pulse length, t_p.

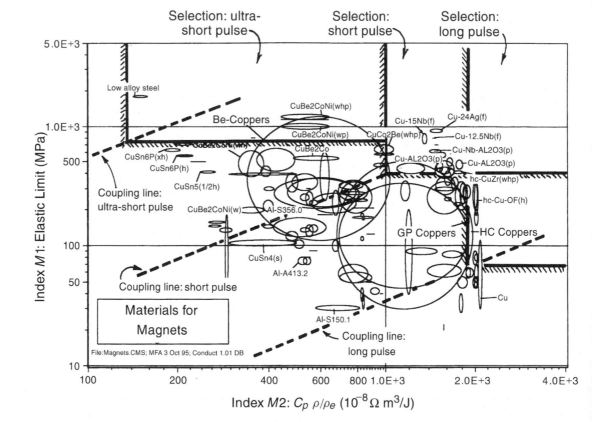

Fig. 10.6 Materials for windings for high-powered magnets, showing the selection for long pulse applications, and for short pulse ultra-high field applications. (Figure created using *CMS* (1995) software.)

Table 10.7 Materials for high field magnet windings

Material	Comment
Continuous and long pulse	
High conductivity coppers	Best choice for low field, long pulse
Pure silver	magnets (heat-limited).
Short pulse	
Copper–AL$_2$O$_3$ composites (Glidcop)	
H–C copper cadmium alloys	Best choice for high field, short pulse
H–C copper zirconium alloys	magnets (heat and strength limited).
H–C copper chromium alloys	
Drawn copper-niobium composites	
Ultra short pulse, ultra high field	
Copper–beryllium–cobalt alloys	Best choice for high field, short pulse
High-strength, low-alloy steels	magnets (strength-limited).

The selection is illustrated in Figure 10.6. Here we have used a database of *conductors*: it is an example of sector-specific database (one containing materials and data relevant to a specific industrial sector, rather than one that is material class-specific). The axes are the two indices M_1 and M_2. Three selections are shown, one for very short-pulse magnets, the other for long pulses. Each selection box is a contour of constant field, B; its corner lies on the coupling line for the appropriate pulse duration. The best choice, for a given pulse length, is that contained in the box which lies farthest up its coupling line. The results are summarized in Table 10.7.

Postscript

The case study, as developed here, is an oversimplification. Magnet design, today, is very sophisticated, involving nested sets of electro and super-conducting magnets (up to 9 deep), with geometry the most important variable. But the selection scheme for coil materials has validity: when pulses are long, resistivity is the primary consideration; when they are very short, it is strength, and the best choice for each is that developed here. Similar considerations enter the selection of materials for very high-speed motors, for bus-bars and for relays.

Further reading

Herlach, F. (1988) The technology of pulsed high-field magnets, *IEEE Transactions on Magnetics*, **24**, 1049.
Wood, J.T., Embury, J.D. and Ashby, M.F. (1995) An approach to material selection for high field magnet design, submitted to *Acta Metal. et Mater.* **43**, 212.

Related case studies

Case Study 10.2: Multiple constraints — con-rods

10.4 Compound objectives — materials for insulation

The objective in insulating a refrigerator (of which that sketched in Figure 10.7 is one class — there are many others) is to minimize the energy lost from it, and thus the running cost. But the insulation

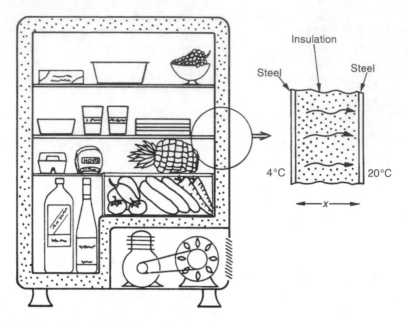

Fig. 10.7 Insulation for refrigerators. The objectives are to minimize heat loss from the interior and to minimize the cost of the insulation itself.

itself has a capital cost associated with it. The most economical choice of material for insulation is that which minimizes the total. There is at least one constraint: an upper limit on the thickness x_{max} of the insulation (Table 10.8).

The model

The first objective is to minimize the cost of the insulation. This cost, per unit area of wall, is

$$C = x_{max} \rho C_m \tag{10.18}$$

Here C_m is the cost/kg of the insulation and ρ is its density.

The second objective is to minimize the energy loss. The heat flux per unit area of wall, Q (W/m^2), assuming steady-state heat flow, is

$$Q = -\lambda \frac{dT}{dx} = \frac{\lambda \, \Delta T}{x_{max}}$$

where λ (W/m K) is its thermal conductivity and ΔT is the temperature difference between the inside and the outside of the insulation layer. If the refrigerator runs continuously, the energy consumed

Table 10.8 Design requirement for refrigerator insulation

Function	Thermal insulation
Objectives	(a) Minimize insulation cost and
	(b) Minimize energy loss, appropriately coupled
Constraint	Thickness $\leq x_{max}$

in time $t(s)$ is

$$H = Qt \, (\text{J/m}^2) \tag{10.19}$$

We identify t with the design life of the refrigerator.

To minimize both objectives in a properly couple way we create a value-function, V,

$$V = -C + E^{\$}H$$

with C given by equation (10.18) and H by (10.19). It contains the exchange constant, $E^{\$}$, relating energy to cost. It can vary widely (Table 9.5). If grid-electricity is available, $E^{\$}$ is low. But in remote areas (requiring power-pack generation), in aircraft (supplementary turbine generator) or in space (solar panels), it can be far higher. (The exchange constant relating value to cost is -1, giving the negative sign.) Inserting equations (10.18) and (10.19) gives

$$V = -x_{\max}[\rho C_m] + E^{\$} \left(\frac{t\Delta T}{x_{\max}} \right) [\lambda] \tag{10.20}$$

Here the material properties are enclosed in square brackets; everything outside these brackets is fixed by the design.

The selection: analytical method

Take the example

$$X_{\max} = 20 \, \text{mm} \qquad \Delta T = 20 \, \text{C} \qquad t = 1 \, \text{year} = 31.5 \times 10^6 \, \text{s}$$

$$E^{\$} = -0.02 \, \text{\$/MJ (grid electricity)}$$

giving, for four candidate foams listed in Table 10.9, the values of V shown in the last column.

The polystyrene foam is the cheapest to buy, but the phenolic has the largest (least negative) value of V. It is the best choice.

The selection: graphical method

Define the indices

$$M_1 = \rho C_m \text{ and } M_2 = \lambda$$

We rewrite equation (10.18) in the form:

$$\tilde{V} = \frac{V}{x_{\max}} = -M_1 + E^{\$} \left[\frac{\Delta T}{x_{\max}^2} t \right] M_2 \tag{10.21}$$

Table 10.9 Value function, V, for thermal insulation

Material	ρ kg/m³	λ W/m K	C_m \$/kg	σ_y MPa	C \$/m²	V $E^{\$} = -0.02$ \$/MJ
Polystyrene foam	30	0.034	2.0	0.2	**1.2**	−22.6
Phenolic foam	35	0.025	4.0	0.2	2.8	**−18.6**
Polymethacrylimide foam	50	0.030	27	0.8	27	−45.9
Polyethersulphone foam	90	0.038	18	0.8	32	−56.0

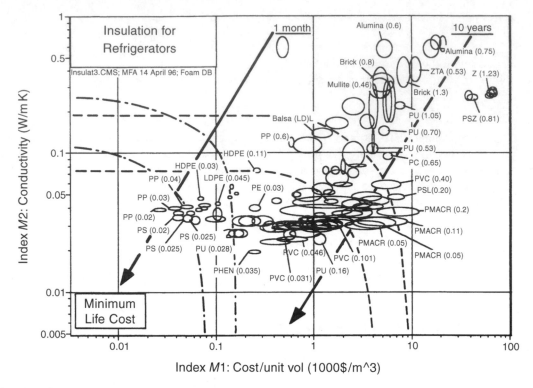

Fig. 10.8 Selection of insulating materials for refrigerators with different design lives. (Figure created using *CMS* (1997) software.)

Everything in the equation is specified except the material groups M_1 and M_2. We seek materials which maximize $\tilde{V}$. Figure 10.8 shows M_1 plotted against M_2. Contours of constant $\tilde{V}$ appear as curved lines; the value of $\tilde{V}$ increases towards the bottom right. Two sets of contours are shown, one for long-term insulation with a design life, t, of 10 years, the other for short-term refrigeration, with a value of t of 1 month. To plot these, we need a value for the term in square brackets. It has been evaluated using $\Delta T = 20°C$, $x_{max} = 20$ mm, and $E^{\$} = -0.02$ \$/MJ, as before. For the shorter design life, the cost of the insulation dominates the value function; then the best choices are simply the cheapest ones: the low density expanded polystyrene EPS (0.03) for instance*. But for the longer design life, the second term on the right of equation (10.21) becomes dominant, and the choice of material changes; the contours shown for $t = 10$ years suggest that low-density phenolics might be a good selection, because their conductivity is lower than that of the polystyrenes. Table 10.10 summarizes the selection.

Postscript

In many insulation applications the foam is bonded to the inner and outer walls of the refrigerator to give stiffness: it performs a mechanical as well as a thermal function. Then the strength, σ_y, may also be relevant. The table includes two high-strength foams.

---

* On Figure 10.6 the letters identify the material; the number in brackets gives the density in Mg/m³. Thus PS(0.03) means 'a polystyrene foam with a density of 0.03 Mg/m³'.

Table 10.10 Materials for refrigerator insulation

Material	Comment
Short design life (t_ℓ = 1 month) Polystyrene (PS) foams, e.g. PS(0.02) or PS(0.025) Polypropylene (PP) foams, e.g. PP(0.02) or PP(0.03)	Cost of insulation dominates the value function; polystyrene and polypropylene foams are the best choice because they are the cheapest.
Long design life (t_ℓ = 10 years) Phenolic (PHEN) foams, e.g. PHEN(0.035) Polyurethane (PU) foams, e.g. PU(0.028) Polystyrene (PS) foams, e.g. PS(0.02) or PS(0.025)	Heat conduction is important in the value function. The more expensive phenolics minimize the value function and are the best choice.

Of the two, the polymethacrylimide foam gives the largest (least negative) value of V.

Related case studies

Case Study 10.5: Compound objectives — disposable coffee cups

10.5 Compound objectives — disposable coffee cups

It is increasingly recognized that the use of materials in engineering carries environmental penalties: pollution of water and air, solid waste, consumption of non-renewable resources and more (collectively called *eco-damage*). One response is to adopt, as a design objective, the minimization of this damage.

Consider, as an example, the replacement of an existing disposable cup (Figure 10.9) by one which is more environmentally benign. The environmental impact it causes is difficult to quantify. One component of impact relates to the *energy content* of the material: many aspects of impact (CO_2 emissions, air-borne particulates) are proportional to this. And energy content *can* be quantified, at least approximately. We shall use it as a measure of environmental impact, to illustrate how it can be balanced against cost.

Disposable cups are not, at present, recycled, so the energy and material they contain are irretrievably lost when they are discarded. To minimize the eco-impact (measured now by energy content), we seek the design which incorporates the least energy to start with. But disposable cups must also be cheap. So we find two conflicting objectives: the environmental goal of minimizing energy content, and the economic one of minimizing cost. There are constraints which must be met: the cup must be sufficiently stiff that it can be picked up without ovalizing severely, and it would be desirable, too, that it also insulates (Table 10.11).

We first write a value function for the cup:

$$V = -C + E^\$ qm \tag{10.22}$$

Here C is the cost of the cup, m is its mass and q the energy content per unit mass of the material of which it is made. The quantity $E^\$$ is the exchange constant: the value associated with one unit

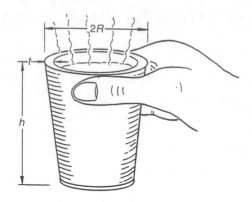

Fig. 10.9 A disposable hot-drink cup. It must be cheap, stiff and of minimum energy-content.

Table 10.11 Design requirements for disposable cup

Function	Disposable hot-drink cup
Objectives	(a) Minimize energy-content and
	(b) Minimize cost, appropriately coupled
Constraints	(a) Stiff enough to be picked up
	(b) Thermally insulating

of environmental damage. Values $E^\$$ are, at present, unknown, but by taking extremes its influence can be explored.

The first term in this equation describes the material cost of the cup. It is the volume of material it contains (thought of as a cylinder of radius R, height h and wall thickness t, closed at one end) times the cost $C_m \rho$ per unit volume (C_m is the material cost per unit weight and ρ the density):

$$C = C_m m \approx (2\pi R h + \pi R^2) t C_m \rho$$
$$= (2\alpha + 1)\pi R^2 t C_m \rho \tag{10.23}$$

where $\alpha = h/R$ the ratio of height-to-radius. The constraint on stiffness requires that ovalization must not become unacceptable when the cup is loaded across a diagonal, as in the figure. This imposes a limit on its stiffness, S:

$$S = \frac{F}{\delta} = \frac{C_1 EI}{R^3} = \frac{\alpha C_1 E t^3}{12 R^2} > S_c \tag{10.24}$$

Here I is the second moment of area of the wall of the cup (proportional to $ht^3/12$ for a wall of uniform thickness, t), E is its Young's modulus, C_1 is a constant and S_c is the critical stiffness required for safe handling. Solving for t gives

$$t = \left(\frac{12 R^2 S_c}{\alpha C_1 E}\right)^{1/3} \tag{10.25}$$

which, when inserted in equation (10.23), gives the *cost* of the cup:

$$C = C_m m = (2\alpha + 1)\pi R^2 C_m \rho \left(\frac{12R^2 S_c}{\alpha C_2 E} \right)^{1/3} \tag{10.26}$$

or

$$C = C_2 \left(\frac{C_m \rho}{E^{1/3}} \right) \tag{10.27}$$

in which the constant C_2 contains the design parameters. By a similar chain of argument, replacing $C_m \rho$ by $q\rho$ (where q is the energy per unit mass of the material), the *energy content* of the cup is

$$qm = C_2 \left(\frac{q\rho}{E^{1/3}} \right) \tag{10.28}$$

If we now associate a cost $E^{\$}$ with environmental impact as measured by energy content (an energy tax, for example, or a pollution tax), environmental impact can be converted to cost, giving:

$$V = C_2 [M_1 + E^{\$} M_2] \tag{10.29}$$

with $M_1 = C_m \rho / E^{1/3}$ and $M_2 = q\rho / E^{1/3}$.

The selection: analytical method

Table 10.12 lists three candidates for the cup: foamed polystyrene (PS), polycarbonate (PC) and high density polyethylene (HDPE), with the relevant properties. The remaining columns list the wall thickness, the cost and the value, taking

$$R = 40\,\text{mm} \qquad \alpha = 4 \qquad C_1 = 24 \qquad S_c = 3\,\text{kN/m}$$

With no penalty on energy ($E^{\$} = 0$), polystyrene has the greatest value. A pollution tax of 0.01 \$/MJ leads to the ranking in the second last column; one of 0.05 \$/MJ gives the values in the last one. With the higher tax, PC becomes more attractive.

We have used numerical values for R, α, C_1 and S_c here, but it was not necessary. It is frequently so that the optimum selection is independent of some or all of the other variables of the design, and this is an example of just that. The variables R, α, C_1 and S_c are all contained in the quantity C_2 of equation (10.29), the value of which does not alter the ranking of the candidates in Table 10.12: ranking by V or by V/C_2.

Table 10.12 Value functions, V, for two values of exchange constant, $E^{\$}$

Material	ρ Mg/m³	E GPa	C_m^* \$/kg	q MJ/kg	t mm	C \$	$V, E^{\$} =$ −0.01 \$/MJ	$V, E^{\$} =$ −0.05 \$/MJ
Expanded PS	0.05	0.03	1.4	180	2.7	0.009	**−0.02**	−0.07
Expanded PC	0.065	0.95	5.0	170	1	0.016	**−0.02**	**−0.04**
Expanded HDPE	0.08	0.006	1.6	150	4.6	0.3	−0.06	−0.17

*Cost of material in shape of cup, when mass produced, is almost the same as that of the material itself.

The selection: graphical method

Figure 10.10 shows M_1 plotted against M_2, allowing the selection of materials to minimize, in a balanced way, both cost and energy content. We will assume that the cups are at present pressed from solid polystyrene (PS) sheet with a density of 1060 kg/m³; it is indicated as a black ellipse on the figure. The contours show the selection 'boundary' for various values of $E^\$$. The materials which lie below the appropriate contour are a better choice than the current material: they give a lower value of V than it does. For small values of $E^\$$, the contours are almost vertical; for large $E^\$$ they are almost horizontal.

Materials in the lower-left quadrant are both cheaper *and* less energy intensive than the current material: they are a better choice than the existing solid PS, regardless of the value of $E^\$$. Among these are a range of polyethylene foams, LDPE, with densities in the range 0.018 to 0.029 Mg/m³ (LDPE (0.018), for instance) and the expanded polystyrenes with densities between 0.02 and 0.05 Mg/m³ (EPS (0.02) or EPS (0.05)). But if the energy tax were high enough — if $E^\$$ were as high as 0.01 \$/MJ, for example — then a range of PVC foams become potential candidates; and if it rose to 0.1 \$/MJ, cups made of cork (!) would become economic. Table 10.13 summarizes the selection.

Further reading

Boustead, I. and Hancock, G.F. (1979) *Handbook of Industrial Energy Analysis*, Wiley, New York.

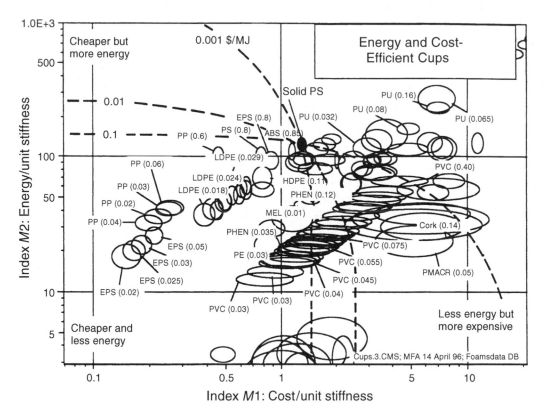

Fig. 10.10 Comparison of polystyrene with competing materials for disposable cups. (Figure created using *CMS* (1995) software.)

Table 10.13 Materials for low energy, cheap coffee cups

Material	Comment
Short design life	
Expanded polystyrene (EPS) foams [e.g. EPS(0.02) to EPS(0.05)]	The best choice: lower cost and energy content than solid PS; good thermal properties.
Polypropylene (PP) foams [e.g. PP(0.02) to PP(0.06)]	A viable alternative to expanded PS.
Polyethylene (LDPE) foams [e.g. LDPE(0.018) to LDPE(0.029)]	Considerably more expensive and more energy intensive than expanded PS.

Related case studies

Case Study 10.4: Compound objectives — materials for insulation

10.6 Summary and conclusions

Most designs are over-constrained: they must simultaneously meet several conflicting require-ments. But although they conflict, an optimum selection is still possible. The 'active constraint' method, developed in Chapter 9, allows the selection of materials which optimally meet two or more constraints. It is illustrated here by two case studies, one of them mechanical, one electro-mechanical.

Greater problems arise when the design must meet two or more conflicting objectives (such as minimizing mass, cost and environmental impact). Here we need a way can be found to express all the objectives in the same units, a 'common currency', so to speak. The conversion factor is called the exchange constant, $E^\$$. Establishing the value of the exchange constant is an important step in solving the problem. With it, a value function V is constructed which combines the objectives. Materials which minimize V meet all the objectives in a properly balanced way. The most obvious common currency is cost itself, requiring an 'exchange rate' to be established between cost and the other objectives. This can be done for energy and for mass, and — at least in principle — for environmental impact. The method is illustrated by two further case studies.

_____

Chapter 11

Materials processing and design

_____

11.1 Introduction and synopsis

A *process* is a method of shaping, finishing or joining a material. *Sand casting*, *injection moulding*, *polishing* and *fusion welding* are all processes; there are hundreds of them. It is important to choose the right process-route at an early stage in the design before the cost-penalty of making changes becomes large. The choice, for a given component, depends on the material of which it is to be made, on its size, shape and precision, and on how many are to be made — in short, on the *design requirements*. A change in design requirements may demand a change in process route.

Each process is characterized by a set of *attributes*: the materials it can handle, the shapes it can make and their precision, complexity and size. The intimate details of processes make tedious reading, but have to be faced: we describe them briefly in the following section, using Process Selection Charts to capture their attributes. *Process selection* is the act of finding the best match between process attributes and design requirements.

Methods for doing this are developed in the remaining sections of this chapter. In using them, one should not forget that material, shape and processing interact (Figure 11.1). Material properties and shape limit the choice of process: ductile materials can be forged, rolled and drawn; those which are brittle must be shaped in other ways. Materials which melt at modest temperatures to low-viscosity liquids can be cast; those which do not have to be processed by other routes. Slender shapes can be made easily by rolling or drawing but not by casting. High precision is possible by machining but not by forging, and so on. And processing affects properties. Rolling and forging change the texture of metals and align the inclusions they contain, enhancing strength and ductility. Composites acquire their properties during processing by control of lay-up; for these the interactions between function, material, shape and process are particularly strong.

Like the other aspects of design, process selection is an iterative procedure. The first iteration gives one or more possible processes-routes. The design must then be re-thought to adapt it, as far as possible, to ease of manufacture by the most promising route. The final choice is based on a comparison of *process cost*, requiring the use of cost models developed later in this chapter, and on *supporting information*: case histories, documented experience and examples of process-routes used for related products.

11.2 Processes and their influence on design

Now for the inevitable catalogue of manufacturing processes. It will be kept as concise as possible; details can be found in the numerous books listed in Further reading at the end of this chapter.

Manufacturing processes can be classified under the nine headings shown in Figure 11.2. *Primary processes* create shapes. The first row lists five primary forming processes: casting, moulding,

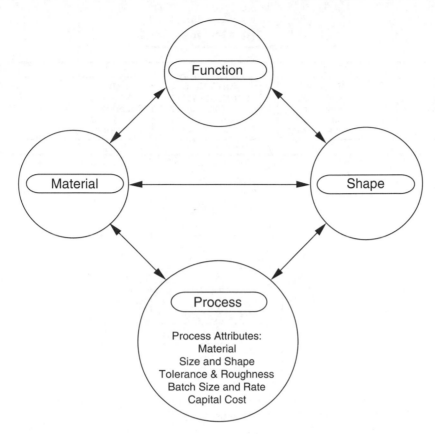

Fig. 11.1 Processing selection depends on material and shape. The 'process attributes' are used as criteria for selection.

deformation, powder methods, methods for forming composites, special methods and rapid proto-typing. *Secondary processes* modify shapes; here they are shown collectively as 'machining'; they add features to an already shaped body. These are followed by *tertiary processes*: like heat treatment, which enhance surface or bulk properties. The classification is completed by *finishing* and *joining*.

(a) In *casting*, a liquid is poured into a mould where it solidifies by cooling (metals) or by reaction (thermosets). Casting is distinguished from moulding, which comes next, by the low viscosity of the liquid: it fills the mould by flow under its own weight (gravity casting, Figure 11.3) or under a modest pressure (centrifugal casting and pressure die casting, Figure 11.4). Sand moulds for one-off castings are cheap; metal dies for making large batches can be expensive. Between these extremes lie a number of other casting methods: shell, investment, plaster-mould and so forth.

Cast shapes must be designed for easy flow of liquid to all parts of the mould, and for progressive solidification which does not trap pockets of liquid in a solid shell, giving shrinkage cavities. Whenever possible, section thicknesses are made uniform (the thickness of adjoining sections should not differ by more than a factor of 2). The shape is designed so that the pattern and the finished casting can be removed from the mould. Keyed-in shapes are avoided because they lead to 'hot tearing' (a tensile creep-fracture) as the solid cools and shrinks. The tolerance and surface finish

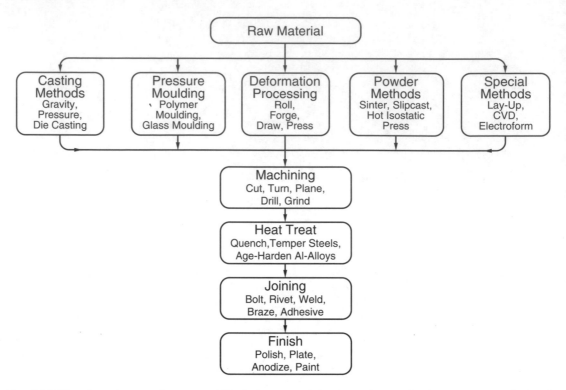

Fig. 11.2 The nine classes of process. The first row contains the primary shaping processes; below lie the secondary shaping, joining and finishing processes.

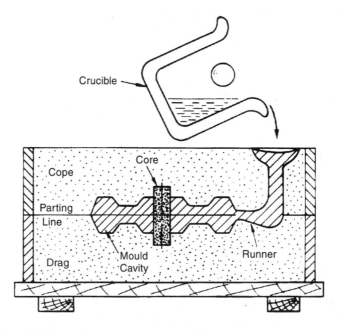

Fig. 11.3 Sand casting. Liquid metal is poured into a split sand mould.

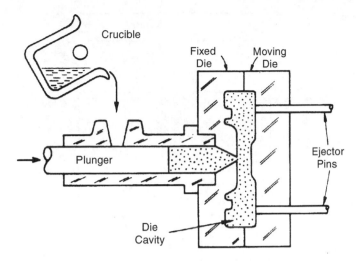

Fig. 11.4 Die casting. Liquid is forced under pressure into a split metal mould.

of a casting vary from poor for cheap sand-casting to excellent for precision die-castings; they are quantified at page 272.

(b) *Moulding* is casting, adapted to materials which are very viscous when molten, particularly thermoplastics and glasses. The hot, viscous fluid is pressed (Figure 11.5) or injected (Figures 11.6 and 11.7) into a die under considerable pressure, where it cools and solidifies. The die must withstand repeated application of pressure, temperature, and the wear involved in separating and removing the part, and therefore is expensive. Elaborate shapes can be moulded, but at the penalty of complexity in die shape and in the way it separates to allow removal.

Blow-moulding (Figure 11.8) uses a gas pressure to expand a polymer or glass blank into a split outer-die. It is a rapid, low-cost process well suited for mass-production of cheap parts like milk bottles.

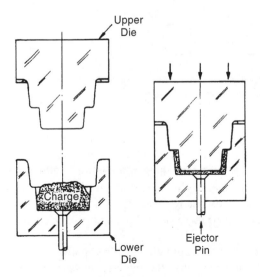

Fig. 11.5 Moulding. A hot slug of polymer or glass is pressed to shape between two dies.

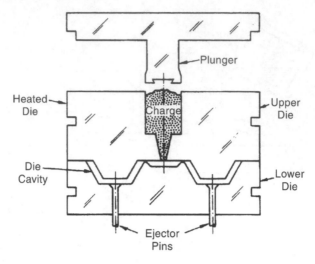

Fig. 11.6 Transfer-moulding. A slug of polymer or glass in a heated mould is forced into the mould cavity by a plunger.

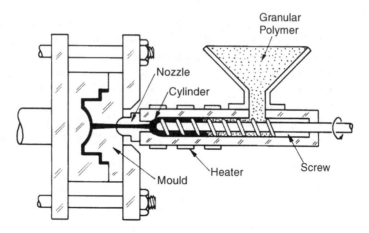

Fig. 11.7 Injection-moulding. A granular polymer (or filled polymer) is heated, compressed and sheared by a screw feeder, forcing it into the mould cavity.

(c) *Deformation processing* (Figures 11.9 to 11.12) can be hot, warm or cold. Extrusion, hot forging and hot rolling ($T > 0.55T_m$) have much in common with moulding, though the material is a true solid not a viscous liquid. The high temperature lowers the yield strength and allows simultaneous recrystallization, both of which lower the forming pressures. Warm working ($0.35T_m < T < 0.55T_m$) allows recovery but not recrystallization. Cold forging, rolling and drawing ($T < 0.35T_m$) exploit work hardening to increase the strength of the final product, but at the penalty of higher forming pressures.

Forged parts are designed to avoid rapid changes in thickness and sharp radii of curvature. Both require large local strains which can cause the material to tear or to fold back on itself ('lapping'). Hot forging of metals allows bigger changes of shape but generally gives a poor surface and

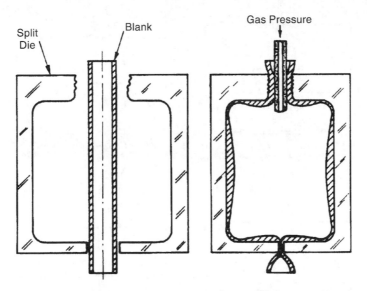

Fig. 11.8 Blow-moulding. A tubular or globular blank of hot polymer or glass is expanded by gas pressure against the inner wall of a split die.

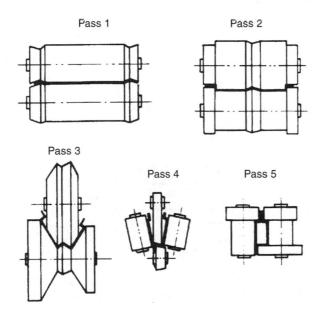

Fig. 11.9 Rolling. A billet or bar is reduced in section by compressive deformation between the rolls. The process can be hot ($T > 0.55T_m$), warm ($0.35T_m < T < 0.55T_m$) or cold ($T < 0.35T_m$).

tolerance because of oxidation and warpage. Cold forging gives greater precision and finish, but forging pressures are higher and the deformations are limited by work hardening.

Sheet metal forming (Figure 11.12) involves punching, bending, and stretching. Holes cannot be punched to a diameter less than the thickness of the sheet. The minimum radius to which a sheet can be bent, its *formability*, is sometimes expressed in multiples of the sheet thickness t: a value

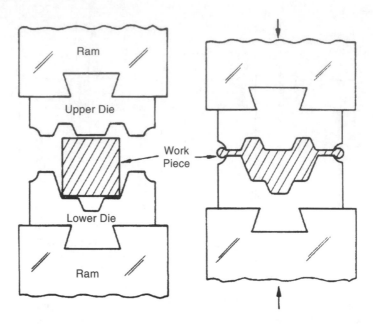

Fig. 11.10 Forging. A billet or blank is deformed to shape between hardened dies. Like rolling, the process can be hot, warm or cold.

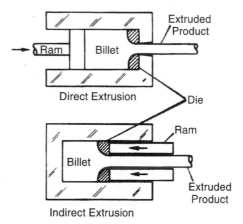

Fig. 11.11 Extrusion. Material is forced to flow through a die aperture to give a continuous prismatic shape. Hot extrusion is carried out at temperatures up to $0.9T_m$; cold extrusion is at room temperature.

of 1 is good; one of 4 is average. Radii are best made as large as possible, and never less than t. The formability also determines the amount the sheet can be stretched or drawn without necking and failing. The *limit forming diagram* gives more precise information: it shows the combination of principal strains in the plane of the sheet which will cause failure. The part is designed so that the strains do not exceed this limit.

(d) *Powder methods* create the shape by pressing and then sintering fine particles of the material. The powder can be cold-pressed and then sintered (heated at up to $0.8T_m$ to give bonding); it can

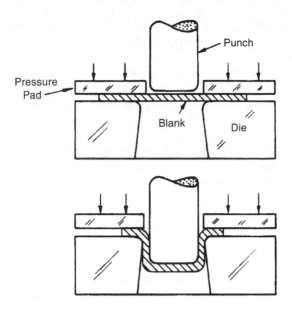

Fig. 11.12 Drawing. A blank, clamped at its edges, is stretched to shape by a punch.

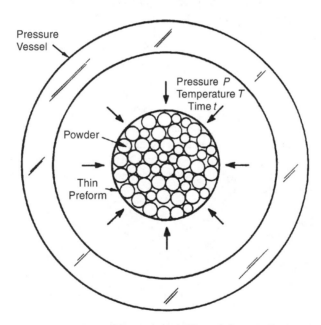

Fig. 11.13 Hot isostatic pressing. A powder in a thin, shaped, shell or preform is heated and compressed by an external gas pressure.

be pressed in a heated die ('die pressing'); or, contained in a thin preform, it can be heated under a hydrostatic pressure ('hot isostatic pressing' or 'HIPing', Figure 11.13). Metals and ceramics which are too high-melting to cast and too strong to deform can be made (by chemical methods) into powders and then shaped in this way. But the processes is not limited to 'difficult' materials; almost any material can be shaped by subjecting it, as a powder, to pressure and heat.

Powder pressing is most widely used for small metallic parts like gears and bearings for cars and appliances, and for fabricating almost all engineering ceramics. It is economic in its use of material, it allows parts to be fabricated from materials that cannot be cast, deformed or machined, and it can give a product which requires little or no finishing.

Since pressure is not transmitted uniformly through a bed of powder, the length of a die-pressed powder part should not exceed 2.5 times its diameter. Sections must be near-uniform because the powder will not flow easily round corners. And the shape must be simple and easily extracted from the die.

(e) *Composite fabrication* methods are adapted to make polymer-matrix composites reinforced with continuous or chopped fibres. Large components are fabricated by filament winding (Figure 11.14) or by laying-up pre-impregnated mats of carbon, glass or Kevlar fibre ('pre-preg') to the required thickness, pressing and curing. Parts of the process can be automated, but it remains a slow manufacturing route; and, if the component is a critical one, extensive ultrasonic testing may be necessary to confirm its integrity. So lay-up methods are best suited to a small number of high-performance, tailor-made, components. More routine components (car bumpers, tennis racquets) are made from chopped-fibre composites by pressing and heating a 'dough' of resin containing the fibres, known as bulk moulding compound (BMC) or sheet moulding compound (SMC), in a mould, or by injection moulding a rather more fluid mixture into a die as in Figures 11.5, 11.6 and 11.7. The flow pattern is critical in aligning the fibres, so that the designer must work closely with the manufacturer to exploit the composite properties fully.

(f) *Special methods* include techniques which allow a shape to be built up atom-by-atom, as in electro-forming and chemical and physical vapour deposition. They include, too, various spray-forming techniques (Figure 11.15) in which the material, melted by direct heating or by injection into a plasma, is sprayed onto a former — processes which lend themselves to the low-number production of small parts, made from difficult materials.

(g) *Machining* almost all engineering components, whether made of metal, polymer or ceramic, are subjected to some kind of machining (Figure 11.16) or grinding (a sort of micro-machining, as in Figure 11.17) during manufacture. To make this possible they should be designed to make gripping and jigging easy, and to keep the symmetry high: symmetric shapes need fewer operations. Metals differ greatly in their *machinability*, a measure of the ease of chip formation, the ability to give a smooth surface, and the ability to give economical tool life (evaluated in a standard test). Poor machinability means higher cost.

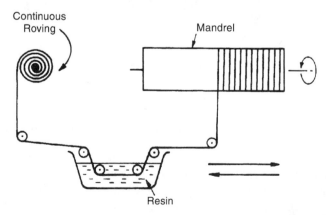

Fig. 11.14 Filament winding. Fibres of glass, Kevlar or carbon are wound onto a former and impregnated with a resin–hardener mix.

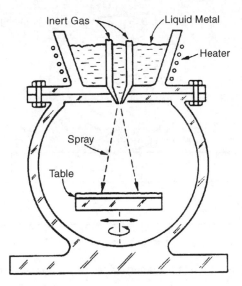

Fig. 11.15 Spray forming. Liquid metal is 'atomized' to droplets by a high velocity gas stream and projected onto a former where it splats and solidifies.

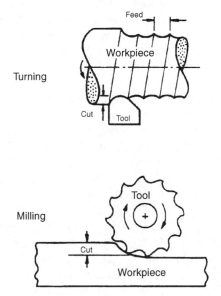

Fig. 11.16 Machining: turning (above left) and milling (below). The sharp, hardened tip of a tool cuts a chip from the workpiece surface.

Most polymers machine easily and can be polished to a high finish. But their low moduli mean that they deflect elastically during the machining operation, limiting the tolerance. Ceramics and glasses can be ground and lapped to high tolerance and finish (think of the mirrors of telescopes). There are many 'special' machining techniques with particular applications; they include electro-machining, spark machining, ultrasonic cutting, chemical milling, cutting by water-jets, sand-jets, electron beams and laser beams.

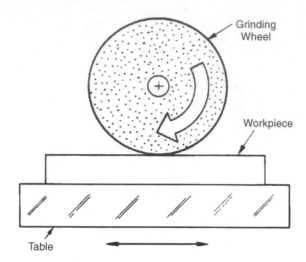

Fig. 11.17 Grinding. The cutting 'tool' is the sharp facet of an abrasive grain; the process is a sort of micro-machining.

Machining operations are often finishing operations, and thus determine finish and tolerance (pp. 271–2). Higher finish and tolerance mean higher cost; overspecifying either is a mistake.

(h) *Heat treatment* is a necessary part of the processing of many materials. Age-hardening alloys of aluminium, titanium and nickel derive their strength from a precipitate produced by a controlled heat treatment: quenching from a high temperature followed by ageing at a lower one. The hardness and toughness of steels is controlled in a similar way: by quenching from the 'austenitizing' temperature (about 800°C) and tempering.

Quenching is a savage procedure; thermal contraction can produce stresses large enough to distort or crack the component. The stresses are caused by a non-uniform temperature distribution, and this, in turn, is related to the geometry of the component. To avoid damaging stresses, the section should be as uniform as possible, and nowhere so large that the quench-rate falls below the critical value required for successful heat treatment. Stress concentrations should be avoided: they are usually the source of quench cracks. Materials which have been moulded or deformed may contain internal stresses which can be removed, at least partially, by stress-relief anneals — another sort of heat treatment.

(i) *Joining* is made possible by a number of techniques. Bolting and riveting (Figure 11.18), welding, brazing and soldering (Figure 11.19) are commonly used for metals. Polymers are joined by snap-fasteners (Figure 11.18 again), and by thermal bonding. Ceramics can be diffusion-bonded to themselves, to glasses and to metals. Improved adhesives give new ways of bonding all classes of materials (Figure 11.20). Friction welding (Figure 11.21) and friction-stir welding rely on the heat and deformation generated by friction to create a bond.

If components are to be welded, the material of which they are made must be characterized by a high *weldability*. Like *machinability*, it measures a combination of basic properties. A low thermal conductivity allows welding with a low rate of heat input, and gives a less rapid quench when the weld torch is removed. Low thermal expansion gives small thermal strains with less risk of distortion. A solid solution is better than an age-hardened alloy because, in the heat-affected zone on either side of the weld, overageing and softening can occur.

Welding always leaves internal stresses which are roughly equal to the yield strength. They can be relaxed by heat treatment but this is expensive, so it is better to minimize their effect by good

Fig. 11.18 Fasteners: (a) bolting; (b) riveting; (c) stapling; (d) push-through snap fastener; (e) push-on snap fastener; (f) rod-to-sheet snap fastener.

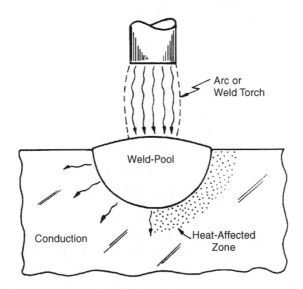

Fig. 11.19 Welding. A torch melts both the workpiece and added weld-metal to give a bond which is like a small casting.

design. To achieve this, parts to be welded are made of equal thickness whenever possible, the welds are located where stress or deflection is least critical, and the total number of welds is minimized.

The large-volume use of fasteners is costly because it is difficult to automate; welding, crimping or the use of adhesives can be more economical.

(j) *Finishing* describes treatments applied to the surface of the component or assembly. They include polishing, plating, anodizing and painting, they aim to improve surface smoothness, protect against corrosion, oxidation and wear, and to enhance appearance.

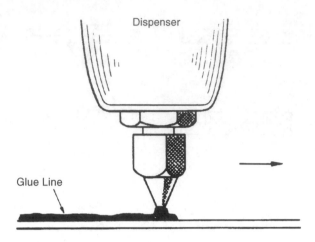

Fig. 11.20 Adhesive bonding. The dispenser, which can be automated, applies a glue-line onto the workpiece against which the mating face is pressed.

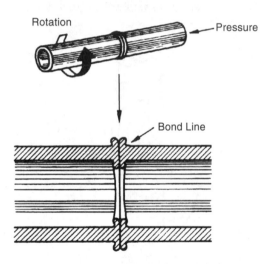

Fig. 11.21 Friction welding. A part, rotating at high speed, is pressed against a mating part which is clamped and stationary. Friction generates sufficient heat to create a bond.

Plating and painting are both made easier by a simple part shape with largely convex surfaces. Channels, crevices and slots are difficult to reach with paint equipment and often inadequately coated by electroplates.

(k) *Rapid prototyping systems* (RPS) allow single examples of complex shapes to be made from numerical data generated by CAD solid-modelling software. The motive may be that of visualization: the aesthetics of an object may be evident only when viewed as a prototype. It may be that of pattern-making: the prototype becomes the master from which moulds for conventional processing, such as casting, can be made. Or — in complex assemblies — it may be that of validating intricate geometry, ensuring that parts fit, can be assembled, and are accessible. All RPS can create shapes of great complexity with internal cavities, overhangs and transverse features, although the precision, at present, is limited to ± 0.3 mm at best.

The methods build shapes layer-by-layer, rather like three-dimensional printing, and are slow (typically 4–40 hours per unit). There are four broad classes of RPS.

(i) The shape is built up from a thermoplastic fed to a single scanning head which extrudes it like a thin layer of toothpaste ('Fused Deposition Modelling' or FDM), exudes it as tiny droplets ('Ballistic Particle Manufacture', BPM, Figure 11.22), or ejects it in a patterned array like a bubble-jet printer ('3-D printing').

(ii) Screen-based technology like that used to produce microcircuits ('Solid Ground Curing' or SGC, Figure 11.23). A succession of screens admits UV light to polymerize a photo-sensitive monomer, building shapes layer-by-layer.

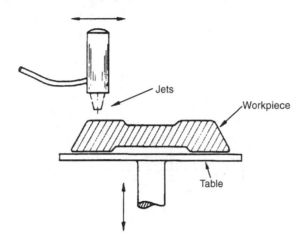

Fig. 11.22 Ballistic particle manufacture (BPM), a rapid prototyping method by which a solid body is created by layer-by-layer deposition of polymer droplets.

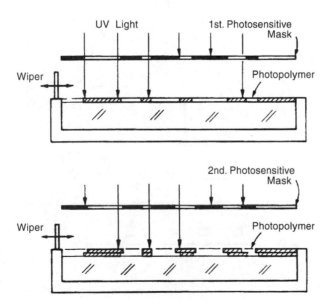

Fig. 11.23 Solid ground curing (SGC), a rapid prototyping method by which solid shapes are created by sequential exposure of a resin to UV light through glass masks.

(iii) Scanned-laser induced polymerization of a photo-sensitive monomer ('Stereo-lithography' or SLA, Figure 11.24). After each scan, the workpiece is incrementally lowered, allowing fresh monomer to cover the surface. Selected laser sintering (SLS) uses similar laser-based technology to sinter polymeric powders to give a final product. Systems which extend this to the sintering of metals are under development.

(iv) Scanned laser cutting of bondable paper elements (Figure 11.25). Each paper-thin layer is cut by a laser beam and heat bonded to the one below.

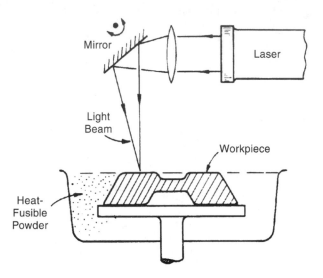

Fig. 11.24 Stereo-lithography (SLA), a rapid prototyping method by which solid shapes are created by laser-induced polymerization of a resin.

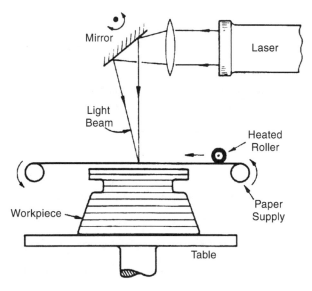

Fig. 11.25 Laminated object manufacture (LOM), a rapid prototyping method by which a solid body is created from layers of paper, cut by a scanning laser beam and bonded with a heat-sensitive polymer.

To be useful, the prototypes made by RPS are used as masters for silicone moul[...] number of replicas to be cast using high-temperature resins or metals.

Enough of the processes themselves; for more detail the reader will have to con[...] reading section.

11.3 Process attributes

The *kingdom* of processes can be classified in the way shown in top half of Figure 11.26. There are the broad *families*: casting, deformation, moulding, machining, compaction of powders, and such like. Each family contains many *classes*: casting contains sand-casting, die-casting, and investment casting, for instance. These in turn have many *members*: there are many variants of sand-casting, some specialized to give greater precision, others modified to allow exceptional size, still others adapted to deal with specific materials.

Each member is characterized by a set of *attributes*. It has *material attributes*: the particular subset of materials to which it can be applied. It has *shape-creating attributes*: the classes of shapes

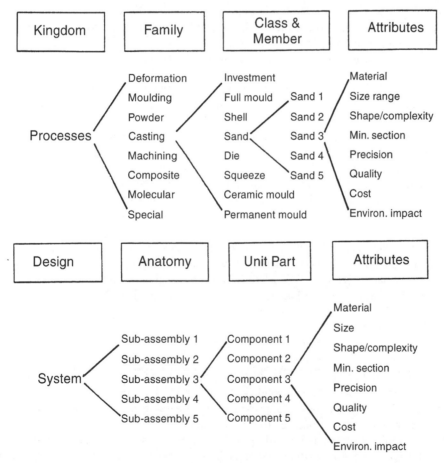

Fig. 11.26 Top: the taxonomy of the kingdom of process, and their attributes; bottom: the design of a component defines a required attribute profile. Process selection involves matching the two.

can make. It has *physical attributes* which relate to the size, precision, finish and quality of its product. It has attributes which relate to the *economics* of its use: its capital cost and running cost, the speed with which it can be set up and operated, the efficiency of material usage. And it has attributes which relate to its impact on the environment: its eco-cost, so to speak.

Process selection is the action of matching process attributes to the attributes required by the design (Figure 11.26, bottom half). The anatomy of a design can be decomposed into *sub-assemblies*; these can be subdivided into *components*; and components have *attributes*, specified by the designer, some relating to material, some to shape, some physical, some economic. The problem, then, is that of matching the attribute-profiles of available processes to that specified by the design.

11.4 Systematic process selection

You need a process to shape a given material to a specified shape and size, and with a given precision. How, from among the huge number of possible processes, are you to choose it? Here is the strategy. The steps parallel those for selecting a material. In four lines:

(a) consider *all processes to be candidates* until shown to be otherwise;
(b) *screen* them, eliminating those which lack the attributes demanded by the design;
(c) *rank* those which remain, using relative cost as the criterion;
(d) *seek supporting information* for the top candidates in the list.

Figure 11.27 says the important things. Start with an open mind: initially, *all* processes are options. The design specifies a material a shape, a precision, a batch size, and perhaps more. The first step — that of *screening* — eliminates the process which cannot meet these requirements. It is done by comparing the attributes specified by the design (material, for instance, or shape or precision) with the attributes of processes, using hard copy or computer-generated Process Selection Charts described in a moment. Here, as always, decisions must be moderated by common sense: some design requirements are absolute, resulting in rejection, others can be achieved by constructing process-chains. As an example, if a process cannot cope with a *material* it must be rejected, but if its *precision* is inadequate, this can be overcome by calling on a secondary process such as machining.

Screening gives the processes which could meet the design requirements. The next step is to *rank* them using economic criteria. There are two ways of doing it. Each process is associated with an 'economic batch size-range' or EBS: it is the range over which that process is found to be cheaper than competing processes. The design specifies a batch size. Processes with an EBS which corresponds to the desired batch size are put at the top of the list. It is not the best way of ranking, but it is quick and simple.

Better is to rank by *relative cost*. Cost, early in the design, can only be estimated in an approximate way, but the cost differences between alternative process routes are often so large that the estimate allows meaningful ranking. The cost of making a component is the sum of the costs of the resources consumed in its production. These resources include materials, capital, time, energy, space and information. It is feasible to associate approximate values of these with a given process, allowing the relative cost of competing processes to be estimated.

Screening and ranking reduce the kingdom of processes to a small subset of potential candidates. We now need *supporting information*. What is known about each candidate? Has it been used before to make components like the one you want? What is its family history? Has it got hidden faults, character defects, so to speak? Such information is found in handbooks, in the data sheets

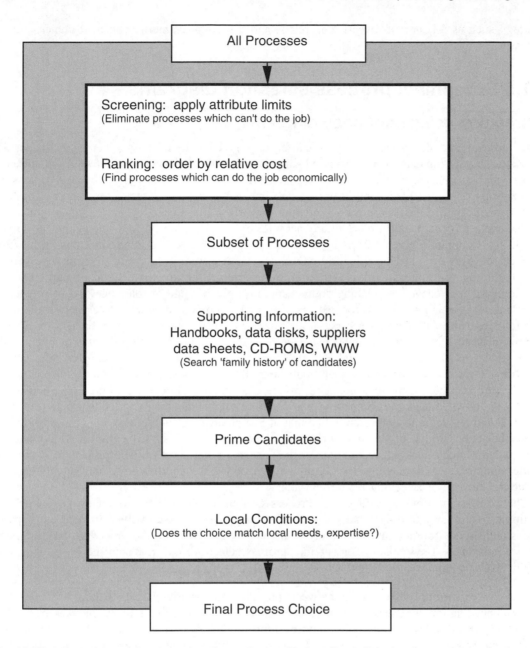

Fig. 11.27 A flow chart of the procedure for process selection. It parallels that for material selection.

of suppliers of process equipment and in documented case studies which, increasingly, appear in electronic format on CD or the World Wide Web.

This is as far as a general strategy can go. In reality there is one more step: it is to examine whether local conditions modify the choice. Available equipment and expertise for one class of process and lack of them for another can, for obvious reasons, bias the selection. But one should be aware that the unbiased choice might, in the long run, be better. That is the value of a systematic strategy such as this one: it reveals the options and their relative merit. The final choice is up to the user.

11.5 Screening: process selection diagrams

Screening using hard copy diagrams

How do you find the processes which can form a given material to a given size, shape, and precision? First eliminate all processes which cannot handle the material; then seek the subset of these which can handle the size, create the shape and achieve the precision you want. Progress can be made by using the hard copy charts shown first in this section. Greater resolution is possible with computer-aided process selection software, described after that.

The axes of a process selection chart are measures of two of the attributes — precision and surface finish, for example. Figure 11.28 is a schematic of such a chart. The horizontal axis is the RMS surface roughness, plotted on a logarithmic scale, running from $10^{-3}\,\mu m$ to $100\,\mu m$. The vertical axis is the tolerance ranging from $\pm 10^{-4}\,mm$ to $\pm 10\,mm$. Each process occupies a particular area of the chart: it is capable of making components in a given range of tolerance and of roughness. Conventional casting processes, for instance, can make components with a tolerance in the range ± 0.1 to $\pm 10\,mm$ (depending on process and size) with a roughness ranging from 5 to $100\,\mu m$; precision casting can improve both by a factor of 10. Machining adds precision: it extends the range down to $T = \pm 10^{-3}\,mm$ and $R = 0.01\,\mu m$. Polymer forming processes give high surface finish but limited tolerance. Lapping and polishing allow the highest precision and finish of all.

Selection is achieved by superimposing on the chart the envelope of attributes specified by the design, as shown in the figure. Sometimes the design sets upper and lower limits on process attributes (here: T and R), defining a closed box like that of Search area 1 of the figure. Sometimes, instead, it prescribes upper limits only, as in Search area 2. The processes which lie within or are bounded by the shaded search envelope are candidates; they are the initial shortlist. The procedure is repeated using similar charts displaying other attributes, narrowing the shortlist to a final small subset of processes capable of achieving the design goal.

There are some obvious difficulties. Process attributes can be hard to quantify: 'shape', for example, is not easy to define and measure. Certain processes have evolved to deal with special needs and do not naturally appear on any of the charts. Despite this, the procedure has the merits that it introduces a systematic element into process selection, and it forms the basis of a more sophisticated computer-based approach, described in a moment.

Material compatibility. The match between process and material is established by the link to material class and by the use of the material compatibility chart of Figure 11.29. Its axes are melting point and hardness. The melting point imposes limits on the processing of materials by conventional casting methods. Low melting point metals can be cast by any one of many techniques. For those which melt above 2000 K, conventional casting methods are no longer viable, and special techniques such as electron-beam melting must be used. Similarly, the yield strength or hardness of a material imposes limitations on the choice of deformation and machining processes. Forging and rolling pressures are proportional to the flow strength, and the heat generated during machining, which limits tool life, also scales with the ultimate strength or hardness. Generally speaking, deformation processing is limited to materials with hardness values below 3 GPa. Other manufacturing methods exist which are not limited either by melting point or by hardness. Examples are: powder methods, CVD and evaporation techniques, and electro-forming.

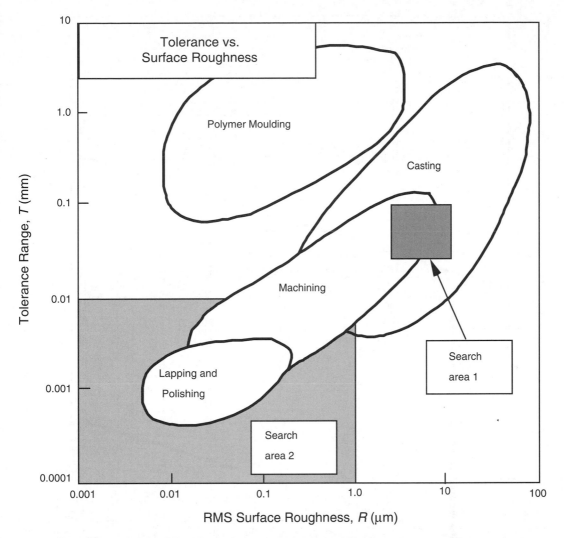

Fig. 11.28 A schematic illustrating the idea of a process-selection chart. The charts have process attributes as axes; a given process occupies a characteristic field. A design demands a certain set of processes attributes, isolating a box ('Search Area 1') or a sub-field ('Search Area 2') of the chart. Processes which overlap the search areas become candidates for selection.

Figure 11.29 presents this information in graphical form. In reality, only part of the space covered by the axes is accessible: it is the region between the two heavy lines. The hardness and melting point of materials are not independent properties: low melting point materials tend to be soft (polymers and lead, for instance); high melting point materials are hard (diamond is the extreme example). This information is captured by the equation

$$0.03 < \frac{H\Omega}{kT_m} < 20 \tag{11.1}$$

where Ω is the atomic or molecular volume and k is Boltzmann's Constant (1.38×10^{-26} J/K). It is this equation which defines the two lines.

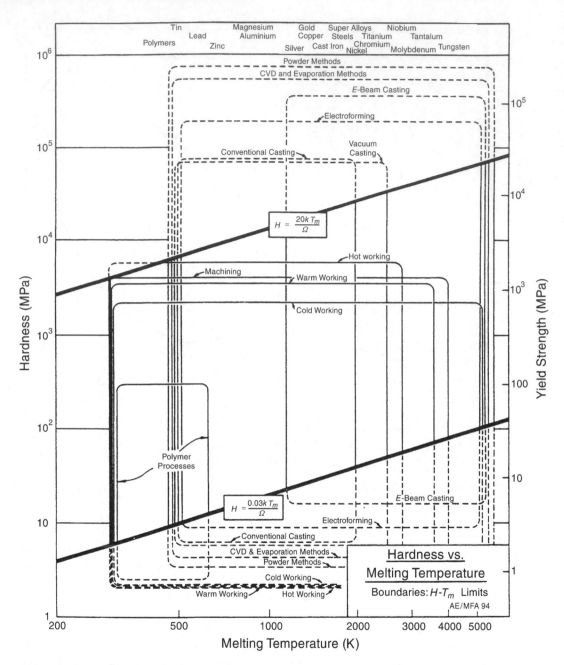

Fig. 11.29 The hardness–melting point chart.

Complexity, and the size–shape chart. Shape and complexity are the most difficult attributes to quantify. Pause for a moment to consider one way of quantifying complexity because it illustrates the nature of the difficulty. It is the idea of characterizing shape and complexity by *information content*. It has two aspects. The obvious one is the number *n* of independent dimensions which must be specified to describe the shape: for a sphere, it is 1 (the radius), for a cylinder, 2; for a tube, 3. A complex casting might have 100 or more specified dimensions. Second there is the precision with which these dimensions are specified. A sphere of radius $r = 10\,\mathrm{m}\ \pm 0.01\,\mathrm{mm}$ is more 'complicated', in a manufacturing sense, than one with a radius $r = 10\,\mathrm{mm}\ \pm 1\,\mathrm{mm}$ because it is harder to make. Both aspects of complexity are captured by the information content

$$C = n \log_2 \left(\frac{\bar{\ell}}{\overline{\Delta\ell}} \right) \tag{11.2}$$

Here $\bar{\ell}$ is the average dimension and $\overline{\Delta\ell}$ is the mean tolerance (see Suh, 1990 for an extensive discussion). It looks as if it makes sense. The information content increases linearly with the number of dimensions, n, and logarithmically with the average relative precision $\bar{\ell}/\overline{\Delta\ell}$. The dimensions cease to have meaning if $\overline{\Delta\ell}$ equals $\bar{\ell}$ because the information content goes to zero.

So far, so good. But now compare a sphere (only one dimension) with a cylinder (with two). Spheres are hard to make, cylinders are not, even though they require twice as much information. Hollow spheres (two dimensions) are harder still, hollow tubes are easy. Information content does not relate directly to the way in which manufacturing processes actually work. Lathes are good at creating axisymmetric shapes (cylinders, tubes); rolling, drawing and extrusion are good at making prismatic ones (sheet, box-sections and the like). Add a single transverse feature and the processing, suddenly, becomes much more difficult. A measure of shape, if it is to be useful here, must recognize the capabilities and limitations of processes.

This directs our thinking towards axial symmetry, translational symmetry, uniformity of section and such like. As mentioned already, turning creates *axisymmetric* shapes; extrusion, drawing and rolling make *prismatic* shapes. Indexing gives shapes with *translational* or *rotational* symmetries, like a gear wheel. Sheet-forming processes make *flat* shapes (stamping) or *dished* shapes (drawing). Certain processes can make three-dimensional shapes, and among these, some can make hollow shapes, whereas others cannot. Figure 11.30 illustrates this classification scheme, building on those of Kusy (1976), Schey (1977) and Dargie *et al.* (1982). The shapes are arranged in the figure in such a way that complexity, defined here as the difficulty of making a shape, increases downwards and to the right.

Shape can be characterized in other ways. One, useful in process selection, is the aspect ratio, or what we call 'slenderness' S. Manufacturing processes vary widely in their capacity to make thin, slender sections. For our purposes, slenderness, S, is measured by the ratio t/ℓ where t is the minimum section and ℓ is the large dimension of the shape: for flat shapes, ℓ is about equal to $\sqrt{A}$ where A is the projected area normal to t. Thus

$$S = \frac{t}{\sqrt{A}} \tag{11.3}$$

Size is defined by the minimum and maximum volumes of which the process is capable. The volume, V, for uniform sections is, within a factor of 2, given by

$$V = At \tag{11.4}$$

Volume can be converted approximately to weight by using an 'average' material density of $5000\,\mathrm{kg/m^3}$; most engineering materials have densities within a factor of 2 of this value. Polymers are the exception: their densities are all around $1000\,\mathrm{kg/m^3}$.

The size–slenderness chart is shown in Figure 11.31. The horizontal axis is the slenderness, S; the vertical axis is the volume, V. Contours of A and t are shown as families of diagonal lines. Casting processes occupy a characteristic field of this space. Surface tension and heat-flow limit

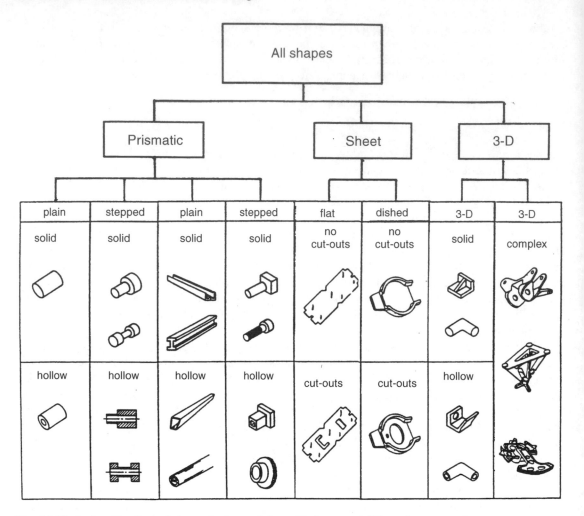

Fig. 11.30 A classification of shape that correlates with the capabilities of process classes.

the minimum section and the slenderness of gravity cast shapes. The range can be extended by applying a pressure, as in centrifugal casting and pressure die casting, or by preheating the mould. But there remain definite upper and lower limits to the size and shape achievable by casting. Deformation processes — cold, warm and hot — cover a wider range. Limits on forging-pressures set a lower limit on thickness and slenderness, but it is not nearly as severely as in casting. Sheet, wire and rod can be made in very great lengths — then the surface area becomes enormous. Machining creates slender shapes by removing unwanted material. Powder-forming methods occupy a smaller field, one already covered by casting and deformation shaping methods, but they can be used for ceramics and very hard metals which cannot be shaped in other ways. Polymer-forming methods — injection moulding, pressing, blow-moulding, etc. — share this regime. Special techniques, which include electro-forming, plasma-spraying, and various vapour-deposition methods, allow very slender shapes. Micro-fabrication technology, in the extreme lower part of the chart, refers to the newest techniques for sub-micron deposition and chemical or electron-beam milling. Joining extends the range further: fabrication allows almost unlimited size and complexity.

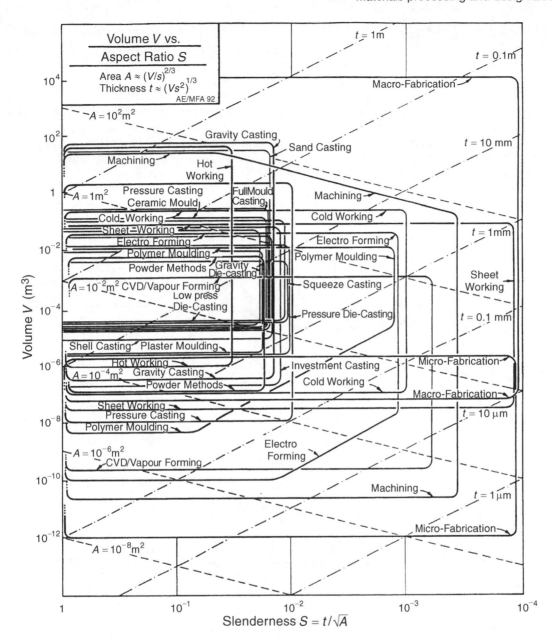

Fig. 11.31 The size–slenderness chart. Diagonal contours give approximate measures of area A and thickness t.

A real design demands certain specific values of S and V, or A and t. Given this information, a subset of possible processes can be read off. Examples are given in the next chapter.

The complexity level vs. *size chart*. Complexity is defined as *the presence of features such as holes, threads, undercuts, bosses, re-entrant shapes, etc., which cause manufacturing difficulty or require additional operations*. So if Figure 11.31 describes the basic shape, complexity describes the additional extra features which are required to produce the final shape. For purposes of comparison, a scale of 1 to 5 is used with 1 indicating the simplest shapes and 5 the most complicated. Each process is given a rating for the maximum complexity of which it is capable corresponding to its proximity to the top left or bottom right shapes in Figure 11.30.

This information is plotted on the complexity level–size chart shown in Figure 11.32. Generally, deformation processes give shapes of limited complexity. Powder routes and composite forming methods are also limited compared with other methods. Polymer moulding does better. Casting processes offer the greatest complexity of all: a cast automobile cylinder block, for instance, is an extremely complicated object. Machining processes increase complexity by adding new features to a component. Fabrication extends the range of complexity to the highest level.

The tolerance–surface roughness chart. No process can shape a part *exactly* to a specified dimension. Some deviation Δx from a desired dimension x is permitted; it is referred to as the *tolerance,*

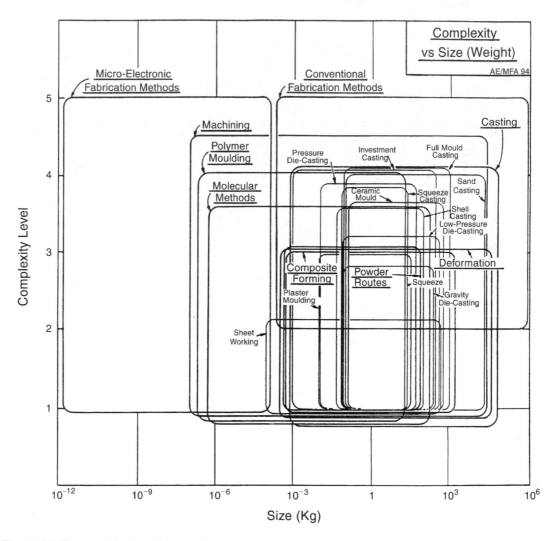

Fig. 11.32 The complexity–size chart.

Table 11.1 Levels of finish

Finish, μm	Process	Typical application
$R = 0.01$	Lapping	Mirrors
$R = 0.1$	Precision grind or lap	High quality bearings
$R = 0.2-0.5$	Precision grinding	Cylinders, pistons, cams, bearings
$R = 0.5-2$	Precision machining	Gears, ordinary machine parts
$R = 2-10$	Machining	Light-loaded bearings, Non-critical components
$R = 3-100$	Unfinished castings	Non-bearing surfaces

T, and is specified as $x = 100 \pm 0.1$ mm, or as $x = 50^{+0.01}_{-0.001}$ mm. Closely related to this is the *surface roughness R*, measured by the root-mean-square amplitude of the irregularities on the surface. It is specified as $R < 100$ μm (the rough surface of a sand casting) or $R < 0.01$ μm (a lapped surface; Table 11.1).

Manufacturing processes vary in the levels of tolerance and roughness they can achieve economically. Achievable tolerances and roughnesses are shown in Figure 11.33. The tolerance is obviously greater than $2R$ (shaded band): indeed, since R is the root-mean-square roughness, the peak roughness is more like $5R$. Real processes give tolerances which range from about $10R$ to $1000R$. Sand casting gives rough surfaces; casting into metal dies gives a better finish. Moulded polymers inherit the finish of the moulds and thus can be very smooth, but tolerances better than ± 0.2 mm are seldom possible because of internal stresses left by moulding and because polymers creep in service. Machining, capable of high dimensional accuracy and smooth surface finish, is commonly used after casting or deformation processing to bring the tolerance or finish to the desired level. Metals and ceramics can be surface-ground and lapped to a high tolerance and smoothness: a large telescope reflector has a tolerance approaching 5 μm over a dimension of a metre or more, and a roughness of about 1/100 of this. But such precision and finish are expensive: processing costs increase almost exponentially as the requirements for tolerance and surface finish are made more severe. The chart shows contours of relative cost: an increase in precision corresponding to the separation of two neighbouring contours gives an increase in cost, for a given process, of a factor of two. It is an expensive mistake to overspecify precision.

Achievable tolerances depend, of course, on dimensions (those given here apply to a 25 mm dimension) and on material. However, for our purposes, typical ranges of tolerance and surface finish are sufficient and discriminate clearly between various processes.

Use of hard copy process selection charts. The charts presented here provide an overview: an initial at-a-glance graphical comparison of the capabilities of various process classes. In a given selection exercise they are not all equally useful: sometimes one is discriminating, another not — it depends on the design requirements. They should not be used blindly, but used to give guidance in selection and engender a feel for the capabilities and limitations of various process types, remembering that some attributes (precision, for instance) can be added later by using secondary processes. That is as far as one can go with hard-copy charts. The number of processes which can be presented on them is obviously limited and the resolution is poor because so many of them overlap. But the procedure lends itself well to computer implementation, overcoming these deficiencies.

Computer-aided screening

If process attributes are stored in a database with an appropriate user-interface, selection charts can be created and selection boxes manipulated with much greater freedom. The Cambridge Process

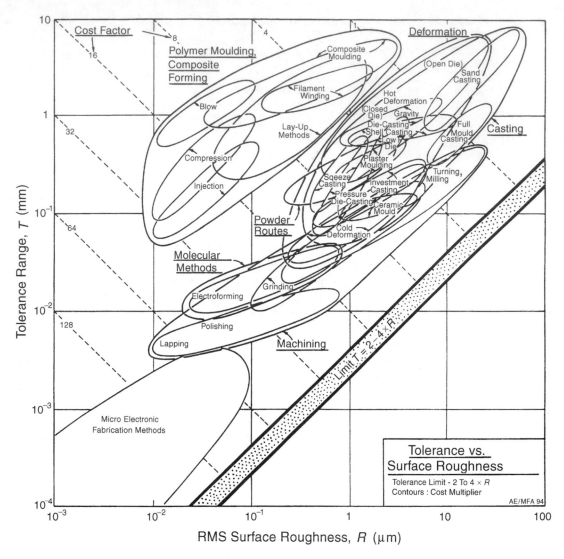

Fig. 11.33 The tolerance–roughness chart.

Selector *(CPS)* is an example of such a system. The way it works is described here; examples of its use are given in Chapter 12. The database contains records, each describing the attributes of a single process. Figure 11.34 shows a typical record: it is that for a particular member of the sand casting class. A schematic indicates how the process works; it is supported by a short description. This is followed by a listing of attributes: the material capability, the attributes relating to shape and physical characteristics, and those which describe economic parameters; the record also contains a brief description of typical uses, references and notes. All the numeric attributes are stored as ranges, indicating the range of capability of the process. The record concludes with a set of references from which the data were drawn and which provide intelligence information, essential in reaching a final selection.

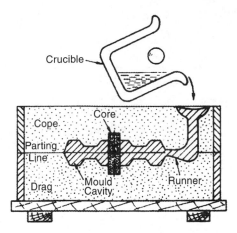

Fig. 11.34 A typical record from a computer-based process selector. It is that for a member of the casting family: CO_2/Silicate sand casting.

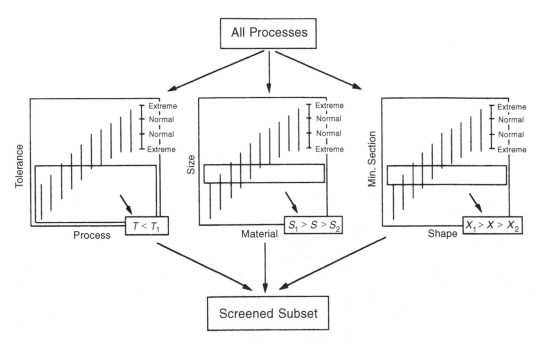

Fig. 11.35 Computer-based screening. The attributes of processes are plotted as bar-charts, isolating processes which can handle a given material class, create a given class of shapes, and meets the design requirements on size, tolerance and minimum section.

The starting point, as in Figure 11.27, is the idea that all processes are potential candidates until shown otherwise. A shortlist of candidates is extracted in two steps: screening to eliminate processes which cannot meet the design specification, and ranking to order the survivors by economic criteria.

A typical three stage screening takes the form shown in Figure 11.35. It shows three bar-charts, on each of which a numeric property is plotted for a selected class property (process class, material

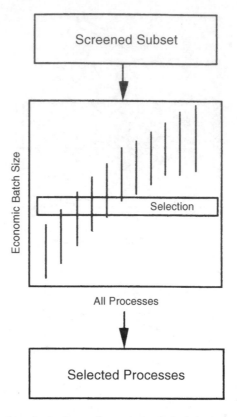

Fig. 11.36 Ranking by economic criteria, here, the economic batch size.

class and shape class). All processes with the selected class attributes appear on the charts. The processes are sorted in order of ascending value of the numeric property which is plotted as a bar to show its range. The left-hand chart selects a process of a given type ('primary', for example) which offers a tolerance better than $\pm T_1$ mm. The second specializes this to those which can shape a chosen class of material ('thermoplastic polymers', for instance) with a size between S_1 and S_2 kg. The third isolates the subset of these which are able to create a given shape (such as '3-D solid, parallel features') with a minimum section thickness as small as X_1 mm. Further stages can be added. The selection is made by placing a selection box onto each chart, identifying the range of tolerance, size, minimum section and so forth specified by the design. The effect is to eliminate the processes which cannot meet the specifications.

The next step is to rank the survivors by economic criteria (Figure 11.36). To do this we need to examine process cost.

11.6 Ranking: process cost

Part of the cost of a component is that of the material of which it is made. The rest is the cost of manufacture, that is, of forming it to a shape, and then of joining it to the other components to give the finished product. Before turning to details, there are three common-sense rules for minimizing cost which the designer should bear in mind.

Keep things standard. If someone already makes the part you want, it will almost certainly be cheaper to buy it than to make it. If nobody does, then it is cheaper to design it to be made from standard stock (sheet, rod, tube) than from non-standard shapes, or special castings or forgings. Try to use standard materials, and as few of them as possible: it reduces inventory costs and the range of tooling the manufacturer needs.

Keep things simple. If a part has to be machined, it will have to be clamped; the cost increases with the number of times it will have to be re-jigged or re-oriented, specially if special tools are necessary. If a part is to be welded or brazed, the welder must be able to reach it with his torch and still see what he is doing. If it is to be cast or moulded or forged, it should be remembered that high (and expensive) pressures are required to make fluids flow into narrow channels, and that re-entrant shapes greatly complicate mould design. All this is pretty obvious, but easily overlooked. Think of making the part yourself: will it be awkward? Could slight re-design make it less awkward?

Do not specify more performance than is needed. Performance must be paid for. High-strength metals are more heavily alloyed with expensive additions; high-performance polymers are chemically more complex; high-performance ceramics require greater quality control in their manufacture. All of these increase material costs. In addition, high-strength materials are hard to fabricate. The forming pressures (whether for a metal or a polymer) are higher; tool wear is greater; ductility is usually less so that deformation processing can be difficult or impossible. This can mean that new processing routes must be used: investment casting or powder forming instead of conventional casting and mechanical working; more expensive moulding equipment operating at higher temperatures and pressures, and so on. The better performance of the high-strength material must be paid for, not only in greater material cost but also in the higher cost of processing. Finally, there are the questions of tolerance and roughness. The 'cost' contours of Figure 11.33 give warning: cost rises exponentially with precision and surface finish. It is an expensive mistake to specify tighter tolerance or smoother surfaces than are necessary. The message is clear. Performance costs money. Do not over specify it.

To make further progress, we must examine the contributions to process costs, and their origins.

Economic criteria for selection. If you have to sharpen a pencil, you can do it with a knife. If, instead, you had to sharpen a thousand pencils, it would pay to buy an electric 'mechanized' sharpener. And if you had to sharpen a million, you might wish to equip yourself with an automatic feeding, gripping and sharpening system. To cope with pencils of different length and diameter, you could go further and devise a microprocessor-controlled system with sensors to measure pencil dimensions, sharpening pressure and so on; that is, you create a system with 'intelligence' which can recognize and adapt to pencil size. The choice of process, then, depends on the number of pencils you wish to sharpen, that is, on the *batch size*. The best choice is that which costs least, per pencil sharpened.

Figure 11.37 is a schematic of how the cost of sharpening a pencil might vary with batch size. A knife does not cost much but it is slow, so the labour cost is high. The other processes involve progressively greater capital investment but do the job more quickly, reducing labour costs. The balance between capital cost and rate gives the shape of the curves. In this figure the best choice is the lowest curve — a knife for up to 100 pencils; mechanization for 10^2 to 10^4, an automatic system for 10^4 to 10^6, and so on.

Economic batch size. Modelling cost may sound easy but it is not. Process cost depends on a large number of independent variables, not all within the control of the modeller. Cost modelling is described in the next section, but — given the disheartening implications of the last sentence — it is

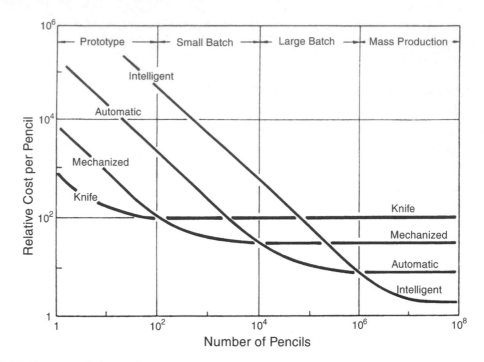

Fig. 11.37 The cost of sharpening a pencil, plotted against batch size, for four processes. The curves all have the form of equation (11.7).

comforting to have an easy, if approximate, way out. The influence of many of the inputs to the cost of a process are captured by a single attribute: the *economic batch size*. A process with an economic batch size with the range B_1-B_2 is one which is found by experience to be competitive in cost when the output lies in that range. The economic batch size is commonly cited for processes. The easy way to introduce economy into the selection is to rank candidate processes by economic batch size and retain those which are economic in the range you want, as illustrated by Figure 11.36. But do not harbour false illusions: many variables cannot be rolled into one without loss of discrimination. It is better to develop a cost model.

Cost modelling. The manufacture of a component consumes resources (Table 11.2). The process cost is the sum of the costs of the resources it consumes. This *resource-based* approach to cost analysis is particularly helpful at the broad level with which we are concerned here since all processes, no matter how diverse, consume the resources listed in the table. Thus the cost of a component of mass m entails the cost C_m ($/kg) of the *material* of which it is made, and it involves the cost of *dedicated tooling*, C_t, which must be amortized by the *batch size*, n. In addition, it requires *time*, chargeable at an *overhead rate* $\dot{C}_L$ (thus with units of $/h or equivalent), *power* $\dot{P}$ (kW) at an *energy cost* C_e ($/kWh), and it requires *space* of area A, incurring a *rental cost* of $\dot{C}_s$ ($/m²h). The cost equation takes the form

$$\text{Material} \quad \text{Tooling} \quad \text{Time} \quad \text{Energy} \quad \text{Space}$$

$$C = [mC_m] + \left[\frac{C_t}{n}\right] + \left[\frac{\dot{C}_L}{\dot{n}}\right] + \left[\frac{\dot{P}C_e}{\dot{n}}\right] + \left[\frac{A\dot{C}_s}{\dot{n}}\right] \tag{11.5}$$

where $\dot{n}$, the *batch rate*, is the number of units produced per hour.

Table 11.2 The resources consumed in production

Resource		Symbol	Unit
Materials:	inc. consumables	C_m	$/kg
Capital:	of equipment	C_c	$
	cost of tooling	C_t	$
Time:	overhead rate	$\dot{C}_L$	$/hr
Energy:	power	$\dot{P}$	kW
	cost of energy	C_e	$/kW h
Space:	area	A	m^2
	cost of space	$\dot{C}_s$	$/m^{2}h
Information:	R & D	C_i	$/yr
	royalty payments		

Where has the *capital cost* C_c of the equipment (as opposed to tooling) gone? A given piece of equipment — a press, for example — is commonly used to make more than one product. It is then usual to convert the capital cost of non-dedicated equipment, and the cost of borrowing the capital itself, into an overhead by dividing it by a capital write-off time, t_c (5 years, say) over which it is to be recovered. Thus the overhead rate becomes

Basic OH rate   Capital write-off

$$\frac{\dot{C}_L}{\dot{n}} = \frac{1}{\dot{n}} \left\{ [\dot{C}_{Lo}] + \left[\frac{C_c}{Lt_c} \right] \right\} \tag{11.6}$$

where $\dot{C}_{Lo}$ is the *basic overhead rate* (labour, etc.) and L is the *load factor* (the fraction of time over which the equipment is productively used).

A detailed analysis breaks cost down further, detailing the contributions of scrap, administration, maintenance, the cost of capital (the interest that must be paid, or could have been earned, on the capital tied up in the equipment) and so on — real cost models can become very complex. Let us, instead, simplify. The terms can be assembled into three groups:

Materials   Tooling   Time   Capital   Energy   Space

$$C = [mC_m] + \frac{1}{n}[C_t] + \frac{1}{\dot{n}} \left[\dot{C}_{Lo} + \frac{C_c}{Lt_c} + \dot{P}C_e + A\dot{C}_s \right]$$

We merge the terms in the final bracket into a single 'gross overhead', $\dot{C}_{L,\text{gross}}$, allowing the equation to be written

Materials   Dedicated cost/unit   Gross overhead/unit

$$C = [mC_m] + \frac{1}{n}[C_t] + \frac{1}{\dot{n}}[\dot{C}_{L,\text{gross}}] \tag{11.7}$$

The equation really says: cost has three types of contributions — one which is independent of batch size and rate, one which varies as the reciprocal of the batch size (n^{-1}), and one which varies as the reciprocal of the batch rate ($\dot{n}^{-1}$). The first — the 'material' costs — includes also material consumed in manufacture. The second — the dedicated capital investment — contains the cost of tooling, dies, jigs and moulds. The last term — the one dependent on time — includes the 'direct' cost of the machine operator plus the 'indirect' or 'overhead' cost associated with administration, maintenance, safety, and so forth. It is sometimes difficult to decide precisely how costs should be assigned between these headings; different companies do it in different ways. But the general point is clear: *material* plus *dedicated capital costs* plus *gross overhead*.

The equation describes a set of curves, one for each process. Each has the shape of the pencil-sharpening curves of Figure 11.37. Figure 11.38 illustrates a second example: the manufacture of an aluminium con-rod by two alternative processes: sand casting and die casting. Sand casting equipment is cheap but the process is slow. Die casting equipment costs much more but it is also much faster. Data for the terms in equation (11.7), for these two processes, are listed in Table 11.3: they show that the capital cost assigned to the die-casting equipment is greater by some 76 times that for sand casting, but that the process is 40 times faster. The material cost (1 unit) and the labour cost per hour (20 units) are, of course, the same for both. Figure 11.38 is a plot of equation (11.7), evaluated with this data for the two processes. The curves intersect at a batch size of 4000. Sand casting is the most economical process for batches less than this; die casting for batches which are larger. Note that, for small batches, the component cost is dominated by that of the process — the material cost hardly matters. But as the batch size grows, the contribution of the second term in the cost equation diminishes; and if the process is fast, the cost falls until it is typically about twice that of the material of which the component is made.

Technical cost modelling. Equation (11.7) is the first step in modelling cost. Greater predictive power is possible with technical cost models which exploit understanding of the way in which the design, the process and cost interact. The capital cost of equipment depends on size and degree of automation. Tooling cost and production rate depend on complexity. These and many other

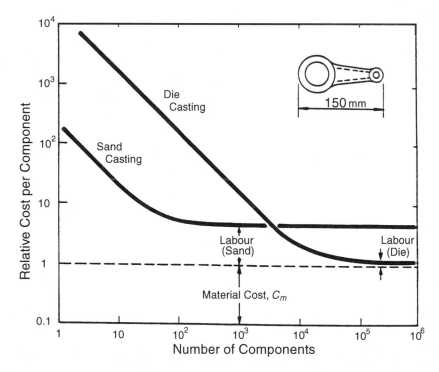

Fig. 11.38 The best choice of casting (or machining or forging) process depends on batch size. Sand casting requires cheap equipment but is labour intensive. Die casting requires more expensive equipment, but is faster. The data shown here are for an automobile connecting rod.

Table 11.3 Data for the cost equation

Relative cost*	Sand Casting	Die Casting	Comment
Material, mC_m	1	1	
Basic overhead $C_{Lo}(h^{-1})$	20	20	Process independent
Capital write-off time t_c (yrs)	5	5	
Dedicated tool cost, C_t	210	16000	
Capital cost C_c	10000	300000	Process dependent
Batch rate, $\dot{n}(h^{-1})$	5	200	

*All costs normalized to the material cost.

dependencies can be captured in theoretical or empirical formulae or look-up tables which can be built into the cost model, giving more resolution in ranking competing processes. For more advanced analyses the reader is referred to the literature listed in the Further reading section of this chapter.

11.7 Supporting information

Systematic screening and ranking based on attributes common to all processes gives a short list of candidates. We now need supporting information — details, case studies, experience, warnings, anything which helps form a final judgement. Where is it to be found?

Start with texts and handbooks — they don't help with systematic selection, but they *are* good on supporting information. They, and other sources are listed in Further reading. Bralla (1986) is particularly good and so is Schey (1977), although it is dated. Many texts and handbooks are specialized to a single class of process, giving more detail. Casting is an example: detailed intelligence is to be found in the ASM *Casing Design Handbook* (1962) and Clegg's (1991) *Precision Casting*.

Next look at the data sheets and design manuals available from the makers and suppliers of process equipment, and, often, from material suppliers. Leading suppliers exhibit at major conferences and exhibitions — these are a useful source of information for small and medium scale industries. Increasingly this sort of supplier-specific information is available on CD, allowing rapid access.

And then there is the World Wide Web. My old mother, were she still here, would have described it as 'a dog's dinner' meaning, I believe, that it contained everything from the best to the worst bits of today's menu. It is certainly mixed, but it *is* today's menu, and that has value. There are an increasing number of web sites which offer information on processes, the best of them very helpful. A selection is given in Chapter 13.

11.8 Summary and conclusions

A wide range of shaping and finishing processes is available to the design engineer. Each has certain characteristics, which, taken together, suit it to the forming of certain materials to certain shapes, but disqualify it for others. Faced with the choice, the designer has, in the past, relied on locally available expertise, or on common practice. Neither of them lead to innovation, nor are they well matched to current design methods. The structured, systematic approach of this chapter provides a way forward. It ensures that potentially interesting processes are not overlooked, and guides the user quickly to processes capable of making the desired shape.

The method parallels that for selection of material, using process selection charts to implement the procedure. The axes of the charts are process attributes: product size, shape, precision, and certain key material properties which influence shaping operations. A product design dictates a certain, known, combination of these attributes. The design requirements are plotted onto the charts, identifying a subset of possible processes.

There is, of course, much more to process selection than this. It is to be seen, rather, as a first systematic step, replacing a total reliance on local experience and past practice. The narrowing of choice helps considerably: it is now much easier to identify the right source for more expert knowledge and to ask of it the right questions. The final choice still depends on local economic and organizational factors which can only be decided on a case-by-case basis.

11.9 Further reading

Alexander, J.M., Brewer, R.C. and Rowe, G.W. (1987) *Manufacturing Technology, Vol. 2: Engineering Processes*, Ellis Horwood, Chichester.

ASM Casting Design Handbook (1962) American Society for Metals, Metals Park, Ohio USA.

Bolz, R.W. (1977) *Production Processes — the Productivity Handbook*, Conquest Publications, NC.

Bralla, J.G. (1986) *Handbook of Product Design for Manufacturing*, McGraw-Hill, New York.

Clegg, A.J. (1991) *Precision Casting Processes*, Pergamon Press, Oxford.

Crane, F.A.A. and Charles, J.A. (1984) *Selection and Use of Engineering Materials*, Butterworth, London, Chapter 13.

Dieter, G.E. (1983) *Engineering Design, A Materials and Processing Approach*, McGraw-Hill, New York, Chapter 7.

Edwards, L. and Endean, M., (eds) (1990) *Manufacturing with Materials*, Materials in Action Series, The Open University, Butterworths, London.

Esawi, A. (1994) 'Systematic Process Selection in Mechanical Design', PhD thesis, Cambridge University Engineering Department, Trumpington Street, Cambridge CB2 1PZ.

Esawi, A. and Ashby, M.F. (1998) 'Computer-based Selection of Manufacturing Processes, Part 1: methods and software; Part 2, case studies', Cambridge University Engineering Department Report TR 50, May 1997.

Esawi, A. and Ashby, M.F. (1998) 'Computer-based selection of manufacturing processes', *Journal of Engineering Manufacture*.

Farag, M.M. (1990) *Selection of Materials and Manufacturing Processes for Engineering Design*, Prentice-Hall, London.

Frost, H.J. and Ashby, M.F. (1982) *Deformation Mechanism Maps*, Pergamon Press, Oxford, Chapter 19, Section 111.6.

Kalpakjian, S. (1984) *Manufacturing Processes for Engineering Materials*, Addison Wesley, London.

Kusy, P.F. (1976) 'Plastic materials selection guide', SAE Technical Paper 760663, Sept. 1976.

Lascoe, O.D. (1988) *Handbook of Fabrication Processes*, ASM International, Metals Park, Columbus, Ohio.

Ludema, K.C., Caddell, R.M. and Atkins, A.G. (1987) *Manufacturing Engineering, Economics and Processes*, Prentice-Hall, Englewood Cliffs, NJ.

Schey, J.A. (1977) *Introduction to Manufacturing Processes*, McGraw-Hill, New York.

Suh, N.P. (1990) *The Principles of Design*, Oxford University Press, Oxford.

Waterman, N. and Ashby, M.F. (1997) *Chapman and Hall Materials Selector*, Chapman and Hall, London, Chapter 1.6.

Yankee, H.W. (1976) *Manufacturing Processes*, Prentice-Hall, Englewood Cliffs, NJ.

Cost modelling

Clark, J.P. and Field, F.R. III (1997) 'Techno-economic issues in materials selection', in *ASM Metals Handbook*, Vol. 20. American Society for Metals, Metals Park, Ohio.

_____

Chapter 12

Case studies: process selection

12.1 Introduction and synopsis

The previous chapter described a systematic procedure for process selection. The inputs are design requirements; the output is a shortlist of processes capable of meeting them. The case studies of this chapter illustrate the method. The first four make use of hard-copy charts; the last two show how computer-based selection works. More details for each are then sought, starting with the texts listed under Further reading for Chapter 11, and progressing to the specialized data sources described in Chapter 13. The final choice evolves from this subset, taking into account local factors, often specific to a particular company, geographical area or country.

The case studies follow a standard pattern. First, we list the *design requirements*: size, minimum section, surface area, shape, complexity, precision and finish, and the *material* and the *processing constraints* that it creates (melting point and hardness). Then we plot these requirements onto the process charts, identifying search areas. The processes which overlap the search areas are capable of making the component to its design specification: they are the candidates. If no one process meets all the design requirements, then processes have to be 'stacked': casting followed by machining (to meet the tolerance specification on one surface, for instance); or powder methods followed by grinding. Computer-based methods allow the potential candidates to be ranked, using economic criteria. More details for the most promising are then sought, starting with the texts listed under Further reading for Chapter 11, and progressing to the specialized data sources described in Chapter 13. The final choice evolves from this subset, taking into account local factors, often specific to a particular company, geographical area or country.

12.2 Forming a fan

Fans for vacuum cleaners are designed to be cheap, quiet and efficient, probably in that order. Case study 6.6 identified a number of candidate materials, among them, aluminium alloys and nylon. Both materials are cheap. The key to minimizing process costs is to form the fan to its final shape in a single operation — that is, to achieve net-shape forming — leaving only the central hub to be machined to fit the shaft with which it mates. This means the selection of a process which can meet the specifications on precision and tolerance, avoiding the need for machining or finishing of the disk or blades.

The design requirements

The pumping rate of a fan is determined by its radius and rate of revolution: it is this which determines its size. The designer calculates the need for a fan of radius 60 mm, with 20 blades of

Table 12.1 Design constraints for the fan

Constraint		Value
Materials	Nylons	$T_m = 550$–573 K
		$H = 150$–270 MPa
		$\rho = 1080$ kg/m^3
	Al-alloys	$T_m = 860$–933 K
		$H = 150$–1500 MPa
		$\rho = 2070$ kg/m^3
Complexity		2 to 3
Min. Section		1.5–6 mm
Surface area		0.01–0.04 m^2
Volume		1.5×10^{-5}–2.4×10^{-4} m^3
Weight		0.03–0.5 kg
Mean precision		± 0.5 mm
Roughness		<1 μm

average thickness 3 mm. The surface area, approximately $2(\pi R^2)$, is 2×10^{-2} m^2. The volume of material in the fan is, roughly, its surface area times its thickness — about 6×10^{-5} m^3, giving a weight in the range 0.03 (nylon) to 0.5 kg (aluminium). If formed in one piece, the fan has a fairly complex shape, though its high symmetry simplifies it somewhat. We classify it as 3-D solid, with a complexity between 2 and 3. In the designer's view, the surface finish is what really matters. It (and the geometry) determine the pumping efficiency of the fan and influence the noise it makes. He specifies a smooth surface: $R < 1$ μm. The design constraints are summarized in Table 12.1.

What processes can meet them?

The selection

We turn first to the size–shape chart, reproduced as Figure 12.1. The surface area and minimum section define the search area labelled 'FAN' — it has limits which lie a factor 2 on either side of the target values. It shows that the fan can be shaped in numerous ways; they include *die-casting* for metals and *injection moulding* for polymers.

Turn next to the complexity–size chart, reproduced in Figure 12.2. The requirements for the fan again define a box. We learn nothing new: the complexity and size of the fan place it in a regime in which many alternative processes are possible. Nor do the material properties limit processing (Figure 12.3); both materials can be formed in many ways.

The discriminating requirement is that for smoothness. The design constraints $R < \pm 1$ μm and $T < 0.5$ mm are shown on Figure 12.4. Any process within the fan search region is a viable choice; any outside is not. Machining from solid meets the specifications, but is not a net-shape process. A number of polymer moulding processes are acceptable, among them, injection moulding. Few metal-casting processes pass — the acceptable choices are pressure die-casting, squeeze casting and investment casting.

The processes which pass all the selection steps are listed in Table 12.2. They include injection moulding for the nylon and die-casting for the aluminium alloy: these can achieve the desired shape, size, complexity, precision and smoothness, although a cost analysis (Case Study 12.5) is now needed to establish them as the best choices.

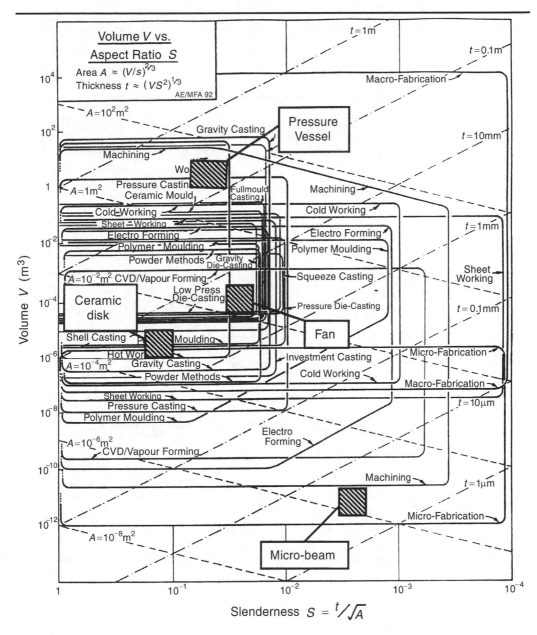

Fig. 12.1 The size–slenderness–area–thickness chart, showing the search areas for the fan, the pressure vessel, the micro-beam and the ceramic tap valve.

Postscript

There are (as always) other considerations. There are the questions of capital investment, batch size and rate, supply, local skills and so forth. The charts cannot answer these. But the procedure has been helpful in narrowing the choice, suggesting alternatives, and providing a background against which a final selection can be made.

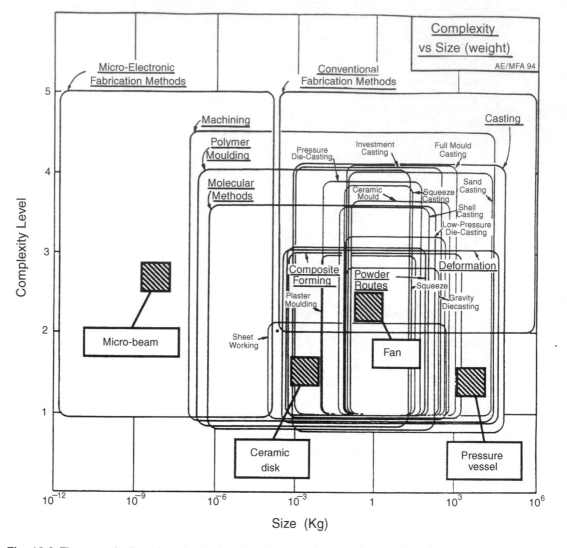

Fig. 12.2 The complexity–size chart, showing the search areas for the fan, the pressure vessel, the micro-beam and the ceramic tap valve.

Related case studies

Case Study 6.7:   Materials for high flow fans
Case Study 14.3: Data for a non-ferrous alloy

12.3 Fabricating a pressure vessel

A pressure vessel is required for a hot-isostatic press or HIP (Figure 11.13). Materials for pressure vessels were the subject of Case Study 6.14; tough steels are the best choice.

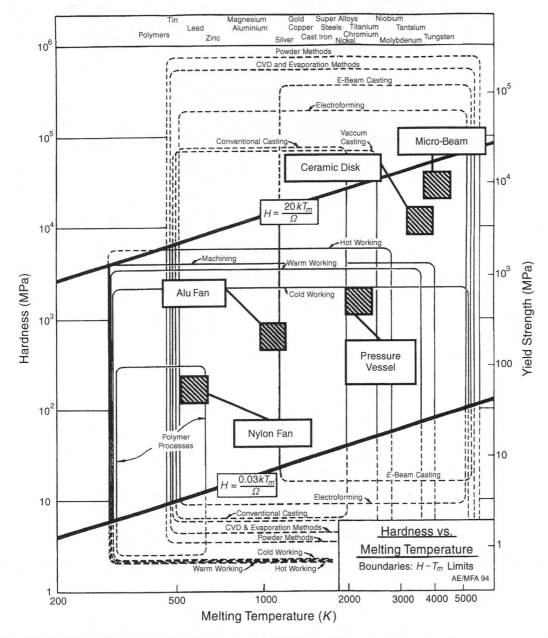

Fig. 12.3 The hardness–melting point chart, showing the search areas for the fan, the pressure vessel, the micro-beam and the ceramic tap valve.

The design requirements

The design asks for a cylindrical pressure vessel with an inside radius R_i of 0.5 m and a height h of 1 m, with removable end-caps (Figure 12.5). It must safely contain a pressure p of 100 MPa. A steel with a yield strength σ_y of 500 MPa (hardness: 1.5 GPa) has been selected. The necessary

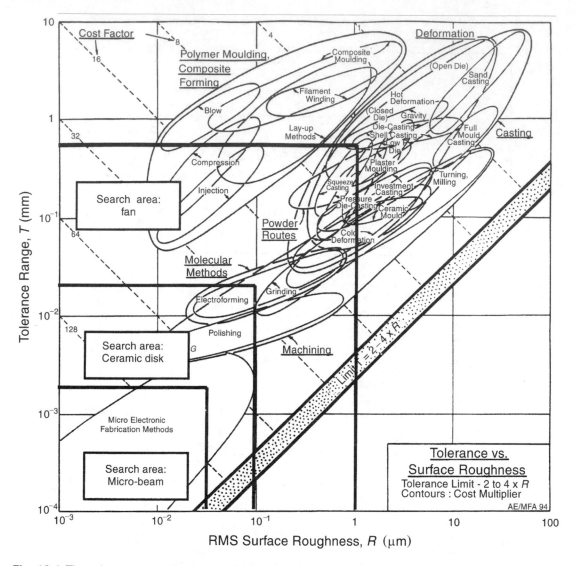

Fig. 12.4 The tolerance–roughness chart, showing the search areas for the fan, the micro-beam and the ceramic tap valve.

wall thickness t is given approximately by equating the hoop stress in the wall, roughly pR/t, to the yield strength of the material of which it is made, σ_y, divided by a safety factor S_f which we will take to be 2:

$$t = \frac{S_f\, pR}{\sigma_y} = 0.2\,\text{m} \tag{12.1}$$

The outside radius R_o is, therefore, 0.7 m. The surface area A of the cylinder (neglecting the end-caps) follows immediately: it is roughly $3.8\,\text{m}^2$. The volume $V = At$ is approximately $0.8\,\text{m}^3$. Lest that sounds small, consider the weight. The density of steel is just under $8000\,\text{kg/m}^3$. The vessel weighs 6 tonnes. The design constraints are shown at Table 12.3.

Table 12.2 Processes for forming the fan

Process	Comment
Machine from solid	Expensive. Not a net-shape process.
Electro-form	Slow, and thus expensive.
Cold deformation	Cold forging meets design constraints.
Investment casting	Accurate but slow.
Pressure die casting	Meets all design constraints.
Squeeze cast	Meets all design constraints.
Injection moulding	Meets all design constraints.
Resin transfer moulding	Meets all design constraints.

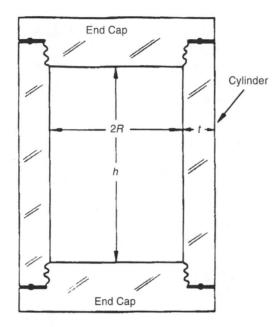

Fig. 12.5 Schematic of the pressure vessel of a hot isostatic press.

Table 12.3 Design constraints for the pressure vessel

Constraint	Value	
Material	Steel	$T_m = 1600\,\text{K}$
		$H = 2000\,\text{MPa}$
		$\rho = 8000\,\text{kg/m}^3$
Complexity	2	
Min. Section	200 mm	
Surface area	3.8 m^2	
Volume	0.8 m^3	
Weight	6000 kg	
Mean precision	±1.0 mm	
Roughness	<1 μm on mating surfaces only	

A range of pressures is envisaged, centred on this one, but with inner radii and pressures which range by a factor of 2 on either side. (A constant pressure implies a constant 'aspect ratio', R/t.) Neither the precision nor the surface roughness of the vessel are important in selecting the primary forming operation because the end faces and internal threads will be machined, regardless of how it is made. What processes are available to shape the cylinder?

The selection

The discriminating requirement, this time, is size. The design requirements of wall thickness and surface area are shown as a labelled box on Figure 12.1. It immediately singles out the four possibilities listed in Table 12.4: the vessel can be machined from the solid, made by hot-working, cast, or fabricated (by welding plates together, for instance).

Complexity and size (Figure 12.2) confirm the choice. Material constraints are worth checking (Figure 12.3), but they do not add any further restrictions. Tolerance and roughness do not matter except on the end faces and threads (where the end-caps must mate) and any ports in the sides — these require high levels of both. The answer here (Figure 12.4) is to machine, and perhaps surface-grind.

Postscript

A 'systematic' procedure is one that allows a conclusion to be reached without prior specialized knowledge. This case study is an example. We can get so far (Table 12.4) systematically, and it is a considerable help. But we can get no further without adding some expertise.

A cast pressure vessel is not impossible, but it would be viewed with suspicion by an expert because of the risk of casting defects; safety might then require elaborate ultrasonic testing. The only way to make very large pressure vessels is to weld them, and here we encounter the same problem: welds are defect-prone and can only be accepted after elaborate inspection. Forging, or machining from a previously forged billet are the best because the large compressive deformation heals defects and aligns oxides and other muck in a harmless, strung-out way.

That is only the start of the expertise. You will have to go to an expert for the rest.

Related case studies

Case Study 6.15: Safe pressure vessels

Table 12.4 Processes for forming pressure vessels

Process	Comment
Machining	Machine from solid (rolled or forged) billet. Much material discarded, but a reliable product. Might select for one-off.
Hot working	Steel forged to thick-walled tube, and finished by machining end faces, ports, etc. Preferred route for economy of material use.
Casting	Cast cylindrical tube, finished by machining end-faces and ports. Casting-defects a problem.
Fabrication	Weld previously-shaped plates. Not suitable for the HIP; use for very large vessels (e.g. nuclear pressure vessels).

12.4 Forming a silicon nitride micro-beam

The ultimate in precision mechanical metrology is the atomic-force microscope; it can measure the size of an atom. It works by mapping, with Angström resolution, the forces near surfaces, and, through these forces, the structure of the surface itself. The crucial component is a micro-beam: a flexible cantilever with a sharp stylus at its tip (Figure 12.6). When the tip is tracked across the surface, the forces acting between it and the sample cause minute deflections of the cantilever which are detected by reflecting a laser beam off its back surface, and are then displayed as an image.

The design requirements

Albrecht and his colleagues (1990) list the design requirements for the micro-beam. They are: minimum thermal distortion, high resonant frequency, and low damping. If these sound familiar, it is perhaps because you have read Case Study 6.19: 'Materials to minimize thermal distortion in precision devices'. There, the requirements of minimum thermal distortion and high resonant frequency led to a shortlist of candidate materials: among them, silicon carbide and silicon nitride.

The demands of sensitivity require beam dimensions which range, by a factor of 2 (depending on material), about those shown in Figure 12.6. The minimum section, t, lies in the range 2 to 8 μm; the surface area is about $10^{-6}\,\text{m}^2$, the volume is roughly $5 \times 10^{-12}\,\text{m}^3$, and the weight approximately $10^{-8}\,\text{kg}$.

Precision is important in a device of this sort. The precision of 1% on a length of order 100 mm implies a tolerance of ± 1 μm. Surface roughness is only important if it interferes with precision, requiring $R < 0.04$ μm.

The candidate materials — silicon carbide and silicon nitride — are, by this time, part of the design specification. They both have very high hardness and melting points. Table 12.5 summarizes the design constraints.

How is such a beam to be made?

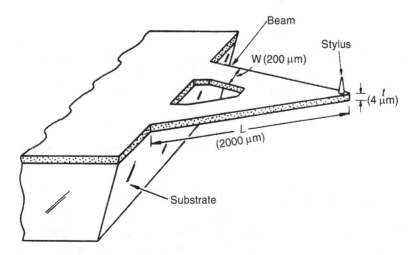

Fig. 12.6 A micro-beam for an atomic-force microscope.

Table 12.5 Design constraints for the micro-beam

Constraint		Value
Materials	Silicon carbide	$T_m = 2973-3200\,\text{K}$
		$H = 30\,000-33\,000\,\text{MPa}$
	Silicon nitride	$T_m = 2170-2300\,\text{K}$
		$H = 30\,000-34\,000\,\text{MPa}$
Complexity	2 to 3	
Min. section	$2-8\,\mu\text{m}$	
Surface area	$5 \times 10^{-7}-2 \times 10^{-6}\,\text{m}^2$	
Volume	$2 \times 10^{-12}-10^{-11}\,\text{m}^3$	
Weight ($\rho = 3000\,\text{kg/m}^3$)	$6 \times 10^{-9}-3 \times 10^{-8}\,\text{kg}$	
Mean precision	± 0.5 to $1\,\mu\text{m}$	
Roughness	$<0.04\,\mu\text{m}$	

The selection

The section and surface area locate the beam on Figure 12.1 in the position shown by the shaded box. It suggests that it may be difficult to shape the beam by conventional methods, but that the methods of micro-fabrication could work. The conclusion is reinforced by Figure 12.2.

Material constraints are explored with the hardness–melting point chart of Figure 12.3. Processing by conventional casting or deformation methods is impossible; so is conventional machining. Powder methods can shape silicon carbide and nitride, but not, Figure 12.3 shows, to anything like the size or precision required here. The CVD and evaporation methods of micro-fabrication look like the best bet.

The dimensions, precision, tolerance and finish all point to micro-fabrication. Silicon nitride can be grown on silicon by gas-phase techniques, standard for micro-electronics. Masking by lithography, followed by chemical 'milling' — selective chemical attack — allows the profile of the beam to be cut through the silicon nitride. A second chemical process is then used to mill away the underlying silicon, leaving the cantilever of silicon nitride meeting the design specifications.

Postscript

Cantilevers with length as small as $100\,\mu\text{m}$ and a thickness of $0.5\,\mu\text{m}$ have been made successfully by this method — they lie off the bottom of the range of the charts. The potential of micro-fabrication for shaping small mechanical components is considerable, and only now being explored.

Related case studies

Case Study 6.20: Materials to minimize thermal distortion in precision devices

12.5 Forming ceramic tap valves

Vitreous alumina, we learn from Case Study 6.20, may not be the best material for a hot water valve — there is evidence that thermal shock can crack it. Zirconia, it is conjectured, could be better. Fine. How are we to shape it?

The design requirements

Each disc of Figure 6.36 has a diameter of 20 mm and a thickness of 5 mm (surface area $\approx 10^{-3}\,\text{m}^2$; volume $1.5 \times 10^{-6}\,\text{m}^3$). They have certain obvious design requirements. They are to be made from zirconia, a hard, high-melting material. Their mating surfaces must be flat and smooth so that they seal well. The specifications for these surfaces are severe: $T \leq \pm 20\,\mu\text{m}$, and $R < 0.1\,\mu\text{m}$. The other dimensions are less critical (constraints are shown in Table 12.6). Any process which will form zirconia to these requirements will do. There aren't many.

The selection

The size is small and the shape is simple: they impose no great restrictions (Figures 12.1 and 12.2). It is the material which is difficult. Its melting point is high (2820 K or 2547°C) and its hardness is too (15 GPa). The chart we want is that of hardness and melting point. The search region for zirconia is shown on Figure 12.3. It identifies a subset of processes, listed in the first column of Table 12.7. Armed with this list, standard texts reveal the further information given in the second column. Powder methods emerge as the only practical way to make the discs.

Powder methods can make the shape, but can they give the tolerance and finish? Figure 12.3, shows that they cannot. The mating face of the disc will have to be ground and polished to give the desired tolerance and smoothness.

Postscript

Here, as in the earlier case studies, the design requirements alone lead to an initial shortlist of processes. Further, detailed, information for these must then be sought. The texts on processing

Table 12.6 Design constraints for the valve

Constraint	Value
Materials	Zirconia $T_m = 2820\,\text{K}$ $\quad\quad\quad\quad H = 15\,000\,\text{MPa}$
Complexity	1–2
Min. Section	5 mm
Surface area	$10^{-3}\,\text{m}^2$
Volume	$1.5 \times 10^{-6}\,\text{m}^3$
Weight ($\rho = 3000\,\text{kg/m}^3$)	$4.5 \times 10^{-3}\,\text{kg}$
Mean precision	$\pm 0.02\,\text{mm}$
Roughness	$<0.1\,\mu\text{m}$

Table 12.7 Processes for shaping the valve

Process	Comment
Powder methods	Capable of shaping the disc, but not to desired precision.
CVD and Evaporation methods	No CVD route available. Other gas-phase methods possible for thin sections.
Electron-beam casting	Difficult with a non-conductor.
Electro-forming	Not practical for an oxide.

(Further reading of Chapter 11) and the material-specific data sources (Chapters 13 and 14) almost always suffice.

Related case studies

Case Study 6.21: Ceramic valves for taps
Case Study 14.5: Data for a ceramic

12.6 Economical casting

Optical benches are required for precision laser-holography. The list of materials thrown up as candidates for precision devices (Case Study 6.20) included aluminium and its alloys. The decision has been taken to cast the benches from Alloy 380, an aluminium–silicon alloy developed for casting purposes (Case Study 14.3).

The design requirements

The designer, uncertain of the market for the benches, asks for advice on the best way to cast one prototype bench, a preliminary run of 100 benches, and (if these succeed) enough benches to satisfy a potential high-school market of about 10 000. The high precision demanded by the design can only be met by machining the working surfaces of the bench, so the tolerance and roughness of the casting itself do not matter. The best choice of casting method is the cheapest.

Process data for four possible casting methods for aluminium alloys are listed in Table 12.8. The costs are given in units of the material cost, C_m, of one bench (that is, $C_m = 1$). In these units, labour costs, C_L, are 20 units per hour. Estimates for the capital cost C_c of setting up each of the four processes come next. Finally, there is the batch rate for each process, in units per hour. Which is the best choice?

The selection

Provided the many components of cost have been properly distributed between C_m, C_L and C_c, the cost of manufacturing one bench is (equation (11 .7))

$$C = C_m + \frac{C_C}{n} + \frac{C_L}{\dot{n}}$$

where n is the batch size and $\dot{n}$ the batch rate. Analytical solutions for the cheapest process are possible, but the most helpful way to solve the problem is by plotting the equation for each of the four casting methods using the data in Table 12.8. The result is shown in Figure 12.7.

Table 12.8 Process costs for four casting methods

Process	Sand casting	Low pressure	Permanent mould	Die casting
Material, C_m	1	1	1	1
Labour, C_L (h^{-1})	20	20	20	20
Capital, C_c	0.9	4.4	700	3000
Rate $\dot{n}$ (h^{-1})	6.25	22	10	50

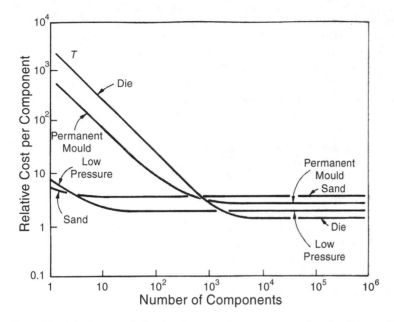

Fig. 12.7 The unit cost/batch size graph for the four casting processes for aluminium alloys.

The selection can now be read off: for one bench, sand casting is marginally the cheapest. But since a production run of 100 is certain, for which low-pressure casting is cheaper, it probably makes sense to use this for the prototype as well. If the product is adopted by schools, die casting becomes the best choice.

Postscript

All this is deceptively easy. The difficult part is that of assembling the data of Table 12.8, partitioning costs between the three heads of material, labour and capital. In practice this requires a detailed, in-house, study of costs and involves information not just for the optical bench but for the entire product line of the company. But when — for a given company — the data for competing processes are known, selecting the cheapest route for a new design can be guided by the method.

Related case studies

Case Study 6.20: Materials to minimize thermal distortion in precision instruments
Case Study 14.3: Data for a non-ferrous alloy

12.7 Computer-based selection — a manifold jacket

The difficulties of using hard-copy charts for process selection will, by now, be obvious: the charts are too cluttered, the overlap too great. They give a helpful overview but they are not the way to get a definitive selection. Computer-based methods increase the resolution.

A computer-based selector (*CPS* 1998) which builds on the method of Chapter 11 is illustrated below. Its database consists of a number of records each containing data for the attributes of one process. These include its *physical attributes* (the ranges of size, tolerance, precision, etc.) and its *economic attributes* (economic batch size, equipment and tooling cost, production rate and so forth). A *material-class menu* allows selection of the subset of process which can shape a given material; a *shape-class menu* allows selection shape (continuous or discrete, prismatic, sheet, 3-D solid, 3-D hollow and the like); and a *process-class menu* allows the choice of process type (primary, secondary, tertiary, etc.).

The best way to use the selector is by creating a sequence of charts with a class attribute on one axis and a physical or economic attribute on the other; superimposed selection boxes define the design requirements, as in Case Studies 12.1 to 12.4. A choice of *Size Range* plotted for processes for which *Material Class = Ferrous Metals*, for instance, gives a bar-chart with bars showing the range of size which lies within the capacity of process which can shape ferrous metals. A selection box positioned to bracket the Size Range between 10 and 15 kg then isolates the subset of processes which can shape ferrous metals to this particular size. The procedure is repeated to select shape, process type, tolerance, economic batch size, and more if required. The output is the subset of processes which satisfy *all* the requirements.

This case study and the next will show how the method works.

The design requirements

The manifold jacket shown in Figure 12.8 is part of the propulsion system of a space vehicle. It is to be made of nickel. It is large, weighing about 7 kg, and very complicated, having a 3D-hollow shape with transverse features and undercuts. The minimum section thickness is between 2 and 5 mm. The requirement on precision is strict (tolerance $< \pm 0.1$ mm). Because of its limited application, only 10 units are to be made. Table 12.9 lists the requirements.

The selection

The output of a computer-based process selector (*CPS*, 1998) is shown in Figures 12.9–12.12. Figure 12.9 shows the first of the selection stages: a bar chart of *mass range* against material class, choosing *non-ferrous metal* from the menu of material classes. The selection box brackets a mass

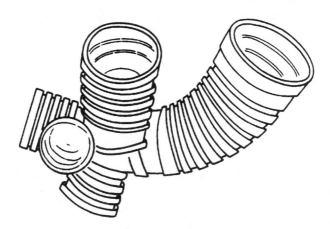

Fig. 12.8 A manifold jacket (source: Bralla, 1986).

Table 12.9 Design requirements for the manifold jacket

Constraint	Value
Material class	Non-ferrous metal: nickel
Process class	Primary, discrete
Shape class	3-D hollow, transverse features
Weight ($\rho = 3000\,\text{kg/m}^3$)	7 kg
Min. section	2 to 5 mm
Tolerance	$< \pm 0.1$ mm
Roughness	$< 10\,\mu$m
Batch size	10

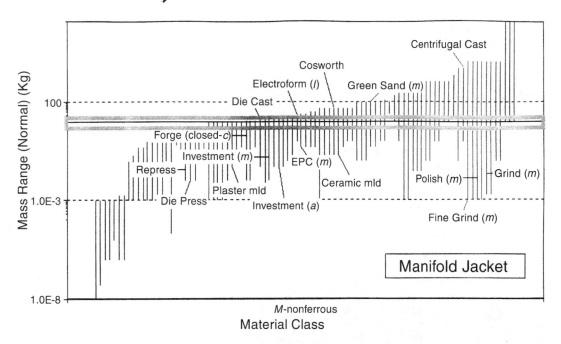

Fig. 12.9 A chart of mass range against material class. The box isolates processes which can shape non-ferrous alloys and can handle the desired mass range.

range of 5–10 kg. Many processes pass this stage, though, of course, all those which cannot deal with non-ferrous metals have been eliminated.

We next seek the subset of processes which can produce the complex shape of the manifold and the desired section thickness, creating a chart of *minimum section thickness* for shapes with *3D-hollow-transverse features*, selected from the menu of shape classes (Figure 12.10). The selection box encloses thicknesses in the range 2 to 5 mm. Again, many processes pass, although any which cannot produce the desired shape fail.

The third selection stage, Figure 12.11, is a bar-chart of *tolerance* against process class selecting *primary processes* (one which creates a shape, rather than one which finishes or joins) from the process class menu. The selection box specifies the tolerance requirement of ± 0.1 mm or better. Very few processes can achieve this precision.

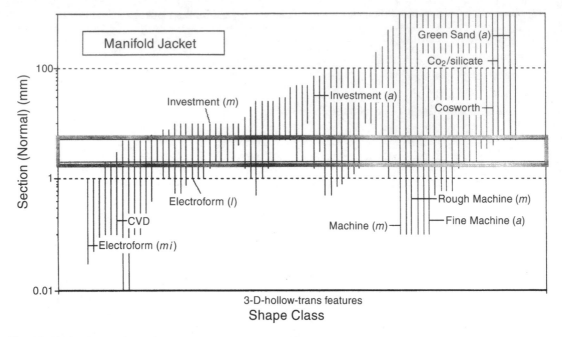

Fig. 12.10 A chart of section thickness range against shape class. The chart identifies processes capable of making 3D-hollow shapes having transverse features with sections in the range 2−5 mm.

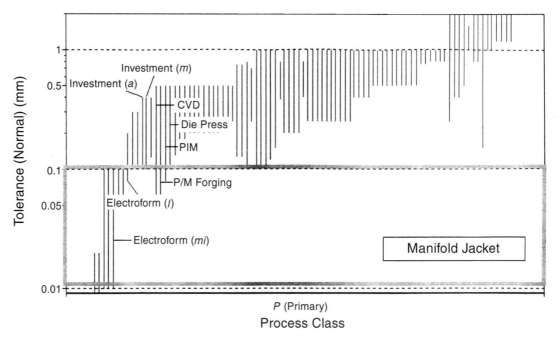

Fig. 12.11 A chart of tolerance against process class. The box isolates primary processes which are capable of tolerance levels of 0.1 mm or better.

The processes which passed all the selection stages so far are listed in Table 12.10. The final step is to rank them. Figure 12.12 shows the economic batch size for discrete processes (selected from the process-class menu), allowing this ranking. It indicates that, for a batch size of 10, automated investment casting is not economic, leaving two processes which are competitive: electro-forming and manual investment casting.

Conclusions and postscript

Electro-forming and *investment casting* emerged as the suitable candidates for making the manifold jacket. A search for further information in the sources listed in Chapter 11 reveals that electro-forming of nickel is established practice and that components as large as 20 kg are routinely made by this process. It looks like the best choice.

Related case studies

Case Study 12.8: Computer-based selection — a spark plug insulator

Table 12.10 Processes capable of making the manifold jacket

Process	Comment
Investment casting (manual)	Practical choice
Investment casting (automated)	Eliminated on economic grounds
Electro-forming	Practical choice

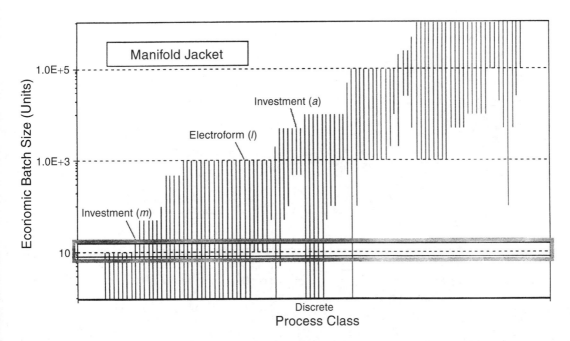

Fig. 12.12 A chart of economic batch size against process class. Three processes have passed all the stages. They are labelled.

12.8 Computer-based selection — a spark plug insulator

This is the second of two case studies illustrating the use of computer-based selection methods.

The design requirements

The anatomy of a spark plug is shown schematically in Figure 12.13. It is an assembly of components, one of which is the insulator. This is to be made of a ceramic, *alumina*, with the shape shown in the figure: an axisymmetric-hollow-stepped shape of low complexity. It weighs about 0.05 kg, has an average section thickness of 2.6 mm and a minimum section of 1.2 mm. Precision is important, since the insulator is part of an assembly; the design specifies a precision of ±0.2 mm and a surface finish of better than 10 μm and, of course, cost should be as low as possible. Table 12.11 summarizes the requirements.

The selection

As in the previous case study, we set up four selection stages. The first (Figure 12.14) combines the requirements of material and mass. Here we have selected the subset of ceramic-shaping processes which can produce components with a mass range of 0.04 to 0.06 kg bracketing that of the insulator. The second stage (Figure 12.15) establishes that the process is a primary one and that it can cope

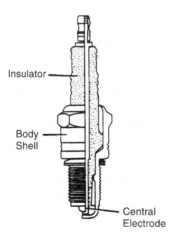

Insulator

Body
Shell

Central
Electrode   **Fig. 12.13** A spark plug.

Table 12.11 Spark plug insulator: design requirements

Constraint	*Value*
Material class	Ceramic (alumina)
Process class	Primary, discrete
Shape class	Prismatic-axisymmetric-hollow-stepped
Weight ($\rho = 3000\,\text{kg/m}^3$)	0.05 kg
Min. section	1.2 mm
Mean precision	$< \pm 0.2$ mm
Roughness	$<10\,\mu\text{m}$
Batch size	100 000

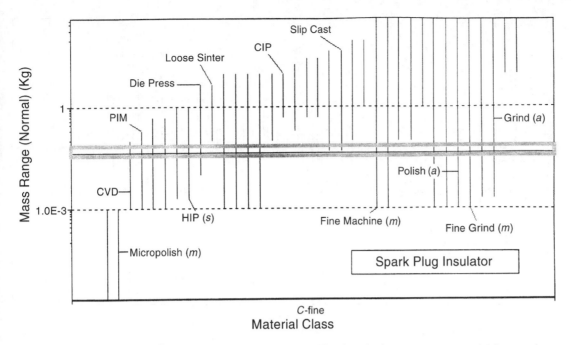

Fig. 12.14 A chart of mass range against material class. The box isolates processes which can shape fine ceramics to the desired mass range.

with the section thickness of the insulator (1 to 4 mm). The third stage (Figure 12.16) deals with shape and precision: processes capable of making 'prismatic-axisymmetric-hollow-stepped' shapes are plotted, and the selection box isolates the ones which can achieve tolerances better than ±0.2 mm.

The three stages allowed the identification of processes which are capable of meeting the design requirements for the insulator. They are listed in Table 12.12: die pressing of powder followed by sintering, powder injection moulding with sintering (PIM) and chemical vapour deposition onto a shaped pre-form (CVD). But this says nothing of the economics of manufacture. A final stage, shown in Figure 12.17, gives an approximate ranking, using the *economic batch size* as the ranking attribute. The first two processes are economic at a batch size of 100 000; the third is not.

Postscript

Insulators are made commercially by die pressing followed by sintering. According to our selection, PIM is a viable alternative and should be investigated further. More detailed cost analysis would be required before a final decision is made. Spark plugs have a very competitive market and, therefore, the cost of manufacturing should be kept low by choosing the cheapest route.

Table 12.12 Processes capable of making the spark plug insulator

Process	*Comment*
Die pressing and sintering	Practical choice
Powder injection moulding (PIM)	Practical choice
Chemical-vapour deposition (CVD)	Eliminated on economic grounds

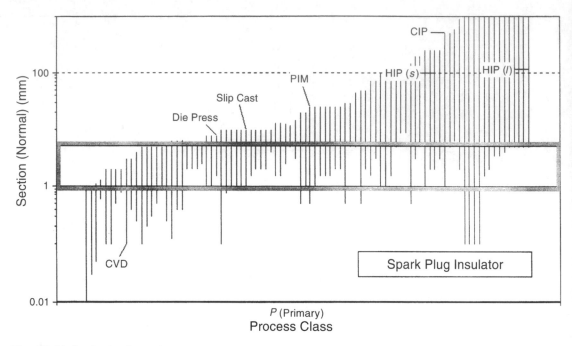

Fig. 12.15 A chart of section thickness range against process class. The chart identifies primary processes capable of making sections in the range 1–4 mm.

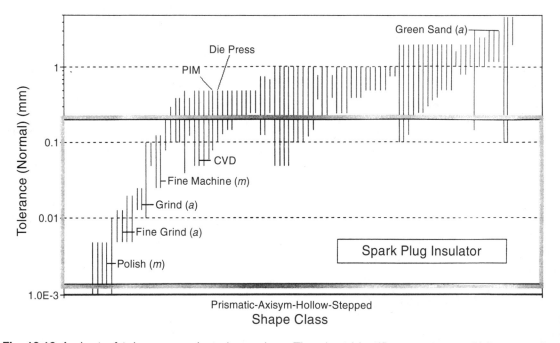

Fig. 12.16 A chart of tolerance against shape class. The chart identifies processes which can make prismatic-axisymmetric-hollow-stepped shapes with a tolerance of 0.2 mm or better.

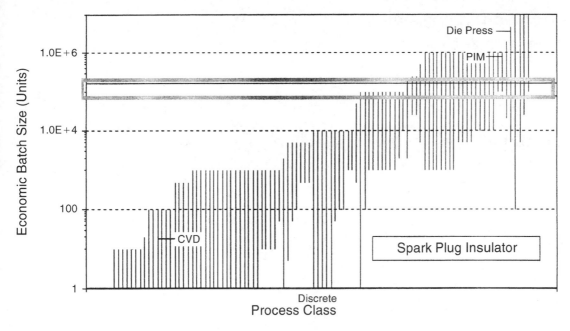

Fig. 12.17 A chart of economic batch size against process class. The three processes which passed the preceding selection stages are labelled. The box isolates the ones which are economic at a batch size of 100 000.

Related case studies

Case Study 12.7: Computer-based selection — a manifold jacket

12.9 Summary and conclusions

Process selection, at first sight, looks like a black art: the initiated know; the rest of the world cannot even guess how they do it. But this — as the chapter demonstrates — is not really so. The systematic approach, developed in Chapter 11 and illustrated here, identifies a subset of viable processes using design information only: size, shape, complexity, precision, roughness and material — itself chosen by the systematic method of Chapter 5. It does not identify the single, best, choice; that depends on too many case-specific considerations. But, by identifying candidates, it directs the user to data sources (starting with those listed in the Further reading of Chapters 11 and 13) which provide the details needed to make a final selection.

The case studies, deliberately, span an exceptional range of size, shape and material. In each, the systematic method leads to helpful conclusions.

12.10 Further reading

Atomic-force microscope design

Albrecht, T.R., Akamine, S., Carver, T.E. and Quate, C.F. (1990) 'Microfabrication of cantilever styli for the atomic force microscope', *J. Vac. Sci. Technol.*, **A8**(4), 3386.

Ceramic-forming methods

Richerson, D.W. (1982) *Modern Ceramic Engineering*, Marcel Dekker, New York.

Economics of manufacture

Kalpakjian, D. (1985) *Manufacturing Processes for Engineering Materials*, Addison Wesley, Reading, MA.

Computer-based process selection

CPS (Cambridge Process Selector) (1998), Granta Design, Trumpington Mews, 40B High Street, Trumpington, Cambridge CB2 2LS, UK.

Esawi, A. and Ashby, M.F. (1998) 'Computer-based selection of manufacturing processes', *J. Engineering Manufacture*.

Esawi, A. and Ashby, M.F. (1998) 'Computer-based selection of manufacturing processes, Part 1: methods and software; Part 2, case studies', Cambridge University Engineering Department Report TR 50, May 1997.

Data sources

13.1 Introduction and synopsis

The engineer, in selecting a material for a developing design, needs data for its properties. Engineers are often conservative in their choice, reluctant to consider material with which they are unfamiliar. One reason is this: that data for the old, well-tried materials are reliable, familiar, easily found; data for newer, more exciting, materials may not exist or, if they do, may not inspire confidence. Yet innovation is often made possible by new materials. So it is important to know where to find material data and how far it can be trusted. This chapter gives information about data sources. Chapter 14, which follows, describes case studies which illustrate data retrieval.

As a design progresses from concept to detail, the data needs evolve in two ways (Figure 13.1). At the start the need is for low-precision data for all materials and processes, structured to facilitate screening. At the end the need is for accurate data for one or a few of them, but with the richness of detail which assists with the difficult aspects of the selection: corrosion, wear, cost estimation and the like. The data sources which help with the first are inappropriate for the second. The chapter surveys data sources from the perspective of the designer seeking information at each stage of the design process. Long-established materials are well documented; less-common materials may be less so, posing problems of checking and, sometimes, of estimation. The chapter proper ends with a discussion of how this can be done.

So much for the text. Half the chapter is contained in the Appendix, Section 13A. It is a catalogue of data sources, with brief commentary. It is intended for reference. When you *really* need data, this is the section you want.

13.2 Data needs for design

Data breadth versus data precision

Data needs evolve as a design develops (Figure 13.1). In the conceptual stage, the designer requires approximate data for the widest possible range of materials. At this stage all options are open: a polymer could be the best choice for one concept, a metal for another, even though the function is the same. Breadth is important; precision is less so. Data for this first-level screening is found in wide-spectrum compilations like the charts of this book, the *Materials Engineering 'Materials Selector'* (1997), and the *Chapman and Hall Materials Selector* (1997).* More effective is software based on these data sources such as the *CMS* and *CPS* (1992, 1998) selection system. The easy access gives the designer the greatest freedom in considering alternatives.

_____

* Details in Further reading.

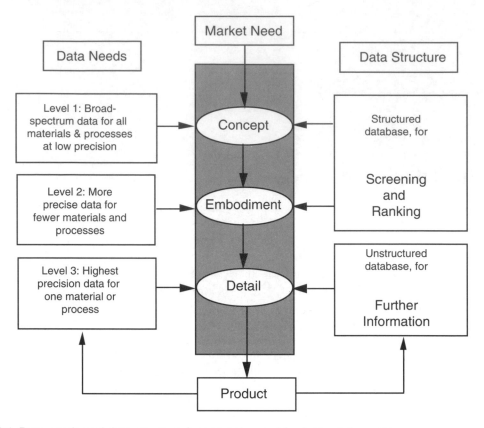

Fig. 13.1 Data needs and data structure for screening and for further information.

The calculations involved in deciding on the scale and lay-out of the design (the embodiment stage) require more complete information than before, but for fewer candidates. Data allowing this second-level screening are found in the specialized compilations which include handbooks and computer databases, and the data books published by associations or federations of material producers. They list, plot and compare properties of closely related materials, and provide data at a level of precision not usually available in the broad, level 1, compilations. And, if they are doing their job properly, they provide further information about processability and possible manufacturing routes. But, because they contain much more detail, their breadth (the range of materials and processes they cover) is restricted, and access is more cumbersome.

The final, detailed design, stage requires data at a still higher level of precision and with as much depth as possible, but for only one or a few materials. They are best found in the data sheets issued by the producers themselves. A given material (low-density polyethylene, for instance) has a range of properties which derive from differences in the way different producers make it. At the detailed-design stage, a supplier should be identified, and the properties of his product used in the design calculation. But sometimes even this is not good enough. If the component is a critical one (meaning that its failure could be disastrous) then it is prudent to conduct in-house tests, measuring the critical property on a sample of the material that will be used to make the component itself. Parts of power-generating equipment (the turbine disc for instance), or aircraft (the wing spar, the landing gear) and nuclear reactors (the pressure vessel) are like this; for

Table 13.1 Material data types

Data type	Example
Numeric point data	Atomic number of magnesium: $N_a = 12$
Numeric range data	Thermal conductivity of polyethylene: $\lambda = 0.28$ to 0.31 W/m K
Boolean (yes/no) data	304 stainless steel can be welded: Yes
Ranked data	Corrosion resistance of alumina in tap water (scale A to E): A
Text	Supplier for aluminium alloys: Alcan, Canada...
Images	

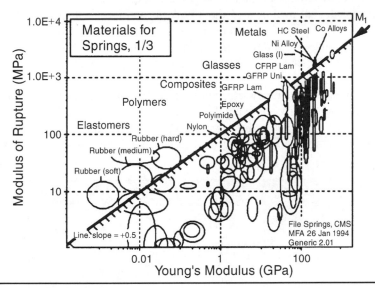

these, every new batch of material is tested, and the batch is accepted or rejected on the basis of the test.

Properties are not all described in the same way. Some, like the atomic number, are described by a single number ('the atomic number of copper = 29'); others, like the modulus or the thermal conductivity are characterized by a range ('Young's modulus for low-density polyethylene = 0.1–0.25 GPa', for instance). Still others can only be described in a qualitative way, or as images. Corrosion resistance is a property too complicated to characterize by a single number; for screening purposes it is ranked on a simple scale: A (very good) to E (very poor), but with further information stored as text files or graphs. The forming characteristics, similarly, are attributes best described by a list ('mild steel can be rolled, forged, or machined'; 'zirconia can be formed by powder methods') with case studies, guidelines and warnings to illustrate how it should be done. The best way to store information about microstructures, or the applications of a material, or the functioning of a process, may be as an image — another data type. Table 13.1 sets out the data types which are typically required for the selection of materials and processes.

13.3 Screening: data structure and sources

Data structure for screening and ranking

To 'select' means: 'to choose'. But from what? Behind the concept of selection lies that of a *kingdom of entities* from which the choice is to be made. The kingdom of materials means: all metals, all polymers, all ceramics and glasses, all composites as in Figure 5.2. If it is *materials* we mean to select, then the kingdom is all of these; leave out part, and the selection is no longer one of materials but of some subset of them. If, from the start, the choice is limited to polymers, then the kingdom becomes a single class of materials, that of polymers. A similar statement holds for processes, based on the kingdom of Figure 11.26.

There is a second implication to the concept of selection; it is that all members of the kingdom must be regarded as candidates — they are, after all, *there* — until (by a series of selection stages) they are shown to be otherwise. From this arises the requirement of a data structure which is *comprehensive* (it includes all members of the kingdom) and the need for characterizing attributes which are *universal* (they apply to all members of the kingdom) and *discriminating* (they have recognizably different values for different members of the kingdom). Similar considerations apply to any selection exercise. We shall use it, in a later chapter, to explore the selection of manufacturing processes.

In the kingdom of materials, many attributes are universal and discriminating: density, bulk modulus and thermal conductivity are examples. Universal attributes can be used for *screening and ranking*, the initial stage of any selection exercise (Figure 13.2, upper half). But if the values of one or more screening attributes are grossly inaccurate or missing, that material is eliminated by

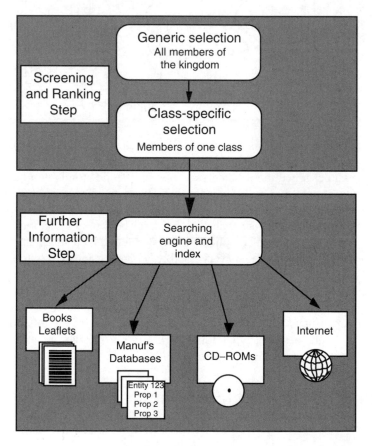

Fig. 13.2 Summary of the selection strategy. The upper box describes screening, the lower one the search for further information.

default. It is important, therefore, that the database be *complete* and be of high *quality*, meaning that the data in it can be trusted. This creates the need for data checking and estimation, tackled by methods described later in this chapter.

The attribute-limits and index methods introduced in Chapters 5 and 11 are examples of the use of attributes to screen, based on design requirements. They provide an efficient way of reducing the vast number of materials in the materials kingdom to a small manageable subset for which further information can be sought.

Data sources for screening (see also the Appendix, Section 13A)

The traditional source of materials data is the handbook. The more courageous of them span all material classes, providing raw data for generic screening. More specialized handbooks and trade-association publications contain data suitable for second-level screening (Figure 13.2) as well as text and figures which help with further information. They are the primary sources, but they are clumsy to use because their data structure is not well suited to screening. Comparison of materials of different classes is possible but difficult because data are seldom reported in comparable formats; there is too much unstructured information, requiring the user to filter out what he needs; and the data tables are almost always full of holes.

Electronic sources for generic screening can overcome these problems. If properly structured, they allow direct comparison across classes and selection by multiple criteria, and it is possible (using methods described in this chapter) to arrange that they have no holes.

Screening, as we have seen, identifies a set of viable candidates. We now need their family history. That is the purpose of the 'further information' step.

13.4 Further information: data structure and sources

Data structure for further information

The data requirements in the further information step differ greatly from those for screening (Figure 13.2, lower half). Here we seek additional details about the few candidates that have already been identified by the screening and ranking step. Typically, this is information about availability and pricing; exact values for key properties of the particular version of the material made by one manufacturer; case studies and examples of uses with cautions about unexpected difficulties (e.g. 'liable to pitting corrosion in dilute acetic acid' or 'material Y is preferred to material X for operation in industrial environments'). It is on this basis that the initial shortlist of candidates is narrowed down to one or a few prime choices.

Sources of further information typically contain specialist information about a relatively narrow range of materials or processes. The information may be in the form of text, tables, graphs, photographs, computer programs, even video clips. The data can be large in quantity, detailed and precise in nature, but there is no requirement that it be comprehensive or that the attributes it contains be universal. The most common media are handbooks, trade association publications and manufacturers' leaflets and catalogues. Increasingly such information is becoming available in electronic form on CD-ROMs and on the Internet. Because the data is in 'free' format, the search strategies differ completely from the numerical optimization procedures used for the screening step. The simplest approach is to use an index (as in a printed book), or a keyword list, or a computerized full text search, as implemented in many hyper-media systems.

Data sources for further information (see also the Appendix, Section 13A)

By 'further information' we mean data sources which, potentially, can contain everything that is known about a material or a process, with some sort of search procedure allowing the user to find and extract the particular details that he seeks. The handbooks and software that are the best sources for screening also contain further information, but because they are edited only infrequently, they are seldom up to date. Trade organizations, listed in the Appendix, Section 13A, do better, providing their members with frequent updates and reports. The larger materials suppliers (Dow Chemical, Ciba-Geigy, Inco, Corning Glass, etc.) publish design guides and compilations of case studies, and all suppliers have data sheets describing their products.

There is an immense resource here. The problem is one of access. It is overcome by capturing the documents on CD-ROM, keyworded and with built-in 'hot-links' to related information, addressed through a search-engine which allows full-text searching on topic strings ('aluminium bronze *and* corrosion *and* sea water', for example).

Expert systems

The main drawback of the simple, common-or-garden, database is the lack of qualification. Some data are valid under all conditions, others are properly used only under certain circumstances. The qualification can be as important as the data itself. Sometimes the question asked of the database is imprecise. The question: 'What is the strength of a steel?' could be asking for yield strength or tensile strength or fatigue strength, or perhaps the least of all three. If the question were put to a materials expert as part of a larger consultation, he would know from the context which was wanted, would have a shrewd idea of the precision and range of validity of the value, and would warn of its limitations. An ordinary database can do none of this.

Expert systems can. They have the potential to solve problems which require reasoning, provided it is based on rules that can be clearly defined: using a set of geometries to select the best welding technique, for instance; or using information about environmental conditions to choose the most corrosion-resistant alloy. It might be argued that a simple checklist or a table in a supplier's data sheet could do most of these things, but the expert system combines qualitative and quantitative information using its rules (the 'expertise'), in a way which only someone with experience can. It does more than merely look up data; it qualifies it as well, allowing context-dependent selection of material or process. In the ponderous words of the British Computer Society: 'Expert systems offer intelligent advice or take intelligent decisions by embodying in a computer the knowledge-based component of an expert's skill. They must, on demand, justify their line of reasoning in a manner intelligible to the user.'

This context-dependent scheme for retrieving data sounds just what we want, but things are not so simple. An expert system is much more complex than a simple database: it is a major task to elicit the 'knowledge' from the expert; it can require massive programming effort and computer power; and it is difficult to update. A full expert system for materials selection is decades away. Success has been achieved in specialized, highly focused applications: guidance in selecting adhesives from a limited set, in choosing a welding technique, or in designing against certain sorts of corrosion. It is only a question of time before more fully developed systems become available. They are something about which to keep informed.

Data sources on the Internet

And today we have the Internet. It contains an expanding spectrum of information sources. Some, particularly those on the World-Wide Web, contain information for materials, placed there by standards organizations, trade associations, material suppliers, learned societies, universities, and individuals — some rational, some eccentric — who have something to say. There is no control over the contents of Web pages, so the nature of the information ranges from useful to baffling, and the quality from good to appalling. The Appendix, Section 13A includes a list of WWW sites which contain materials information, but the rate of change here is so rapid that it cannot be seen as comprehensive.

13.5 Ways of checking and estimating data

The value of a database of material properties depends on its precision and its completeness — in short, on its quality. One way of maintaining or enhancing quality is to subject data to validating procedures. The property ranges and dimensionless correlations, described below, provide powerful tools for doing this. The same procedures fill a second function: that of providing estimates for missing data, essential when no direct measurements are available.

Property ranges

Each property of a given class of materials has a characteristic *range*. A convenient way of presenting the information is as a table in which a low (L) and a high (H) value are stored, identified by the material class. An example listing Young's modulus, E, for the generic material classes is shown in Table 13.2, in which E_L is the lower limit and E_H the upper one.

All properties have characteristic ranges like these. The range becomes narrower if the classes are made more restrictive. For purposes of checking and estimation, described in a moment, it is helpful to break down the class of *metals* into cast irons, steels, aluminium alloys, magnesium alloys, titanium alloys, copper alloys and so on. Similar subdivisions for polymers (thermoplastics, thermosets, elastomers) and for ceramics and glasses (engineering ceramics, whiteware, silicate glasses, minerals) increases resolution here also.

Table 13.2 Ranges of Young's modulus E for broad material classes

Material class	E_L (GPa)	E_H (GPa)
All solids	0.00001	1000
Classes of solid		
Metals: ferrous	70	220
Metals: non-ferrous	4.6	570
Fine ceramics*	91	1000
Glasses	47	83
Polymers: thermoplastic	0.1	4.1
Polymers: thermosets	2.5	10
Polymers: elastomers	0.0005	0.1
Polymeric foams	0.00001	2
Composites: metal-matrix	81	180
Composites: polymer-matrix	2.5	240
Woods: parallel to grain	1.8	34
Woods: perpendicular to grain	0.1	18

*Fine ceramics are dense, monolithic ceramics such as SiC, Al_2O_3, ZrO_2, etc.

Correlations between material properties

Materials which are stiff have high melting points. Solids with low densities have high specific heats. Metals with high thermal conductivities have high electrical conductivities. These rules-of-thumb describe correlations between two or more material properties which can be expressed more quantitatively as limits for the values of *dimensionless property groups*. They take the form

$$C_L < P_1 P_2^n < C_H \qquad (13.1)$$

or
$$C_L < P_1 P_2^n P_3^m < C_H \qquad (13.2)$$

(or larger groupings) where P_1, P_2, P_3 are material properties, n and m are simple powers (usually -1, $-1/2$, $1/2$ or 1), and C_L and C_H are dimensionless constants — the lower and upper limits between which the values of the property-group lies. The correlations exert tight constraints on the data, giving the 'patterns' of property envelopes which appear on the material selection charts. An example is the relationship between expansion coefficient, α (units: K^{-1}), and the melting point, T_m (units: K) or, for amorphous materials, the glass temperature T_g:

$$C_L \leq \alpha T_m \leq C_H \qquad (13.3a)$$

$$C_L \leq \alpha T_g \leq C_H \qquad (13.3b)$$

— a correlation with the form of equation (13.1). Values for the dimensionless limits C_L and C_H for this group are listed in Table 13.3 for a number of material classes. The values span a factor to 2 to 10 rather than the factor 10 to 100 of the property ranges. There are many such correlations. They form the basis of a hierarchical data checking and estimating scheme (one used in preparing the charts in this book), described next.

Data checking

The method is shown in Figure 13.3. Each datum is associated with a material class, or, at a higher level of checking, with a sub-class. It is first compared with the range limits L and H for that class and property. If it lies within the range limits, it is accepted; if it does not, it is flagged for checking.

Table 13.3 Limits for the group αT_m and αT_g for broad material classes*

Correlation $C_L < \alpha T_m < C_H$	$C_L (\times 10^{-3})$	$C_H (\times 10^{-3})$
All solids	0.1	56
Classes of solid		
Metals: ferrous	13	27
Metals: non-ferrous	2	21
Fine ceramics*	6	24
Glasses	0.3	3
Polymers: thermoplastic	18	35
Polymers: thermosets	11	41
Polymers: elastomers	35	56
Polymeric foams	16	37
Composites: metal-matrix	10	20
Composites: polymer-matrix	0.1	10
Woods: parallel to grain	2	4
Woods: perpendicular to grain	6	17

*For amorphous solids the melting point T_m is replaced by the glass temperature T_g.

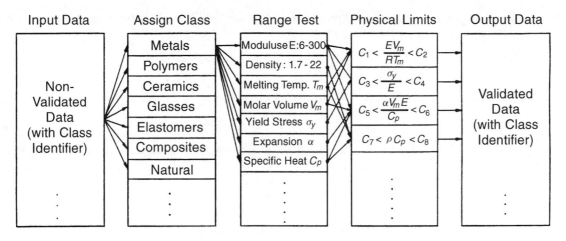

Fig. 13.3 The checking procedure. Range checks catch gross errors in all properties. Checks using dimensionless groups can catch subtler errors in certain properties. The estimating procedure uses the same steps, but in reverse order.

Why bother with such low-level stuff? It is because in compilations of material or process properties, the commonest error is that of a property value which is expressed in the wrong units, or is, for less obvious reasons, in error by one or more orders of magnitude (slipped decimal point, for instance). Range checks catch errors of this sort. If a demonstration of this is needed, it can be found by applying them to the contents of almost any standard reference data books; none among those we have tried has passed without errors.

In the second stage, each of the dimensionless groups of properties like that of Table 13.3 is formed in turn, and compared with the range bracketed by the limits C_L and C_H. If the value lies within its correlation limits, it is accepted; if not, it is checked. Correlation checks are more discerning than range checks and catch subtler errors, allowing the quality of data to be enhanced further.

Data estimation

The relationships have another, equally useful, function. There remain gaps in our knowledge of material properties. The fracture toughness of many materials has not yet been measured, nor has the electric breakdown potential; even moduli are not always known. The absence of a datum for a material would falsely eliminate it from a selection which used that property, even though the material might be a viable candidate. This difficulty is avoided by using the correlation and range limits to estimate a value for the missing datum, adding a flag to alert the user that they are estimates.

In estimating property values, the procedure used for checking is reversed: the dimensionless groups are used first because they are the more accurate. They can be surprisingly good. As an example, consider estimating the expansion coefficient, α, of polycarbonate from its glass temperature T_g. Inverting equation (13.3) gives the estimation rule:

$$\frac{C_L}{T_g} \leq \alpha \leq \frac{C_H}{T_g} \tag{13.4}$$

Inserting values of C_L and C_H from Table 13.3, and the value $T_g = 420\,\text{K}$ for a particular sample of polycarbonate gives the mean estimate

$$\bar{\alpha} = 63 \times 10^{-3}\,\text{K}^{-1} \qquad\qquad (13.5)$$

The reported value for polycarbonate is

$$\alpha = 54 - 62 \times 10^{-3}\,\text{K}^{-1}$$

The estimate is within 9% of the mean of the measured values, perfectly adequate for screening purposes. That it is an estimate must not be forgotten, however: if thermal expansion is crucial to the design, better data or direct measurements are essential.

Only when the potential of the correlations is exhausted are the property ranges invoked. They provide a crude first estimate of the value of the missing property, far less accurate than that of the correlations, but still useful in providing guide-values for screening.

13.6 Summary and conclusions

The systematic way to select materials or processes (or anything else, for that matter) is this.

(a) Identify the taxonomy of the *kingdom* from which the selection is to be made; its *classes*, *subclasses* and *members*.
(b) Identify the *attributes* of the members, remembering that they should be *universal* and *discriminating* within this kingdom; resolution is increased by defining second-level 'sub-kingdoms' allowing an expanded set of attributes, universal within the sub-kingdom.
(c) Assess the *quality* and *completeness* of the data sources for the attributes; both can be increased by techniques of checking and estimation described in the previous section.
(d) Reduce the large population of the kingdom to a shortlist of potential candidates by *screening* on attributes in the first and second-level kingdoms.
(e) Identify sources of *further information* for the candidates: texts, design guides, case studies, suppliers' data sheets or (better) searchable electronic versions of these, including the Internet.
(f) Compare full character profiles of the candidates with requirements of the design, taking into account local constraints (preferences, experience, compatibility with other activities, etc.).

To do all this you need to know where to find data, and you need it at three levels of breadth and precision. Conceptual design requires a broad survey at the low accuracy offered by the selection charts of Chapters 4 and 11, and by other broad-spectrum data tabulations. Embodiment design needs more detail and precision, of the kind found in the handbooks and computer databases listed in the Appendix, Section 13A. The final, detailed, phase of design relies on the yet more precise (and attributable) information contained in material suppliers' data sheets.

The falling cost and rising speed of computing makes databases increasingly attractive. They allow fast retrieval of data for a material or a process, and the selection of the subset of them which have attributes within a specified range. Commercially available databases already help enormously in selection, and are growing every year. Some of those currently available are reviewed in the Appendix, Section 13A.

Expert systems lurk somewhere in the future. They combine a database with a set of rules for reasoning to permit simple, logical deductions to be made by the computer itself, allowing it to retrieve relevant information which the operator did not know or forgot to ask for. They combine the data of a handbook with some of the expertise of a materials consultant. They are difficult to create and demand much computer power, but the selection process lends itself well to expert-systems programming; they will, sooner or later, be with us.

Don't leave this chapter without at least glancing at the compilation of data sources in the next section. It is probably the most useful bit.

13.7 Further reading

Ashby, M.F. (1998) 'Checks and estimates for material properties', Cambridge University Engineering Department, *Proc. Roy. Soc. A* **454**, 1301–1321.

Bassetti, D., Brechet, Y. and Ashby, M.F. (1998) 'Estimates for material properties: the method of multiple correlations', *Proc. Roy. Soc. A* **454**, 1323–1336.

Cebon, D. and Ashby, M.F. (1992) 'Computer-aided selection for mechanical design', *Metals and Materials*, January, 25–30.

Cebon, D. and Ashby, M.F. (1996) 'Electronic material information systems', *I. Mech. E. Conference on Electronic Delivery of Design Information*, October, 1996, London, UK.

CMS (Cambridge Materials Selector) (1992), Granta Design, Trumpington Mews, 40B High Street, Trumpington, Cambridge CB2 2LS, UK.

CPS (Cambridge Process Selector) (1998), Granta Design, Trumpington Mews, 40B High Street, Trumpington, Cambridge CB2 2LS, UK.

The Copper Development Association (1994) *Megabytes on Coppers*, Orchard House, Mutton Lane, Potters Bar, Herts EN6 3AP, UK; and Granta Design Limited, 20 Trumpington St., Cambridge CB2 1PZ, UK, 1994.

13A Appendix: Data sources for material and process attributes

13A.1 Introduction

Background

This appendix tells you where to look to find material property data. The sources, broadly speaking, are of three sorts: hard copy, software and the Internet. The hard copy documents listed below will be found in most engineering libraries. The computer databases are harder to find: the supplier is listed, with address and contact number, as well as the hardware required to run the database. Internet sites are easy to find but can be frustrating to use.

Section 13A.2 lists sources of information about database structure and functionality. Sections 13A.3 catalogues hard-copy data sources for various classes of material, with a brief commentary where appropriate. Selection of material is often linked to that of processing; Section 13A.4 provides a starting point for reading on processes. Section 13A.5 gives information about the rapidly growing portfolio of software for materials and process data and information. Section 13A.6 — the last — lists World-wide Web sites on which materials information can be found.

13A.2 General references on databases

Waterman, N.A., Waterman, M. and Poole, M.E. (1992) 'Computer based materials selection systems', *Metals and Materials* **8**, 19–24.

Sargent, P.M. (1991) *Materials Information for CAD/CAM*, Butterworths-Heinemann, Oxford. A survey of the way in which materials data-bases work. No data.

Demerc, M.Y. (1990) *Expert System Applications in Materials Processing and Manufacture*. TMS Publications, 420 Commonwealth Drive, Warrendale, Penn. 15086, USA.

13A.3 Hard-copy data sources

Data sources, all materials

Few hard-copy data sources span the full spectrum of materials and properties. Six which, in different ways, attempt to do so are listed below.

Materials Selector (1997), Materials Engineering, Special Issue. Penton Publishing, Cleveland, Ohio, USA. Tabular data for a broad range of metals, ceramics, polymers and composites, updated annually. Basic reference work.

The Chapman and Hall 'Materials Selector' (1996), edited by N.A. Waterman and M.F. Ashby. Chapman and Hall, London, UK. A 3-volume compilation of data for all materials, with selection and design guide. Basic reference work.

ASM Engineered Materials Reference Book, 2nd edition (1994), editor Bauccio, M.L., ASM International, Metals Park, Ohio 44073, USA. Compact compilation of numeric data for metals, polymers, ceramics and composites.

Materials Selector and Design Guide (1974), Design Engineering, Morgan-Grampian Ltd, London. Resembles the *Materials Engineering 'Materials Selector'*, but less detailed and now rather dated.

Handbook of Industrial Materials (1992) 2nd edition, Elsevier, Oxford, UK. A compilation of data remarkable for its breadth: metals, ceramics, polymers, composites, fibres, sandwich structures, leather...

Materials Handbook (1986) 12th edition, editors Brady, G.S. and Clauser, H.R., McGraw-Hill, New York, USA. A broad survey, covering metals, ceramics, polymers, composites, fibres, sandwich structures and more.

Handbook of Thermophysical Properties of Solid Materials (1961) Goldsmith, A., Waterman, T.E. and Hirschhorn, J.J. MacMillan, New York, USA. Thermophysical and thermochemical data for elements and compounds.

Guide to Engineering Materials Producers (1994) editor Bittence, J.C. ASM International, Metals Park, Ohio 44037, USA. A comprehensive catalog of addresses for material suppliers.

Data sources, all metals

Metals and alloys conform to national and (sometimes) international standards. One consequence is the high quality of data. Hard copy sources for metals data are generally comprehensive, well-structured and easy to use.

ASM Metals Handbook (1986) 9th Edition, and (1990) 10th Edition. ASM International, Metals Park, Ohio, 44073 USA. The 10th Edition contains Vol. 1: Irons and Steels; Vol 2: Non-ferrous Alloys; Vol 3: Heat Treatment; Vol 4: Friction, Lubrication and Wear; Vol 5: Surface Finishing and Coating; Vol 6: Welding and Brazing; Vol 7: Microstructural Analysis; more volumes are planned for release in 1992/93. Basic reference work, continuously upgraded and expanded.

ASM Metals Reference Book, 3rd edition (1993) ed. M.L. Bauccio, ASM International, Metals Park, Ohio 44073, USA. Consolidates data for metals from a number of ASM publications. Basic reference work.

Brandes, E.A. and Brook, G.B. (1997) *Smithells Metals Reference Book*, 7th edition, Butterworth-Heinemann, Oxford. A comprehensive compilation of data for metals and alloys. Basic reference work.

Metals Databook (1990), Colin Robb. The Institute of Metals, 1 Carlton House Terrace, London SW1Y 5DB, UK. A concise collection of data on metallic materials covered by the UK specifications only.

ASM Guide to Materials Engineering Data and Information (1986). ASM International, Metals Park, Ohio 44073, USA. A directory of suppliers, trade organizations and publications on metals.

The Metals Black Book, Volume 1, Steels (1992) ed. J.E. Bringas, Casti Publishing Inc. 14820–29 Street, Edmonton, Alberta T5Y 2B1, Canada. A compact book of data for steels.

The Metals Red Book, Volume 2, Nonferrous Metals (1993) ed. J.E. Bringas, Casti Publishing Inc. 14820–29 Street, Edmonton, Alberta T5Y 2B1, Canada.

Data sources, specific metals and alloys

In addition to the references listed under Section 13A.2, the following sources give data for specific metals and alloys

Pure metals

Most of the sources listed in the previous section contain some information on pure metals. However, the publications listed below are particularly useful in this respect.

Winter, M. 'WebElements', http://www.shef.ac.uk/~chem/web-elements/, University of Sheffield. A comprehensive source of information on all the elements in the Periodic Table. If it has a weakness, it is in the definitions and values of some mechanical properties.

Emsley, J. *The Elements*, Oxford University Press, Oxford, UK (1989). A book aimed more at chemists and physicists than engineers with good coverage of chemical, thermal and electrical properties but not mechanical properties. A new edition is expected early in 1997.

Brandes, E.A. and Brook, G.B. (eds) *Smithells Metals Reference Book* (7th edition), Butterworth-Heinemann, Oxford (1997). Data for the mechanical, thermal and electrical properties of pure metals.

Goodfellow Catalogue (1995–6), Goodfellow Cambridge Limited, Cambridge Science Park, Cambridge, CB4 4DJ, UK. Useful though patchy data for mechanical, thermal and electrical properties of pure metals in a tabular format. Free.

Alfa Aesar Catalog (1995–96) Johnson Matthey Catalog Co. Inc., 30 Bond Street, Ward Hill, MA 01835–8099, USA. Coverage similar to that of the Goodfellow Catalogue. Free.

Samsonov, G.V. (ed.) *Handbook of the Physiochemical Properties of the Elements*, Oldbourne, London (1968). An extensive compilation of data from Western and Eastern sources. Contains a number of inaccuracies, but also contains a large quantity of data on the rarer elements, hard to find elsewhere.

Gschneidner, K.A. 'Physical properties and interrelationships of metallic and semimetallic elements', *Solid State Physics*, **16**, 275–426 (1964). Probably the best source of its time, this reference work is very well referenced, and full explanations are given of estimated or approximate data.

Non-ferrous metals

Aluminium alloys

Aluminium Standards and Data, The Aluminium Association Inc., 900, 19th Street N.W., Washington, DC 20006, USA (1990).

The Properties of Aluminium and its Alloys, The Aluminium Federation, Broadway House, Calthorpe Road, Birmingham, B15 1TN, UK (1981).

Technical Data Sheets, ALCAN International Ltd, Kingston Research and Development Center, Box 8400, Kingston, Ontario, Canada KL7 4Z4, and Banbury Laboratory, Southam Road, Banbury, Oxon., UK, X16 7SP (1993).

Technical Data Sheets, ALCOA, 1501 Alcoa Building, Pittsburg, PA 15219, USA (1993).

Technical Data Sheets, Aluminium Pechiney, 23 Bis, rue Balzac, Paris 8, BP 78708, 75360 Paris Cedex 08, France (1994).

Babbitt metal

The term 'Babbitt metal' denotes a series of lead–tin–antimony bearing alloys, the first of which was patented in the USA by Isaac Babbitt in 1839. Subsequent alloys are all variations on his original composition.

ASTM Standard B23-83: 'White Metal Bearing Alloys (Known Commercially as 'Babbitt Metal')', *ASTM Annual Book of Standards*, Vol. 02.04.

Beryllium

Designing with Beryllium, Brush Wellman Inc, 1200 Hana Building, Cleveland, OH 44115, USA (1996).
Beryllium Optical Materials, Brush Wellman Inc, 1200 Hana Building, Cleveland, OH 44115, USA (1996).

Cadmium

International Cadmium Association, *Cadmium Production, Properties and Uses*, ICdA, London, UK (1991).

Chromium

ASTM Standard A560-89: 'Castings, Chromium-Nickel Alloy', *ASTM Annual Book of Standards*, Vol. 01.02.

Cobalt alloys

Betteridge, W. *Cobalt and its alloys*, Ellis Horwood, Chichester, UK (1982). A good general introduction to the subject.

Columbium alloys: see Niobium alloys

Copper alloys

ASM Metals Handbook 10th edition, ASM International, Metals Park, Ohio, USA (1990).
The Selection and Use of Copper-based Alloys, E.G. West, Oxford University Press, Oxford, UK (1979).
Copper Development Association Data Sheets, 26 (1988), 27 (1981), 31 (1982), 40 (1979), and Publication 82 (1982), Copper Development Association Inc., Greenwich Office, Park No. 2, Box 1840, Greenwich CT 06836, USA, and The Copper Development Association, Orchard House, Mutton Lane, Potters Bar, Herts, EN6 3AP, UK.
Megabytes on Coppers CD-ROM, Granta Design Ltd., 20 Trumpington Street, Cambridge CB2 1QA, UK (1994).
Smithells Metals Reference Book, 7th edition, eds E.A. Brandes and G.B. Brook, Butterworth-Heinemann, Oxford, UK (1992).

Dental alloys

O'Brien, W.J. 'Biomaterial Properties Database', http://www.lib.umich.edu/libhome/Dentistry.lib/Dental_tables, School of Dentistry, Univ. of Michigan, USA. An extensive source of information, both for natural biological materials and for metals used in dental treatments.
Jeneric Pentron Inc., 'Casting Alloys', http://www.jeneric.com/casting, USA. An informative commercial site.
ISO Standard 1562:1993, 'Dental casting gold alloys', International Standards Organization, Switzerland.
ISO Standard 8891:1993, 'Dental casting alloys with Noble metal content of 25% up to but not including 75%', International Standards Organization, Switzerland.

Gold

Rand Refinery Limited, http://www.bullion.org.za/associates/rr.htm, Chamber of Mines Web-site, SOUTH AFRICA. Contains useful information on how gold is processed to varying degrees of purity.
See also the section on Dental alloys, above.

Indium

The Indium Info Center, http://www.indium.com/metalcenter.html, Indium Corp. of America.

Lead

ASTM Standard B29-79: 'Pig Lead', *ASTM Annual Book of Standards*, Vol. 02.04.
ASTM Standard B102-76: 'Lead- and Tin-Alloy Die Castings', *ASTM Annual Book of Standards*, Vol. 02.04.
ASTM Standard B749-85: 'Lead and Lead Alloy Strip, Sheet, and Plate Products', *ASTM Annual Book of Standards*, Vol. 02.04.
Lead Industries Association, *Lead for Corrosion Resistant Applications*, LIA Inc., New York, USA.
ASM Metals Handbook, 9th edition, Vol. 2, pp. 500–510 (1986).
See also Babbitt metal (above)

Magnesium alloys

Technical Data Sheets, Magnesium Elektron Ltd., PO Box 6, Swinton, Manchester, UK (1994).
Technical Literature, Magnesium Corp. of America, Div. of Renco, Salt Lake City, UT, USA (1994).

Molybdenum

ASTM Standard B386-85: 'Molybdenum and Molybdenum Alloy Plate, Sheet, Strip and Foil', *ASTM Annual Book of Standards*, Vol. 02.04.
ASTM Standard B387-85: 'Molybdenum and Molybdenum Alloy Bar, Rod and Wire', *ASTM Annual Book of Standards*, Vol. 02.04.

Nickel

A major data source for Nickel and its alloys is the Nickel Development Institute (NIDI), a global organization with offices in every continent except Africa. NIDI freely gives away large quantities of technical reports and data compilations, not only for nickel and high-nickel alloys, but also for other nickel-bearing alloys, e.g. stainless steel.

ASTM Standard A297-84, 'Steel Castings, Iron-Chromium and Iron-Chromium-Nickel, Heat Resistant, for General Application', *ASTM Annual Book of Standards*, Vol. 01.02;
ASTM Standard A344-83, 'Drawn or Rolled Nickel-Chromium and Nickel-Chromium-Iron Alloys for Electrical Heating Elements', *ASTM Annual Book of Standards*, Vol. 02.04.
ASTM Standard A494-90, 'Castings, Nickel and Nickel Alloy', *ASTM Annual Book of Standards*, Vol. 02.04.
ASTM Standard A753-85, 'Nickel-Iron Soft Magnetic Alloys', *ASTM Annual Book of Standards*, Vol. 03.04.
Betteridge, W., 'Nickel and its alloys', Ellis Horwood, Chichester, UK (1984). A good introduction to the subject.
INCO Inc., 'High-Temperature, High-Strength Nickel Base Alloys', Nickel Development Institute (1995). Tabular data for over 80 alloys.
Elliott, P. *Practical Guide to High-Temperature Alloys*, Nickel Development Institute, Birmingham, UK (1990).
INCO Inc., *Heat & Corrosion Resistant Castings*, Nickel Development Institute, Birmingham, UK (1978).
INCO Inc., *Engineering Properties of some Nickel Copper Casting Alloys*, Nickel Development Institute, Birmingham, UK (1969).
INCO Inc., *Engineering Properties of IN-100 Alloy*, Nickel Development Institute, Birmingham, UK (1968).
INCO Inc., *Engineering Properties of Nickel-Chromium Alloy 610 and Related Casting Alloys*, Nickel Development Institute, Birmingham, UK (1969).
INCO Inc., *Alloy 713C: Technical Data*, Nickel Development Institute, Birmingham, UK (1968).
INCO Inc., *Alloy IN-738: Technical Data*, Nickel Development Institute, Birmingham, UK (1981).
INCO Inc., *36% Nickel-Iron Alloy for Low Temperature Service*, Nickel Development Institute, Birmingham, UK (1976).
ASTM Standard A658 (Discontinued 1989) 'Pressure Vessel Plates, Alloy Steel, 36 Percent Nickel', *ASTM Annual Book of Standards*, pre-1989 editions.
ASM Metals Handbook, 9th ed., Vol. 3, pp. 125–178 (1986).
Carpenter Technology Corp. Website, http//www.cartech.com/

Niobium (columbium) alloys

ASTM Standard B391-89: 'Niobium and Niobium Alloy Ingots', *ASTM Annual Book of Standards*, Vol. 02.04;

ASTM Standard B392-89: 'Niobium and Niobium Alloy Bar, Rod and Wire', *ASTM Annual Book of Standards*, Vol. 02.04.

ASTM Standard B393-89: 'Niobium and Niobium Alloy Strip, Sheet and Plate', *ASTM Annual Book of Standards*, Vol. 02.04.

ASTM Standard B652-85: 'Niobium-Hafnium Alloy Ingots', *ASTM Annual Book of Standards*, Vol. 02.04.

ASTM Standard B654-79: 'Niobium-Hafnium Alloy Foil, Sheet, Strip and Plate', *ASTM Annual Book of Standards*, Vol. 02.04.

ASTM Standard B655-85: 'Niobium-Hafnium Alloy Bar, Rod and Wire', *ASTM Annual Book of Standards*, Vol. 02.04.

Husted, R, http://www-c8.lanl.gov/infosys/html/periodic/41.html, Los Alamos National Laboratory, USA. An overview of Niobium and its uses.

Palladium

ASTM Standard B540-86: 'Palladium Electrical Contact Alloy', *ASTM Annual Book of Standards*, Vol. 03.04.

ASTM Standard B563-89: 'Palladium-Silver-Copper Electrical Contact Alloy', *ASTM Annual Book of Standards*, Vol. 03.04.

ASTM Standard B589-82: 'Refined Palladium' *ASTM Annual Book of Standards*, Vol. 02.04.

ASTM Standard B683-90: 'Pure Palladium Electrical Contact Material', *ASTM Annual Book of Standards*, Vol. 03.04.

ASTM Standard B685-90: 'Palladium-Copper Electrical Contact Material', *ASTM Annual Book of Standards*, Vol. 03.04.

ASTM Standard B731-84: '60% Palladium-40% Silver Electrical Contact Material', *ASTM Annual Book of Standards*, Vol. 03.04.

Jeneric Pentron Inc., 'Casting Alloys', http://www.jeneric.com/casting, USA. An informative commercial site, limited to dental alloys.

Platinum alloys

ASTM Standard B684-81: 'Platinum-Iridium Electrical Contact Material', *ASTM Annual Book of Standards*, Vol. 03.04;

'Elkonium Series 400 datasheets', CMW Inc., Indiana, USA.

ASM Metals Handbook, 9th edition, Vol. 2, pp. 688–698 (1986).

Silver alloys

ASTM Standard B413-89: 'Refined Silver', *ASTM Annual Book of Standards*, Vol. 02.04.

ASTM Standard B 617-83: 'Coin Silver Electrical Contact Alloy', *ASTM Annual Book of Standards*, Vol. 03.04.

ASTM Standard B 628-83: 'Silver-Copper Eutectic Electrical Contact Alloy', *ASTM Annual Book of Standards*, Vol. 03.04.

ASTM Standard B 693-87: 'Silver-Nickel Electrical Contact Materials', *ASTM Annual Book of Standards*, Vol. 03.04.

ASTM Standard B742-90: 'Fine Silver Electrical Contact Fabricated Material', *ASTM Annual Book of Standards*, Vol. 03.04.

ASTM Standard B 780-87: '75% Silver, 24.5% Copper, 0.5% Nickel Electrical Contact Alloy', *ASTM Annual Book of Standards*, Vol. 03.04.

Elkonium Series 300 datasheets, CMW Inc., 70 S. Gray Street, PO Box 2266, Indianapolis, Indiana, USA (1996).

Elkonium Series 400 datasheets, CMW Inc., 70 S. Gray Street, PO Box 2266, Indianapolis, Indiana, USA (1996).

Jeneric Pentron Inc., 'Casting Alloys', http://www.jeneric.com/casting, USA. An informative commercial site, limited to dental alloys.

Tantalum alloys

ASTM Standard B365-86: 'Tantalum and Tantalum Alloy Rod and Wire', *ASTM Annual Book of Standards*, Vol. 02.04.
ASTM Standard B521-86: 'Tantalum and Tantalum Alloy Seamless and Welded Tubes', *ASTM Annual Book of Standards*, Vol. 02.04.
ASTM Standard B560-86: 'Unalloyed Tantalum for Surgical Implant Applications', *ASTM Annual Book of Standards*, Vol. 13.01.
ASTM Standard B708-86: 'Tantalum and Tantalum Alloy Plate, Sheet and Strip', *ASTM Annual Book of Standards*, Vol. 02.04.
Tantalum Data Sheet, The Rembar Company Inc., 67 Main St., Dobbs Ferry, NY 10522, USA (1996).
ASM Handbook, 9th edn., Vol. 3, pp. 323–325 & 343–347 (1986).

Tin alloys

ASTM Standard B32-89: 'Solder Metal', *ASTM Annual Book of Standards*, Vol. 02.04.
ASTM Standard B339-90: 'Pig Tin', *ASTM Annual Book of Standards*, Vol. 02.04.
ASTM Standard B560-79: 'Modern Pewter Alloys', *ASTM Annual Book of Standards*, Vol. 02.04.
Barry, B.T.K. and Thwaites, C.J., *Tin and its alloys and compounds*, Ellis Horwood, Chichester, UK (1983).
ASM Metals Handbook, 9th edition, Vol. 2, pp. 613–625.
See also Babbitt metal (above)

Titanium alloys

Technical Data Sheets, Titanium Development Association, 4141 Arapahoe Ave., Boulder, Colorado, USA (1993).
Technical Data Sheets, The Titanium Information Group, c/o Inco Engineered Products, Melbourne, UK (1993).
Technical Data Sheets, IMI Titanium Ltd. PO Box 704, Witton, Birmingham B6 7UR, UK (1995).

Tungsten alloys

ASTM Standard B777-87, 'Tungsten Base, High-Density Metal', *ASTM Annual Book of Standards*, Vol. 02.04.
Yih, S.W.H. and Wang, C.T., *Tungsten*, Plenum Press, New York (1979).
ASM Metals Handbook, 9th edition, Vol. 7, p. 476 (1986).
Tungsten Data Sheet (1996), The Rembar Company Inc., 67 Main St., Dobbs Ferry, NY 10522, USA.
Royal Ordnance Speciality Metals datasheet, British Aerospace Defence Ltd., PO Box 27, Wolverhampton, West Midlands, WV10 7NX, UK(1996).
CMW Inc. Datasheets, CMW Inc., 70 S. Gray Street, PO Box 2266, Indianapolis, Indiana, USA (1996).

Vanadium

Teledyne Wah Chang, 'Vanadium Brochure', TWC, Albany, Oregon, USA (1996).

Zinc

ASTM Standard B6-87: 'Zinc', *ASTM Annual Book of Standards*, Vol. 02.04, ASTM, USA.
ASTM Standard B69-87: 'Rolled Zinc', *ASTM Annual Book of Standards*, Vol. 02.04, ASTM, USA.
ASTM Standard B86-88: 'Zinc-Alloy Die Castings', *ASTM Annual Book of Standards*, Vol. 02.02, ASTM, USA.
ASTM Standard B418-88: 'Cast and Wrought Galvanic Zinc Anodes', *ASTM Annual Book of Standards*, Vol. 02.04, ASTM, USA.
ASTM Standard B791-88: 'Zinc-Aluminium Alloy Foundry and Die Castings', *ASTM Annual Book of Standards*, Vol. 02.04, ASTM, USA.
ASTM Standard B792-88: 'Zinc Alloys in Ingot Form for Slush Casting', *ASTM Annual Book of Standards*, Vol. 02.04, ASTM, USA.

ASTM Standard B793-88: 'Zinc Casting Alloy Ingot for Sheet Metal Forming Dies', *ASTM Annual Book of Standards*, Vol. 02.04, ASTM, USA.

Goodwin, F.E. and Ponikvar, A.L. (eds), *Engineering Properties of Zinc Alloys* (3rd edition), International Lead Zinc Research Organization, North Carolina, USA (1989). An excellent compilation of data, covering all industrially important zinc alloys.

Chivers, A.R.L., *Zinc Diecasting*, Engineering Design Guide no. 41, OUP, Oxford, UK (1981). A good introduction to the subject.

ASM Metal Handbook, 'Properties of Zinc and Zinc Alloys', 9th edition, Vol. 2, pp. 638–645.

Zirconium

ASTM Standard B350-80: 'Zirconium and Zirconium Alloy Ingots for Nuclear Application', *ASTM Annual Book of Standards*, Vol. 02.04, ASTM, USA.

ASTM Standard B352-85, B551-83 and B752-85: 'Zirconium and Zirconium Alloys', *ASTM Annual Book of Standards*, Vol. 02.04, ASTM, USA.

Teledyne Wah Chang, 'Zircadyne: Properties & Applications', TWC, Albany, OR 97231, USA (1996).

ASM Metals Handbook, 9th edition, Vol. 2, pp. 826–831 (1986).

Ferrous metals

Ferrous metals are probably the most thoroughly researched and documented class of materials. Nearly every developed country has its own system of standards for irons and steels. Recently, continental and worldwide standards have been developed, which have achieved varying levels of acceptance. There is a large and sometimes confusing literature on the subject. This section is intended to provide the user with a guide to some of the better information sources.

Ferrous metals, general data sources

Bringas, J.E. (ed.) *The Metals Black Book — Ferrous Metals*, 2nd edition, CASTI Publishing, Edmonton, Canada (1995). An excellent short reference work.

ASM Metals Handbook, 10th edition, Vol. 1 (1990), ASM International, Metals Park, Cleveland, Ohio, USA. Authoritative reference work for North American irons and steels.

ASM Metals Handbook, Desk edition, (1985), ASM International, Metals Park, Cleveland, Ohio, USA. A summary of the multi-volume ASM Metals Handbook.

Wegst, C.W., *Stahlschlüssel* (in English: *Key to Steel*), Verlag Stahlschlüssel Wegst GmbH, D-1472 Marbach, Germany. Published every 3 years, in German, French and English. Excellent coverage of European products and manufacturers.

Woolman, J. and Mottram, R.A., *The Mechanical and Physical Properties of the British Standard En Steels*, Pergamon Press, Oxford (1966). Still highly regarded, but is based around a British Standard classification system that has been officially abandoned.

Brandes, E.A. and Brook, G.R. (eds) *Smithells Metals Reference Book*, 7th edition, Butterworth-Heinemann, Oxford, UK (1992). An authoritative reference work, covering all metals.

Chapman and Hall 'Materials Selector', Waterman, N.A. and Ashby, M.F. (eds), Chapman and Hall, London, UK (1996). Covers all materials — Irons and steels are in Vol. 2.

Sharpe, C. (ed.) *Kempe's Engineering Year-Book*, 98th edition (1993), Benn, Tonbridge, Kent, UK. Updated each year — has good sections on irons and steels.

Iron and steels standards

Increasingly, national and international standards organizations are providing a complete catalogue of their publications on the World-Wide Web. Two of the most comprehensive printed sources are listed below.

Iron and Steel Specifications, 9th edition (1998), British Iron and Steel Producers Association (BISPA), 5 Cromwell Road, London, SW7 2HX. Comprehensive tabulations of data from British Standards on irons and steels, as well as some information on European and North American standards. The same information is available on searchable CD.

ASTM Annual Book of Standards, Vols 01.01 to 01.07, The most complete set of American iron and steel standards. Summaries of the standards can be found on the WWW at http://www.astm.org/stands.html.

Cross-referencing of similar international standard grades

It is difficult to match, even approximately, equivalent grades of iron and steel between countries. No coverage of this subject can ever be complete, but the references listed below are helpful.

Gensure, J.G. and Potts, D.L., *International Metallic Materials Cross Reference*, 3rd edition, Genium Publishing, New York (1988). Comprehensive worldwide coverage of the subject, well indexed.

Bringas, J.E. (ed.) *The Metals Black Book — Ferrous Metals*, 2nd edition, CASTI Publishing, Edmonton, Canada (1995). Easy-to-use tables for international cross-referencing. (See General section for more information.)

Unified Numbering System for Metals and Alloys, 2nd edition, Society of Automotive Engineers, Pennsylvania (1977). An authoritative reference work, providing a unifying structure for all standards published by US organizations. No coverage of the rest of the world.

Iron and Steel Specifications, 7th edition (1989), British Steel, 9 Albert Embankment, London, SE1 7SN. Lists 'Related Specifications' for France, Germany, Japan, Sweden, UK and USA.

Cast irons

Scholes, J.P., *The Selection and Use of Cast Irons*, Engineering Design Guides, OUP, Oxford, UK (1979).
Angus, H.T., *Cast Iron: Physical and Engineering Properties*, Butterworths, London (1976).
Gilbert, G.N.J., *Engineering Data on Grey Cast Irons* (1977).
Gilbert, G.N.J., *Engineering Data on Nodular Cast Irons* (1986).
Gilbert, G.N.J., *Engineering Data on Malleable Cast Irons* (1983).
Smith, L.W.L., Palmer, K.B. and Gilbert, G.N.J., *Properties of Modern Malleable Irons* (1986).
Palmer, K.B., *Mechanical & Physical Properties of Cast Irons at Sub-zero Temperatures* (1988).
Palmer, K.B., *Mechanical & Physical Properties of Cast Irons up to 500°C* (1986).

Irons, American Standards

These can all be found in the *Annual Book of ASTM Standards*, Vol. 01.02
ASTM A220M-88: 'Pearlitic Malleable Iron'.
ASTM A436-84: 'Austenitic Gray Iron Castings'.
ASTM A532: 'Abrasion-Resistant Cast Irons'.
ASTM A602-70: (Reapproved 1987) 'Automotive Malleable Iron Castings'.

Cast irons, International Standards

These are available from ISO Central Secretariat, 1, rue de Varembe, Case postale 56, CH-1211 Geneve 20, Switzerland.

ISO 185:1988 'Grey cast iron — classification'
ISO 2892:1973 'Austenitic Cast Iron'
ISO 5922:1981 'Malleable cast iron'

Cast irons, British Standards

Compared with steels, there are relatively few standards on cast iron, which makes it feasible to list them all. Standards are available from BSI Customer Services, 389 Chiswick High Road, London, W4 4AL, UK.

BS 1452:1990 'Flake graphite cast iron'.
BS 1591:1975 'Specification for corrosion resisting high silicon castings'.
BS 2789:1985, 'Iron Castings with spheroidal or nodular graphite'.
BS 3468:1986 'Austenitic cast iron'.
BS 4844:1986 'Abrasion resisting white cast iron'.
BS 6681:1986 'Specification for malleable cast iron'.

Carbon and low alloy steels

ASM Metals Handbook, 10th edition, Vol. 1 (1990), ASM International, Metals Park, Cleveland, Ohio, USA. Authoritative reference work for North American irons and steels.
Fox, J.H.E., *An Introduction to Steel Selection: Part 1, Carbon and Low-Alloy Steels*, Engineering Design Guide no. 34, Oxford University Press.

Stainless steels

ASM Metals Handbook, 10th edition, Vol. 1 (1990), ASM International, Metals Park, Cleveland, Ohio, USA. Authoritative reference work for North American irons and steels.
Elliott, D. and Tupholme, S.M., *An Introduction to Steel Selection: Part 2, Stainless Steels*, Engineering Design Guide no. 43, Oxford University Press (1981).
Peckner, D. and Bernstein, I.M., *Handbook of Stainless Steels*, McGraw-Hill, New York (1977).
Design Guidelines for the Selection and Use of Stainless Steel, Designers' Handbook Series no. 9014, Nickel Development Institute (1991).

(The Nickel Development Institute (NIDI) is a worldwide organization that gives away a large variety of free literature about nickel-based alloys, including stainless steels. NIDI European Technical Information Centre, The Holloway, Alvechurch, Birmingham, B48 7QB, ENGLAND.)

Ferrous metals, World-wide Web sites

Details of a few general sites are given below.

http://www.steelnet.org/: Steel Manufacturers Association, which claims to be North America's largest steel trade group. Contains links to the homepages of many US steel companies, but these are currently more likely to provide business information than data on material properties.
http://www.asm-intl.org/: ASM International (American Society of Materials). Linked to a wide range of useful sites.
http://www.astm.org/: American Society of Testing and Materials, publisher of a wide range of American standards.
http://www.iso.ch/: International Standards Organization — has links to all national standards organizations that have a presence on the WWW.
http://www.iso.ch/cate/77.html: Section 77 of the ISO catalogue, which includes descriptions of all their standards on ferrous metals, plus ordering information.

Polymers and elastomers

Polymers are not subject to the same strict specification as metals. Data tend to be producer-specific. Sources, consequently, are scattered, incomplete and poorly presented. Saechtling is the best; although no single hard-copy source is completely adequate, all those listed here are worth consulting. See also Databases as Software, Section 13A.5; some (Plascams, CMS) are good on polymers.

Saechtling: International Plastics Handbook, editor Dr Hansjurgen Saechtling, MacMillan Publishing Co (English edition), London, UK (1983). The most comprehensive of the hard-copy data-sources for polymers.

Polymers for Engineering Applications, R.B. Seymour. ASM International, Metals Park, Ohio 44037, USA (1987). Property data for common polymers. A starting point, but insufficient detail for accurate design or process selection.

New Horizons in Plastics, a Handbook for Design Engineers, editor J. Murphy, WEKA Publishing, London, UK.

ASM Engineered Materials Handbook, Vol. 2. Engineering Plastics (1989). ASM International, Metals Park, Ohio 44037, USA (1991).

Handbook of Plastics and Elastomers, editor C.A. Harper, McGraw-Hill, New York, USA (1975).

International Plastics Selector, Plastics, 9th edition, Int. Plastics Selector, San Diego, CA, USA (1987).

Die Kunststoffe and Ihre Eigenschaften, editor Hans Domininghaus, VDI Verlag, Dusseldorf, Germany (1992).

Properties of Polymers, 3rd edition, D.W. van Krevelen, Elsevier, Amsterdam, Holland, (1990). Correlation of properties with structure; estimation from molecular architecture.

Handbook of Elastomers, A.K. Bhowmick and H.L. Stephens. Marcel Dekker, New York, USA (1988).

ICI Technical Service Notes, ICI Plastics Division, Engineering Plastics Group, Welwyn Garden City, Herts, UK (1981).

Technical Data Sheets, Malaysian Rubber Producers Research Association, Tun Abdul Razak Laboratory, Brickendonbury, Herts. SG13 8NL (1995). Data sheets for numerous blends of natural rubber.

Ceramics and glasses

Sources of data for ceramics and glasses, other than the suppliers data sheets, are limited. Texts and handbooks such as the ASM's (1991) *Engineered Materials Handbook Vol. 4*, Morell's (1985) compilations, Neville's (1996) book on concrete, Boyd and Thompson (1980) *Handbook on Glass* and Sorace's (1996) treatise on stone are useful starting points. The *CMS Ceramics Database* contains recent data for ceramics and glasses. But in the end it is the manufacturer to whom one has to turn: the data sheets for their products are the most reliable source of information.

Ceramics and ceramic-matrix composites

ASM Engineered Materials Handbook, Vol. 4, Ceramics and Glasses. ASM International, Metals Park, Ohio 44073, USA (1991).

Chapman and Hall 'Materials Selector', editors N. Waterman and M.F. Ashby, Chapman and Hall, London UK (1996).

Concise Encyclopedia of Advanced Ceramic Materials editor R.J. Brook, Pargamon Press, Oxford, UK (1991).

Creyke, W.E.C., Sainsbury, I.E.J. and Morrell, R., *Design with Non Ductile Materials*, Applied Science, London, UK (1982).

Handbook of Ceramics and Composites, 3 Vols, editor N.P. Cheremisinoff, Marcel Dekker Inc., New York, USA (1990).

Handbook of Physical Constants, Memoir 97, editor S.P. Clark, Geological Society of America (1966).

Handbook of Structural Ceramics, editor M.M. Schwartz, McGraw-Hill, New York, USA (1992). Lots of data, information on processing and applications.

Kaye, G.W.C. and Laby, T.H., *Tables of Physical & Chemical Constants*, 15th edition, Longman, New York, USA (1986).

Kingery, W.D., Bowen, H.K. and Uhlmann, D.R., *Introduction to Ceramics*, 2nd edition, New York, Wiley (1976).

Materials Engineering 'Materials Selector', Penton Press, Cleveland, Ohio, USA (1992).

Morrell, R., *Handbook of Properties of Technical & Engineering Ceramics*, Parts I and II, National Physical Laboratory, Her Majesty's Stationery Office, London, UK (1985).

Musikant, S., *What Every Engineer Should Know About Ceramics*, Marcel Dekker Inc (1991). Good on data.

Richerson, D.W., *Modern Ceramic Engineering*, 2nd edition, Marcel Dekker, New York, USA (1992).

Smithells Metals Reference Book, 7th edition, editors E.A. Brandes and G.B. Brook, Butterworth-Heinemann, Oxford (1992).

Glasses

ASM Engineered Materials Handbook, Vol. 4, Ceramics and Glasses. ASM International, Metals Park, Ohio 44073, USA (1991).

Boyd, D.C. and Thompson, D.A., 'Glass', Reprinted from Kirk-Othmer: *Encyclopedia of Chemical Technology*, Volume 11, third edition, pp. 807–880, Wiley (1980).

Engineering Design Guide 05: The Use of Glass in Engineering, Oliver, D.S. Oxford University Press, Oxford, UK (1975).

Handbook of Glass Properties, Bansal, N.P. and Doremus, R.H., Academic Press, New York, USA (1966).

Cement and concrete

Cowan, H.J. and Smith, P.R., *The Science and Technology of Building Materials*, Van Nostrand-Reinhold, New York USA (1988).

Illston, J.M., Dinwoodie, J.M. and Smith, A.A., *Concrete, Timber and Metals*, Van Nostrand-Reinhold, New York USA (1979).

Neville, A.M., *Properties of Concrete*, 4th edition, Longman Scientific and Technical (1996). An excellent introduction to the subject.

Composites: PMCs, MMCs and CMCs

The fabrication of composites allows so many variants that no hard-copy data source can capture them all; instead, they list properties of matrix and reinforcement, and of certain generic lay-ups or types. The *Engineers Guide* and the *Composite Materials Handbook*, listed first, are particularly recommended.

Composite, general

Engineers Guide to Composite Materials, edited by Weeton, J.W., Peters, D.M. and Thomas, K.L. ASM International, Metals Park, Ohio 44073, USA (1987). The best starting point: data for all classes of composites.

Composite Materials Handbook, 2nd edition, editor Schwartz, M.M., McGraw-Hill, New York, USA (1992). Lots of data on PMCs, less on MMCs and CMCs, processing, fabrication, applications and design information.

ASM Engineered Materials Handbook, Vol. 1: Composites. ASM International, Metals Park, Ohio 44073, USA (1987).

Reinforced Plastics, Properties and Applications, R.B. Seymour. ASM International, Metals Park, Ohio 44073, USA (1991).

Handbook of Ceramics and Composites, Volumes 1–3, editor N.P. Cheremisinoff, Marcel Dekker Inc., New York, USA (1990).

Concise Encyclopedia of Composited Materials, editor A. Kelly, Pergamon Press, Oxford, UK (1989).

Middleton, D.H., *Composite Materials in Aircraft Structures*, Longman Scientific and Technical Publications, John Wiley, New York, USA (1990).

Smith, C.S., *Design of Marine Structures in Composite Materials*, Elsevier Applied Science, London, UK (1990).

Metal matrix composites

See, first, the sources listed under 'All Composite Types', then, for more detail, go to:

Technical Data Sheets, Duralcan USA, 10505 Roselle Street, San Diego, CA 92121, USA (1995).

Technical Data Sheets, 3M Company, 3M Xenter, Building 60-1N-001, St Paul MN 55144–1000, USA (1995).

Foams and cellular solids

Many of the references given in 13A.3 for polymers and elastomers mention foam. The references given here contain much graphical data, and simple formulae which allow properties of foams to be estimated from its density and the properties of the solid of which it is made, but in the end it is necessary to contact suppliers. See also Databases as software (Section 13A.5); some (Plascams, CMS) are good on foams. For Woods and wood-based composites, see below.

Cellular Polymers (a Journal), published by RAPRA Technology, Shrewsbury, UK (1981–1996)
Encyclopedia of Chemical Technology, Vol. 2, 3rd edition, pp. 82–126. Wiley, New York, USA (1980).
Encyclopedia of Polymer Science and Engineering, Vol. 3, 2nd edition, Section C, Wiley, New York, USA (1985).
Gibson, L.J. and Ashby, M.F., *Cellular Solids*, Cambridge University Press, Cambridge, UK (1997). Basic text on foamed polymers, metals, ceramics and glasses, and natural cellular solids.
Handbook of Industrial Materials, 2nd edition, pp. 537–556, Elsevier Advanced Technology, Elsevier, Oxford, UK (1992).
Low Density Cellular Plastics — Physical Basis of Behaviour, edited by Hilyard, N.C. and Cunningham, A. Chapman and Hall, London, UK (1994). Specialized articles on aspects of polymer-foam production, properties and uses.
Plascams (1995), Version 6, Plastics Computer-Aided Materials Selector, RAPRA Technology Limited, Shawbury, Shrewsbury, Shropshire, SY4 4NR, UK.
Saechtling (1983): *International Plastics Handbook*, editor Dr Hans Jurgen, Saechtling, MacMillan Publishing Co. (English edition), London, UK.
Seymour, R.P. (1987) *Polymers for Engineering Applications*, ASM International, Metals Park, Ohio 44037, USA.

Stone, rocks and minerals

There is an enormous literature on rocks and minerals. Start with the handbooks listed below; then ask a geologist for guidance.

Atkinson, B.K., *The Fracture Mechanics of Rock*, Academic Press, UK (1987).
Handbook of Physical Constants, editor S.P. Clark, Jr, Memoir 97, The Geological Society of America, 419 West 117 Street, New York, USA (1966). Old but trusted compilation of property data for rocks and minerals.
Handbook on Mechanical Properties of Rocks, Volumes 1–4, editors R.E. Lama and V.S. Vutukuri, Trans Tech Publications, Clausthal, Germany (1978).
Rock Deformation, editors Griggs, D. and Handin, J., Memoir 79, The Geological Society of America, 419 West 117 Street, New York, USA (1960).
Sorace, S., 'Long-term tensile and bending strength of natural building stones' *Materials and Structures*, **29**, 426–435 (August/September 1996).

Woods and wood-based composites

Woods, like composites, are anisotropic; useful sources list properties along and perpendicular to the grain. The US Forest Products Laboratory 'Wood Handbook' and Kollmann and Côté's 'Principles of Wood Science, and Technology' are particularly recommended.

Woods, general information

Bodig, J. and Jayne, B.A. (1982) *Mechanics of Wood and Wood Composites*, Van Nostrand Reinhold Company, New York, USA.
Dinwoodie, J.M. (1989) *Wood, Nature's Cellular Polymeric Fibre Composite*, The Institute of Metals, London, UK.
Dinwoodie, J.M., *Timber, its Nature and Behaviour*, Van Nostrand-Reinhold, Wokingham, UK (1981). Basic text on wood structure and properties. Not much data.
Gibson, L.J. and Ashby, M.F. (1997) *Cellular Solids*, 2nd edition, Cambridge University Press, Cambridge, UK.
Jane, F.W. (1970) *The Structure of Wood*, 2nd edition, A. and C. Black, Publishers, London, UK.
Kollmann, F.F.P. and Côté, W.A. Jr. (1968) *Principles of Wood Science and Technology*, Vol. 1 (Solid Wood), Springer-Verlag, Berlin, Germany. The bible.

Kollmann, F., Kuenzi, E. and Stamm, A. (1968), *Principles of Wood Science and Technology*, Vol. 2 (Wood Based Materials), Berlin: Springer-Verlag.

Schniewind, A.P. (ed.) (1989) *Concise Encyclopedia of Wood and Wood-Based Materials*, Pergamon Press, Oxford, UK.

Woods, data compilations

BRE (1996) 'BRE Information Papers', Building Research Establishment (BRE), Garston, Watford, WD2 7JR, UK.

Forest Products Laboratory (1989), Forest Service, US. Department of Agriculture, *Handbook of Wood and Wood-based Materials*, New York: Hemisphere Publishing Corporation. A massive compilation of data for North-American woods.

Informationsdienst Holz (1996), Merkblattreihe Holzarten, Verein Deutscher Holzeinfuhrhäuser e.V., Heimbuder Strabe 22, D-20148 Hamburg, Germany.

TRADA (1978/1979) *Timbers of the World, Volumes 1–9*, Timber Research and Development Association, High Wycombe, UK.

TRADA (1991) Information Sheets, Timber Research and Development Association, High Wycombe, UK.

Wood and wood-composite standards

Great Britain

British Standards Institution (BSI), 389 Chiswick High Road, GB-London W4 4AL, UK (Tel: +44 181 996 9000; Fax: +44 181 996 7400; e-mail: info@bsi.org.uk).

Germany

Deutsches Institut für Normung (DIN), Burggrafenstrasse 6, D-10772, Berlin, Germany (Tel +49 30 26 01-0; Fax: +49 30 26 01 12 31; e-mail: postmaster@din.de.

USA

American Society for Testing and Materials (ASTM), 1916 Race Street, Philadelphia, Pennsylvania 19103-1187 (Tel 215 299 5400; Fax: 215 977 9679).

ASTM European Office, 27–29 Knowl Piece, Wilbury Way, Hitchin, Herts SG4 0SX, UK (Tel: +44 1462 437933; Fax: +44 1462 433678; e-mail: 100533.741@compuserve.com).

Woods, software data sources

CMS WOODS DATABASE, Granta Design, Trumpington Street, Cambridge CB2 1QA, UK. A database of the engineering properties of softwoods, hardwoods and wood-based composites. PC format, Windows environment.

PROSPECT (Version 1.1) (1995), Oxford Forestry Institute, Department of Plant Sciences, Oxford University, South Parks Road, Oxford OX1 3RB, UK. A database of the properties of tropical woods of interest to a wood user; includes information about uses, workability, treatments, origins. PC format, DOS environment.

WOODS OF THE WORLD (1994), Tree Talk, Inc., 431 Pine Street, Burlington, VT 05402, USA. A CD-ROM of woods, with illustrations of structure, information about uses, origins, habitat etc. PC format, requiring CD drive; Windows environment.

Natural fibres and other materials

Houwink, R., *Elasticity, Plasticity and Structure of Matter*, Dover Publications, Inc, New York, USA (1958).

Handbook of Industrial Materials, 2nd edition, Elsevier, Oxford, UK (1992). A compilation of data remarkable for its breadth: metals, ceramics, polymers, composites, fibres, sandwich structures, leather....

Materials Handbook, 2nd edition, editors Brady, G.S. and Clauser, H.R., McGraw-Hill, New York, USA (1986). A broad survey, covering metals, ceramics, polymers, composites, fibres, sandwich structures, and more.

13A.4 Data for manufacturing processes

Alexander, J.M., Brewer, R.C. and Rowe, G.W., *Manufacturing Technology, Vol. 2: Engineering Processes*, Ellis Horwood Ltd., Chichester, UK (1987).

Bralla, J.G., *Handbook of Product Design for Manufacturing*, McGraw-Hill, New York, USA (1986).

Chapman and Hall Materials Selector, Waterman, N.A. and Ashby, M.F. (eds), Chapman and Hall, London, UK (1996).

CPS Cambridge Process Selector (1995), Granta Design Ltd, 20, Trumpington Street, Cambridge CB2 1QA, UK (Phone: +44-1223-334755; Fax: +44-1223-332797) Software for process selection.

Dieter, G.E., *Engineering Design, A Materials and Processing Approach*, McGraw-Hill, New York, USA, Chapter 7 (1983).

Kalpakjian, S., *Manufacturing Processes for Engineering Materials*, Addison Wesley, London, UK (1984).

Lascoe, O.D., *Handbook of Fabrication Processes*, ASM International, Metals Park, Columbus, Ohio, USA (1989).

Schey, J.A., *Introduction to Manufacturing Processes*, McGraw-Hill, New York, USA (1977).

Suh, N.P., *The Principles of Design*, Oxford University Press, Oxford, UK (1990).

13A.5 Databases and expert systems in software

The number and quality of computer-based materials information systems is growing rapidly. A selection of these, with comment and source, is given here. There has been consumer resistance to on-line systems; almost all recent developments are in PC-format. The prices vary widely. Five price groups are given: free, cheap (less than $200 or £125), modest (between $200 or £125 and $2000 or £1250), expensive (between $2000 or £1250 and $10 000 or £6000) and very expensive (more than $10 000 or £6000). The databases are listed in alphabetical order.

Active Library on Corrosion, ASM International, Metals Park, Ohio 44073, USA. PC format requiring CD ROM drive. Graphical, numerical and textual information on corrosion of metals. Price modest.

Alloy Digest (1997) ASM International, Metals Park, Ohio 44073, USA. PC format requiring CD ROM drive. 3500 datasheets for metals and alloys, regularly updated. Price: modest/expensive.

Alloy Finder, 2nd edition (1997), ASM International, Metals Park, Ohio 44073, USA. PC format requiring CD ROM drive. Lists 70 000 alloys by trade name, composition and designation. Price modest.

ALUSELECT P1.0: Engineering Property Data for Wrought Aluminium Alloys (1992). European Aluminium Association, Königsallee 30, P.O. Box 1012, D-4000 Dusseldorf 1, Germany (Tel: 0211-80871; fax 0211-324098). PC format, DOS environment. Mechanical, thermal, electrical and environmental properties of wrought aluminium alloys. Price cheap.

CAMPUS: Computer Aided Material Preselection by Uniform Standards (1995): Published separately by eight polymer producers:

Bayer UK Ltd. Bayer House, Strawberry Hill, Newbury, Berks, RG13 1JA, UK.

Hoechst Aktiengesellschaft, Marketing Technische Kunststoffe, D-65926, Frankfurt am Main, Germany (Tel: 06172-87-2755; fax: 01672-87-2761).

DuPont UK Ltd, Maylands Ave., Hemel Hempstead, Herts HP2 7DP, UK.

BASF UK Ltd. PO Box 4, Earl Road, Cheadle Hulme, Cheshire SK8 6QG, UK (Tel: 0161-485-6222; fax: 0161-486-0225).

EMS Grilon UK Ltd., Polymers Division, Walton Manor, Milton Keynes, Bucks MK7 7AJ, UK.

PC format DOS environment. A collection of four databases of Hoechst, BASF, Bayer DuPont, and Dow thermoplastic polymers, containing information on modulus, strength, viscosity and thermal properties. Regularly updated, but limited in scope. Free.

CETIM-EQUIST II: Centre Technique des Industries Mécaniques, (1997), BP 67, 60304 Senlis Cedex, France. PC format, DOS environment. Compositions and designations of steels.

CETIM-Matériaux: Centre Technique des Industries Mécaniques, (1997), BP 67, 60304 Senlis Cedex, France. On-line system. Compositions and mechanical properties of materials.

CETIM-SICLOP: Centre Technique des Industries Mécaniques, (1997), BP 67, 60304 Senlis Cedex, France. On-line system. Mechanical properties of steels.

CMS **Cambridge Materials Selector** (1995), Granta Design Ltd, 20, Trumpington Street, Cambridge CB2 1QA, UK (Phone: +44-1223-334755; Fax: +44-1223-332797) All materials, PC format, Windows environment. It implements the selection procedures developed in this book, allowing successive application of up to 16 selection stages. System includes a hierarchy of databases, with a 'Generic' database supported by detailed databases for ferrous metals, non-ferrous metals, polymers and composites, ceramics and glasses, and woods. Modest price.

CPS **Cambridge Process Selector** (1995), Granta Design Ltd, 20, Trumpington Street, Cambridge CB2 1QA, UK (Phone: +44-1223-334755; Fax: +44-1223-332797) All process classes, PC format, Windows environment. It implements the selection procedures developed in this book, allowing successive application of up to 16 selection stages. Modest price.

CopperSelect: Computerized System for Selecting Copper Alloys: Copper Development Association Inc, Greenwich Office, Park No. 2, Box 1840, Greenwich CT 06836, USA (Tel: 203-625-8210; Fax: 203-625-0174). PC-format, DOS environment. A database of properties and processing information for wrought and cast-copper alloys. All these and much more are also contained in *Megabytes on Copper*. Free.

CUTDATA: Machining Data System; Metcut Research Associates Inc, Manufacturing Technology Division, 11240 Cornell Park Drive, Cincinnati, Ohio 45242 USA (Tel: 513-489-6688). A PC-based system which guides the choice of machining conditions: tool materials, geometries, feed rates, cutting speeds, and so forth. Modest price.

EASel: Engineering Adhesives Selector Program (1986): The Design Centre, Haymarket, London SW1Y 4SU, UK. PC and Mac formats. A knowledge-based program to select industrial adhesives for joining surfaces. Modest price.

SF-CD (replacing **ELBASE**), Metal Finishing/Surface Treatment Technology (1992): Metal Finishing Information Services Ltd, PO Box 70, Stevenage, Herts SG1 4DF, UK (Tel: 01438-745115; fax: 01438-364536). PC format. Comprehensive information on published data related to surface treatment technology. Regularly updated. Modest price.

EPOS Engineering Plastics On Screen (1989): ICI Engineering, Plastics Sales Office, PO Box 90, Wilton, Middlesborough, Cleveland TS6 8JE, UK (Tel: 0642-454144 or 0707-337852). PC format, DOS environment. The software lists general and electrical properties of ICI polymer products, with a search facility. Updated periodically. Free.

FUZZYMAT 2.0: Software to Assist the Selection of Materials (1995). SNC Bassetti et Isaac, 91 bis, rue General Mangin, 38100 Grenoble, France (Tel: 04 76 23 35 44, Fax: 04 76 23 35 49). PC format, Windows environment. Materials selection by weight-factors, using fuzzy logic methods. Uses CMS databases and methodology (see *CMS*) and can be customized to meet special needs. Manual, screen display in French. Modest price.

IMAMAT Institute of Metals and Materials, Australasia, PO Box 19, Parkville 3052, Vic, Australia (Tel: 03-347-2544, Fax: 03-348-1208). Price and functionality not known.

M-VISION (1990): PDA Engineering, 2975 Redhill Avenue, Costa Mesa, CA 92626, USA (Tel: 714-540-8900; Fax: 714-979-2990). Requires a workstation. An ambitious image and database, with flexible selection procedures. Data for aerospace alloys and composites. Very expensive.

MAPP (replacing **Mat.DB** and, before that, **METSEL 2**): Materials Data-base; ASM International, Metals Park, Ohio 44073, USA (Tel: 216-338-5151; Fax: 216-338-4634). PC format, Windows environment and Mac. A Windows-based materials data source using the old Mat.DB data files, with an improved user interface. Data files for settles, aluminium alloys, titanium alloys, magnesium alloys, copper alloys, and a limited range of polymers. Selection based on user-defined target values. Expensive if a full suite of databases is wanted.

MATUS Materials User Service: Engineering Information Company Ltd, 23, Cardiff House, Peckham Park Road, London SE15 6TT, UK (Tel +44 0171-538-0096). Formerly an on-line data bank of UK material suppliers, trade names and properties for metals, polymers and ceramics, using data from suppliers' catalogues and data sheets. Much of the information is now available on PC format floppy disks, cheap or free.

Megabytes on Coppers: Information on copper and copper alloys, The Copper Development Association (1994), Orchard House, Mutton Lane, Potters Bar, Hertfordshire, EN6 3AP, UK (Tel: 01707-650711; Fax: 01707-642769). A CD-ROM with Windows search engine, containing all the current publications as well as interactive programs published by CDA on topics of electrical energy efficiency, cost effectiveness and corrosion resistance. Cheap.

PAL II Permabond Adhesives Locator (1996): Permabond, Woodside Road, Eastleigh, Hants SO5 4EX, UK (Tel: +44 703-629628; Fax: +44 703-629629). A knowledge-based, PC-system (DOS environment) for adhesive selection among Permabond adhesives. An impressive example of an expert system that works. Modest price.

PLASCAMS Version 6 Plastics Computer-aided Materials Selector (1995), RAPRA Technology Ltd, Shawbury, Shrewsbury, Shropshire SY4 4NR, UK (Tel: +44-1939-250383; Fax: +44-1939-251118). PC format, Windows environment. Polymers only. Mechanical and processing properties of polymers, thermoplastics and thermosets. Easy to use for data retrieval, with much useful information. Selection procedure cumbersome and not design-related. Modest initial price plus annual maintenance fee. Updated regularly.

Polymerge Modern Plastics International (1997), Emil-von-Behring-Str 2, D-60439 Frankfurt am Main, Germany (Tel: +49-69-5801-135; fax +49-69-5801-104). Allows CAMPUS discs to be merged, allowing comparison.

PROSPECT (Version 1.1) (1995), Oxford Forestry Institute, Department of Plant Sciences, Oxford University, South Parks Road, Oxford OX1 3RB, UK. A database of the properties of tropical woods of interest to a wood user; includes information about uses, workability, treatments, origins. PC format, DOS environment.

SMF Special Metals Fabrication (1996). See MATUS Data Publications, Engineering Information Co. Ltd. PC format, DOS environment. Properties of refractoriy metals for corrosion resistant and high-temperature applications in the chemical, aerospace, electronic and furnace industries. Useful compilation of data and applications. Free.

Stainless Steels A Guide to Stainless Steels (1994). Nickel Development Institute, 214 King Street West, Suite 510, Toronto Canada M5H 3S6 (Tel: 416 591-7999; Fax: 416 591-7987). PC format, DOS environment. Free.

teCal Steel Heat-Treatment Calculations; ASM International, Metals Park, Ohio 44073, USA (Tel: 216-338-5151; Fax: 216-338-4634). PC format, DOS environment. Computes the properties resulting from defined heat-treatments of low-alloy steels, using the composition as input. Modest price.

Stahlschlüssel 17th edition (1997), Wegst, C.W. *Stahlschlüssel* (in English: *Key to Steel*), Verlag Stahlschlüssel Wegst GmbH, D-1472 Marbach, Germany. CD ROM, PC format. Excellent coverage of European products and manufacturers.

SOFINE PLASTICS Société CERAP (1997) 27, Boulevard du 11 november 1918, BP 2132, 69603 Villeurbanne Cedex, France (Tel 04-72-69-58-30; fax: 04-78-93-15-56). Database of polymer properties. Environment and price unknown.

STRAIN Plastic Properties of Materials; Rob Bailey, Lawrence Livermore Laboratory, Materials Laboratory, PO Box 808, Livermore, Ca 94550, USA (Tel: 415-422-8512). PC format, DOS environment. Very simple but useful compilation of room-temperature mechanical properties of ductile materials. Free.

TAPP 2.0 Thermochemical and Physical Properties, (1994). ES Microware, 2234 Wade Court, Hamilton, OH 45013 USA (Tel: 513 738-4773, Fax: 513 738-4407, e-mail: ESMicro@aol.com). PC format, CD ROM, Windows environment. A database of thermochemical and physical properties of solids, liquids and gases, including phase diagrams neatly packaged with good user manual. Modest price.

THERM Thermal Properties of Materials; Rob Bailey, Lawrence Livermore Laboratory, Materials Laboratory, PO Box 808, Livermore, Ca 94550, USA (Tel: 415-422-8512). PC format, DOS environment. Very simple but useful compilation of thermal data for materials: specific heat, thermal conductivity, density and melting point. Free.

Titanium: Titanium Information Group, (1994), See MATUS Data Publications, Engineering Information Co. Ltd. PC format, DOS environment. Properties of titanium alloys. Useful compilation of data and applications. Free.

UNSearch Unified Metals and Alloys Composition Search; ASTM, 1916 Race Street, Philadelphia, PA 19103, USA. PC format, DOS environment. A database of information about composition, US designation and specification of common metals and alloys. Modest price.

WOODS OF THE WORLD (1994), Tree Talk, Inc., 431 Pine Street, Burlington, VT 05402, USA. A CD-ROM of woods, with illustrations of structure, information about uses, origins, habitat, etc. PC format, requiring CD drive; Windows environment.

13A.6 World-wide Web sites

The number of WWW sites carrying information about materials increases every week. There is almost no control of Web-site contents, which can vary enormously in nature and quality. The best are genuinely useful, establishing the Web as a potent 'further information' source. The sites listed below were accessible at the time of writing (August 1998), but at the time of reading some will have changed, and new sites will have appeared.

Sites of materials suppliers and producers

Carpenter Technology Home Page	http://www.cartech.com/
Cerac Incorporated	http://www.cerac.com/index.html
CMW Inc. Home Page	http://www.cmwinc.com/cmw
Copper Page	http://www.copper.org/
Elemental Carbon Information	http://fozzie.chem.wisc.edu/curriculum_development/CurrRef/BDGTopic/BDGtext/BDGtoc.html
Engelhard Corporation–Electro Metallics Department	http://www.engelhardemd.com/
GE Home Page	http://www.ge.com/index.htm
Gold information	http://www.interfaceweb.com/alpine/gold.htm
Indium Corporation of America	http://www.indium.com/
Jeneric/Pentron	http://www.jeneric.com/index.html
Materials (aeronautics): Lockheed Martin -	http://lmtas.com/AMMC/
Materials (high performance): MatTech	http://www.mat-tech.com/
Materials (research): Alfa Aesar	http://www.alfa.com/
Materials Preparation Center	http://www.ameslab.gov/mat_ref/mpc.html
Nickel: INCO Web Site	http://www.incoltd.com/toc-inct.htm
Niobium	http://www-c8.lanl.gov/infosys/html/periodic/41.html
Rare earths: Pacific Industrial Development Corp.	http://pidc.com/home.html
Rare Metals: Stanford Materials Inc.	http://www.stanfordmaterials.com/
Refractory Metals: Teledyne Wah Chang	http://www.twca.com/
Special Metals Corp.	http://www.specialmetals.com/
Steel & nickel based alloys	http://www.specmp.co.uk/
Steel: Bethlehem Steel's Web Site	http://www.bethsteel.com/
Steel: British Steel Home Page	http://www.britishsteel.co.uk/
Steel: Arcus Stainless Steel	http://194.178.135.1/
Steel: Automotive Steel Library	http://www.autosteel.org/home/steelmakers/members/
Steel: Great Plains Stainless	http://www.gpss.com/
Titanium.NET	http://www.titanium.org/
Tungsten: North American Tungsten	http://www.canvest.com/tungsten/strategic.shtml
Uranium Information Centre, Australia	http://www.uic.com.au/
Uranium Institute, London	http://www.uilondon.org/
Zinc Industrias Nacionales S.A.-Peru	http://www.zinsa.com/espanol.htm
Zinc: Eastern Alloys	http://www.eazall.com/ahome.html

Sites that list other sites

Granta Design Site Catalog	http://www.granta.co.uk
Automotive engineering	http://www.mlc.lib.mi.us/~stewarca/auto.html
CASTI Publishing Site Catalog	http://www.casti-publishing.com/intsite.htm
Directory of e-Conferences	http://www.n2h2.com/KOVACS/
Potter's Science Gems — Engineering	http://www-sci.lib.uci.edu/SEP/engineer.html
IndustryLink Homepage	http://www.industrylink.com/
TWI: WWW sites	http://www.twi.co.uk/links.html#tag14
Industry and Trade Associations	http://www.amm.com/ref/trade.HTM
Martindale's: Physics Web Pages	http://133.28.55.52:10080/=@=:www-sci.lib.uci.edu:80/HSG/GradPhysics.html
Materials-related sites	http://www.infodex.com/mlinks.html
Materials science	http://www.mlc.lib.mi.us/~stewarca/materials.html
Metallurgy	http://www.mlc.lib.mi.us/~stewarca/metallurgy.html
New sci.engr.: WWW sources	http://members.aol.com/RonGraham1/www.html
Product Data Management Info Ctr.	http://www.pdmic.com/
Scientific Web Resources	http://boris.qub.ac.uk/edward/index.html
SHAREWARE.COM	http://www.shareware.com/
Steelmaking	http://www.mlc.lib.mi.us/~stewarca/steelmaking.html
Forest Products and Wood Science	http://weber.u.washington.edu/~esw/fpm.htm
Thomas Register–Home Page	http://www.thomasregister.com:8000/login.cgi
Top 50 US/Canadian Metal Co.s	http://www.amm.com/ref/top50.HTM
Tree Talk & The Forest Partnership	http://www.woodweb.com/~treetalk/home.html
Tribology	http://www.mlc.lib.mi.us/~stewarca/tribology.html

Materials database

Elements Info on the Web	http://www.shef.ac.uk/~chem/web-elements/
Biomaterials Properties — TOC	http://www.lib.umich.edu/libhome/Dentistry.lib/Dental_tables/toc.html
CenBASE Materials on WWW	http://www.sgi.com/Works/iii/index.html
Chemscope — Medical Materials	http://chemscope.com/mat.htm
Corrosion relevant databases	http://www.clihouston.com/dbase.html
Electronic Products from ASM International	http://www.asm-intl.org/www-asm/e-prod/top.htm
F*A*C*T 2.1 — COMPOUND–Web	http://www.crct.polymtl.ca/fact/web/compweb.htm
Granta Design Limited	http://www.granta.co.uk
IDEMAT, Environmental Materials Database	http://www.io.tudelft.nl/research/mpo/general.htm
Japanese material database directory	http://fact.jicst.go.jp/~ritu/mdb/mdb.html
Periodic Table	http://steele.isgs.uiuc.edu/isgsroot/geochem/analytic/pt/ptable.html
Single Element Standards	http://radian.com/standards/el-singl.htm
STN Database Catalog	http://info.cas.org/ONLINE/CATALOG/descript.html

Metals prices and economic reports

American Metal Market On-line	http://www.amm.com/
Business Communications Company	http://www.vyne.com/bcc/
CRU International	http://www.cru-int.com/cruint/index.html#top
Daily Economic Indicators	http://www.bullion.org.za/prices.htm
Kitco Inc Gold & Precious Metal Prices	http://www.kitco.com/gold.live.html#ourtable
London Metal Exchange	http://www.lme.co.uk/
Mineral-Resource	http://minerals.er.usgs.gov/USGSminCommodSpecs.html
Precious Metal & Bonds	http://lab.busfac.calpoly.edu/pub/quoter/commoditiesgraph.html
Precious metals 5F page	http://www.ccn.cs.dal.ca/~an388/Precious.html#price
Rand Refinery	http://www.bullion.org.za/associates/rr.htm#goldprice
Roskill Reports from TMS	http://www.tms.org/pubs/Books/Roskill.html
The Precious Metal and Gem Connection	http://www.pm-connect.com/images/right.map
Trelleborg Metals Prices	http://akropolis.malmo.trab.se/trellgroup/PRI.html
Ux Jan 96 Uranium Indicator Update	http://www.uxc.com/ux_u3o8_ind.html

Government organizations and professional societies

ASM International	http://www.asm-intl.org/
ASME International	http://asme.web.aol.com/index.html
ASTM Web Site	http://www.astm.org/
Commonwealth Scientific and Industrial Research Organization(Australia)	http://www.csiro.au
DIN Deutsches Institut für Normung e.V	http://www.din.de/frames/Welcome.html
Institute of Materials, London, UK	http://www.instmat.co.uk/
International Standards Organization	http://www.iso.ch/
Japanese Learned Societies on the Web	http://wwwsoc.nacsis.ac.jp/index-e.html
National Academy Press, USA	http://www.nap.edu/
National Institute for Standards and Technology (USA) - Home Page	http://www.nist.gov/
National Standards Authority of Ireland	http://www.nsai.ie/
Society of Automotive Engineers (SAE)	http://www.sae.org/
The Minerals, Metals & Materials Society	http://www.tms.org/TMSHome.html
UK Government Information Service	http://www.open.gov.uk/
US Govt. Info on Minerals	http://minerals.er.usgs.gov:80/minerals/pubs/mcs/

Miscellaneous

Common unit of measure conversion	http://www.conweb.com/tblefile/conver.shtml
Conversion Factors, Material Properties and Constants Related to Telescope Design	http://www.apo.nmsu.edu/Telescopes/SDSS/eng.papers/ 19950926_ConversionFactors/19950926.html#aa1
Information for Physical Chemists	http://www.liv.ac.uk/Electrochem/online.html
K & K associate's thermal connection	http://www.kkassoc.com/~takinfo/
Thermal data	http://www.csn.net/~takinfo/prop-top.html

Case studies: use of data sources

14.1 Introduction and synopsis

Screening requires data sources with one structure, further information, sources with another. This chapter illustrates what they look like, what they can do and what they cannot.

The procedure follows the flow-chart of Figure 13.2, exploring the use of handbooks, databases, trade-association publications, suppliers data sheets, the Internet, and, if need be, in-house tests. Examples of the use of all of these appear in the case studies which follow. In each we seek detailed data for one of the materials short-listed in various of the case studies of earlier chapters. Not all the steps are reproduced, but the key design data and some indication of the level of detail, reliability and difficulty are given. They include examples of the output of software data sources, of suppliers data sheets and of information retrieved from the World-wide Web.

Data retrieval sounds a tedious task, but when there is a goal in mind it can be fun, a sort of detective game. The problems in Appendix B at the end of this book suggests some to try.

14.2 Data for a ferrous alloy — type 302 stainless steel

An easy one first: finding data for a standard steel. A spring is required to give a closing torque for the door of a dishwasher. The spring is exposed to hot, aerated water which may contain food acids, alkalis and salts. The performance indices for materials for springs

$$M_1 = \frac{\sigma_f^2}{E} \qquad \text{(small springs)}$$

or

$$M_2 = \frac{\sigma_f^2}{E C_{R\rho}} \qquad \text{(cheap springs)}$$

A screening exercise using the appropriate charts, detailed in Case Study 6.8, led to a shortlist which included elastomers, polymers, composites and metals. Elastomers and polymers are eliminated here by the additional constraint on temperature. Although composites remain a possibility, the obvious candidates are metals. Steels make good springs, but ordinary carbon steels would corrode in the hot, wet, chemically aggressive environment. Screening shows that stainless steels can tolerate this.

The detailed design of the spring requires data for the properties that enter M_1 or M_2, — the strength σ_f (in the case of a metal, the yield strength σ_y), the modulus E, the density ρ and the cost C_m — and data for the resistance to corrosion. The handbooks are the place to start.

Table 14.1 Data for hard drawn type 302 stainless steels*

Property	Source A*	Source B*	Source C*
Density (Mg/m^3)	7.8	7.9	7.86
Modulus E (GPa)	210	215	193
0.2% Strength σ_y (MPa)	965	1000	—
Tensile strength (MPa)	1280	1466	1345
Elongation (%)	9	6	—
Corrosion resistance	'Good'	'Highly resistant'	No information
Cost	No information	No information	No information

*Source A: *ASM Metals Handbook*, 10th Edition, Vol. 1 (1990); Source B: Smithells (1987); Source C: http.//www.matweb.com. All data have been converted to SI units.

Source A, the *ASM Metals Handbook* and Source B Smithells (1987) both have substantial entries listing the properties of some 15 stainless steels. Hard-drawn Type 302 has a particularly high yield strength, promising attractive values of the indices M_1 and M_2. Information for Type 302 is abstracted in Table 14.1. Both handbooks give further information on composition, heat treatment and applications. The *ASM Metals Handbook* adds the helpful news: 'Type 302 has excellent spring properties in the fully hard or spring-temper condition, and is readily available'. The World-wide Web yields Source C, broadly confirming what we already know.

No problems here: the mechanical-property data from three quite different sources are in substantial agreement; the discrepancies are of order 2% in density and modulus, and 10% in strength, reflecting the permitted latitude in specification on composition and treatment. To do better than this you have to go to suppliers data sheets.

One piece of information is missing: cost. Handbooks are reluctant to list it because, unlike properties, it varies. But a rough idea of cost would be a help. We turn to the databases. MatDB is hopelessly cumbersome and gives no help. The *CMS* gives the property profile shown in Figure 14.1; it includes the information: 'Price: Range 1.4 to 1.6 £/kg' (or 1.1 to 1.3 $/lb). Not very precise, but enough to be going on with.

Postscript

We are dealing here with a well-bred material with a full pedigree. Unearthing information about it is straightforward. That given above is probably sufficient for the dishwasher design. If more is wanted it must be sought from the steel company or the local supplier of the material itself, who will advise on current availability and price.

Related case studies

Case Study 6.9: Materials for springs

14.3 Data for a non-ferrous alloy — Al–Si die-casting alloys

Candidate materials determined in Case Study 6.6 for the fan included aluminium alloys. Processing charts (Chapter 12) establish that the fan could be made with adequate precision and smoothness by die casting. To proceed with detailed design we now need data for density, ρ, and strength σ_f;

Name: Wrought austenitic stainless steel, AISI 302

State: HT grade D
Composition Fe/<.15C/17-19Cr/8-11Ni/<2Mn/<1Si/<.045P/<.03S

Similar Standards
UK (BS): 302S25; UK (former BS): En 58A; ISO: 683/XIII Type 12; USA (UNS): S30200; Germany (W.-Nr.): 1.4300; Germany (DIN): X12 CrNi 18 8; France (AFNOR): Z12 CN 18.10; Italy (UNI): X15 CrNi 18 09; Sweden (SIS): 2332; Japan (JIS): SUS 302;

General

Density	7.81	—	8.01	Mg/m^3
Price	1.75	—	2.55	£/kg

Mechanical

Bulk Modulus	134	—	146	GPa
Compressive Strength	760	—	900	MPa
Ductility	0.05	—	0.2	
Elastic Limit	760	—	900	MPa
Endurance Limit	436	—	753	MPa
Fracture Toughness	68	—	185	$MPa\,m^{1/2}$
Hardness	3.50E+3	—	5.70E+3	MPa
Loss Coefficient	2.90E-4	—	4.80E-4	
Modulus of Rupture	760	—	900	MPa
Poisson's Ratio	0.265	—	0.275	
Shear Modulus	74	—	78	GPa
Tensile Strength	1.03E+3	—	2.24E+3	MPa
Young's Modulus	189	—	197	GPa

Thermal

Latent Heat of Fusion	260	—	285	kJ/kg
Maximum Service Temperature	1.02E+3	—	1.20E+3	K
Melting Point	1.67E+3	—	1.69E+3	K
Minimum Service Temperature	1	—	2	K
Specific Heat	490	—	530	J/kg K
Thermal Conductivity	15	—	17	W/m K
Thermal Expansion	16	—	20	$10^{-6}/K$

Electrical

Resistivity	65	—	77	10^{-8} ohm m

Typical uses
Exhaust parts; internal building fasteners; sinks; trim; washing-machine tubs; water tubing, springs

References
Elliot, D. and Tupholme, S.M. 'An Introduction to Steel, Selection: Part 2, Stainless Steels', OUP (1981);
'Iron & Steel Specifications', 8th edition (1995), BISPA, 5 Cromwell Road, London, SW7 2HX;
Brandes, E.A. and Brook, G.R. (eds.) 'Smithells Metals Reference Book' 7th Edition (1992), Butterworth-Heinemann, Oxford, UK.
ASM Metals Handbook (9th edition), Vol. 3, ASM International, Metals Park, Ohio, USA (1980);
'Design Guidelines for the Selection and Use of Stainless Steel', Designers' Handbook Series no.9014, Nickel Development Institute (1991);

Fig. 14.1 Part of the output of the PC-format database *CMS* for Type 302 stainless steel. Details of this and other databases are given in the Appendix to Chapter 13, Section 13A.5.

in this case we might interpret σ_f as the fatigue strength. Prudence suggests that we should check the yield and ultimate strengths too.

Aluminium alloys, like steels, have a respectable genealogy. Finding data for them should not be difficult. It isn't. But there *is* a problem: a lack of harmony in specification. We reach for the handbooks again. Volume 2 of the *ASM Metals Handbook* reveals that 85% of all aluminium die-castings are made of Alloy 380, a highly fluid (i.e. castable) alloy containing 8% silicon with a little iron and copper. It gives the data listed under Source A in Table 14.2.

So far so good. But when we turn to Smithells (1987) we find no mention of Alloy 380, or of any other with the same composition. Among die-casting alloys, Alloy LM6 (alias 3L33 and LM20) features. It contains 11.5% silicon, and, not surprisingly, has properties which differ from those of Alloy 380. They are listed under Source B in Table 14.2. The density and modulus of the two alloys are the same, but the fatigue strength of LM6 is less than half that of Alloy 380.

This leaves us vaguely discomforted. Are they really so different? Are the data to be trusted at all? Before investing time and money in detailed design, we need corroboration of the data. A third handbook — the *Chapman and Hall Materials Selector* — gives data for LM6 (Source C, Table 14.2); it fully corroborates Smithells. This looks better, but just to be sure we seek help from the Trade Federations: the Aluminium Association in the US; the Aluminium Federation (ALFED) in the UK. We are at this moment in the UK — we contact ALFED — they mail their publication *The Properties of Aluminium and its Alloys*. It contains everything we need for LM6, including its seven equivalent names in Europe, Russia and Australasia. The data for moduli and strength are identical with those of Source C in the Table — Mr Chapman and Ms Hall got their data from ALFED, a sensible thing to have done. A similar appeal to the US Aluminium Association reveals a similar story — their publication was the origin of the ASM data of Source A.

So there is nothing wrong with the data. It is just that die-casters in the US use one alloy; those in Europe prefer another. But what about cost? None of the handbooks help. A quick scan through the WWW sites listed in Chapter 13 directs us to the London Metal Exchange http://www.metalprice.com./. Todays quoted price for aluminium alloy is Al-alloy 1.408 to 1.43 $/kg.

Postscript

Discord in standards is a common problem. Committees charged with the task of harmonization sit late into the EU night, and move slowly towards a unifying system. In the case of both steels and aluminium alloys, the US system of specification, which has some reason and logic to it, is likely to become the basis of the standard.

Table 14.2 Data for aluminium alloys 380 and LM6

Property	Source A*	Source B*	Source C*
Density (Mg/m^3)	2.7	2.65	2.65
Modulus (GPa)	71	70.6	71
0.2% Yield strength (MPa)	165	77	80
Ultimate strength (MPa)	330	216	200
Fatigue strength (MPa)	145	62	68
Elongation (%)	3	10	13

*Source A: *ASM Metals Handbook*, 10th Edition, Volume 2 (1990); Source B: Smithells (1987); Source C: *Chapman and Hall Materials Selector* (1997) and ALFED (1981). All data have been converted to SI units.

Related case studies

Case Study 6.7:   Materials for high-flow fans
Case Study 12.2: Forming a fan
Case Study 12.6: Economical casting

14.4 Data for a polymer — polyethylene

Now something slightly less clear cut: the selection of a polymer for the elastic seal analysed in Case Study 6.10. One candidate was low-density polyethylene (LDPE). The performance index

$$M = \frac{\sigma_f}{E}$$

required data for modulus and for strength; we might reasonably ask, additionally, for density, thermal properties, corrosion resistance and cost.

Start, as before, with the handbooks. The *Chapman and Hall Materials Selector* compares various grades of polyethylene; its data for LDPE are listed in Table 14.3 under Source A. The *Engineered Materials Handbook, Vol. 2, Plastics*, leaves us disappointed. The *Polymers for Engineering Applications* (1987) is rather more helpful, but gives values for strength and thermal properties which differ by a factor of 2 from those of Source A, and no data at all for the modulus. The *Handbook of Polymers and Elastomers* (1975), after some hunting, gives the data listed under Source B — big discrepancies again. The *Materials Engineering 'Materials Selector'* (Source C) does much the same. None give cost. Things are not wholly satisfactory: we could do this well by simply reading data off the charts of Chapter 4. We need something better.

How about computer databases? The PLASCAMS and the *CMS* systems both prove helpful. We load PLASCAMS. Some 10 keystrokes and two minutes later, we have the data shown in Figure 14.2. They include a modulus, a strength, cost, processing information and applications: we are reassured to observe that these include gaskets and seals. The same database also contains the address and phone number of suppliers who will, on request, send data sheets. All much more satisfactory.

Table 14.3 Data for low-density polyethylene (LDPE)

Property	Source A*	Source B*	Source C*
Density (Mg/m^3)	0.92	0.91–0.93	0.92
Modulus (GPa)	0.25	0.1–0.2	0.2
Heat deflection temp (°C)	50	43	—
Max service temp (°C)	50	82	69
T-expansion (10^{-6} K^{-1})	200	100–200	160–198
T-conductivity (W/m K)	—	0.33	0.33
Tensile strength (MPa)	9	4–15	13
Rockwell hardness	D48	D41–50	D50
Corrosion in water/dilute acid	satisfactory	resistant	excellent

*Source A: *Chapman and Hall Materials Selector* (1997); Source B: *Handbook of Polymers and Elastomers* (1975); Source C: *Materials Engineering Materials Selector* (1997). All data have been converted to SI units.

Material: 119 LDPE

Resin type: TP S.Cryst.   Cost/tonne: 600   S.G. 0.92

Max. Operating Temp	°C	50	Surface hardness		SD48
Water absorption	%	0.01	Linear expansion	E-5	20
Tensile strength	MPa	10	Flammability	UL94	HB
Flexural modulus	GPa	0.25	Oxygen index	%	17
Elongation at break	%	40	Vol. Resist.	$\log \Omega\,cm$	16
Notched Izod	kJ/m	1.06+	Dielect. strength	MV/m	27
HDT @ 0.45 MPa	°C	50	Dielect. const. 1kHz		2.3
HDT @ 1.80 MPa	°C	35	Dissipation Fact. 1kHz		0.0003
Matl. drying	hrs @ °C	NA	Melt temp. range	°C	220–260
Mould shrinkage	%	3	Mould temp. range	°C	20–40

ADVANTAGES   Cheap, good chemical resistance. High impact strength at low temperatures. Excellent electrical properties.

DISADVANTAGES   Low strength and stiffness. Susceptible to stress cracking. Flammable.

APPLICATIONS   Chemically resistant fittings, bowls, lids, gaskets, toys, containers packaging film, film liners, squeeze bottles. Heat-seal film for metal laminates. Pipe, cable covering, core in UHF cables.

Fig. 14.2 Part of the output of PLASCAMS, a PC database for engineering polymers, for low-density polyethylene. It also gives trade names and addresses of UK suppliers. Details of this and other databases are given in the Appendix to Chapter 13, Section 13A.5.

But is it up to date? Not, perhaps, as much so as the World-wide Web. A search reveals company-specific web sites of polymer manufacturers (GE, Hoechst, ICI, Bayer and more). It also guides us to sites which collect and compile data from suppliers data sheets. One such is http://www.matweb.com./ from which Figure 14.3 was downloaded.

Postscript

There are two messages here. The first concerns the properties of polymers: they vary from supplier to supplier much more than do the properties of metals. And the way they are reported is quirky: a flexural modulus but no Young's modulus; a Notched Izod number instead of a fracture toughness, and so on. These we have to live with for the moment. The second concerns the relative ease of use of handbooks and databases: when the software contains the information you need, it surpasses, in ease, speed and convenience, any handbook. But software, like a book, has a publication date. The day after it is published it is, strictly speaking, out of date. The World-wide Web is dynamic; a well maintained site yields data which has not aged.

Related case studies

Case Study 6.10: Elastic hinges
Case Study 6.11: Materials for seals

Polyethylene, Low Density; Molded/Extruded

Polymer properties are subject to a wide variation, depending on the grade specified.

Physical Properties	Values	Comments
Density, g/cc	0.91	0.910–0.925 g/cc
Linear Mold Shrinkage, cm/cm	0.03	1.5–5% ASTM D955
Water Absorption, %	1.5	in 24 hours per ASTM D570
Hardness, Shore D	44	41–46 Shore D

Mechanical Properties	Values	Comments
Tensile Strength, Yield, MPa	10	4–16 MPa; ASTM D638
Tensile Strength, Ultimate, MPa	25	7–40 MPa
Elongation %; break	400	100–800%; ASTM D638
Modulus of Elasticity, GPa	0.2	0.07–0.3 GPa; In Tension; ASTM D638
Flexural Modulus, GPa	0.4	0–0.7 GPa; ASTM D790
Izod Impact in J, J/cm, or J/cm^2	999	No Break; Notched

Thermal Properties	Values	Comments
CTE, linear 20°C, μm/m-°C	30	20–40 μm/m-°C; ASTM D696
HDT at 0.46 MPa, °C	45	40–50°C
Processing Temperature, °C	200	150–320°C
Melting Point, °C	115	
Maximum Service Temp, Air, °C	70	60–90°Cv
Heat Capacity, J/g-°C	2.2	2.0–2.4 J/g-°C; ASTM C351
Thermal Conductivity, W/m-K	0.3	ASTM C177

Electrical Properties	Values	Comments
Electrical Resistivity, Ohm-cm	1E+16	ASTM D257
Dielectric Constant	2.3	2.2–2.4; 50–100 Hz; ASTM D150
Dielectric Constant, Low Frequency	2.3	2.2–2.4; 50–100 Hz; ASTM D150
Dielectric Strength, kV/mm	19	18–20 kV/mm; ASTM D149
Dissipation Factor	0.0005	Upper Limit; 50–100 Hz; ASTM D150
Dissipation Factor, Low Frequency	0.0005	Upper Limit; 50–100 Hz; ASTM D150

Fig. 14.3 Data for low-density polyethylene from the web site http://www.matweb.com.

14.5 Data for a ceramic — zirconia

Now a challenge: data for a novel ceramic. The ceramic valve of the tap examined in Case Study 6.20 failed, it was surmised, because of thermal shock. The problem could be overcome by choosing a ceramic with a greater thermal shock resistance. Zirconia (ZrO_2) emerged as a possibility. The performance index

$$M = \frac{\sigma_t}{E\alpha}$$

contains the tensile strength, σ_t, the modulus E and the thermal expansion coefficient α. The design will require data for these, together with hardness or wear resistance, fracture toughness, and some indication of availability and cost.

Table 14.4 Data for zirconia

Properties	Source A*	Source B*	Source C*	Source D*	Source E*
Density (Mg/m^3)	5.0–5.8	5.4	—	6.0	5.65
Modulus (GPa)	200	150	150	200	200
Tensile strength (MPa)	—	240	—	—	—
Modulus of rupture (MPa)	—	83	—	400–800	550
Hardness (MPa)	12 000	11 000	6000	12 000	11 000
Fracture toughness (MPa m$^{1/2}$)	2.5–5	7.6	4.7	4.5	8.4
T-expansion (10^{-6} K^{-1})	8–9	4.9	7	8–9	7
T-conductivity (W/m K)	1.8	2.4	1.8	1.7–2.0	1.67

*Source A: Morrell, *Handbook of Properties of Technical and Engineering Ceramics* (1985); Source B: *ASM Engineered Materials Reference Book* (1989); Source C: *Handbook of Ceramics and Composites* (1990); Source D: *Chapman and Hall 'Materials Selector'* (1997); Source E: http.//matweb.com./. All data have been converted to SI units.

After some hunting, entries are found in four of the handbooks; the best they can offer is listed in Table 14.4. One (the *ASM Engineered Materials Reference Book*), supplies the further information that zirconia 'has low friction coefficient, good wear and corrosion resistance, good thermal shock resistance, and high fracture toughness'. Sounds promising; but the numeric data show alarming divergence and have unpleasant gaps. No cost data, of course.

There are large discrepancies here. It is not unusual to find that samples of ceramics which are chemically identical can be as strong as steel or as brittle as a biscuit. Ceramics are not yet manufactured to the tight standards of metallic alloys. The properties of a zirconia from one supplier can differ, sometimes dramatically, from those of material from another. But the problem with Source B, at least, is worse: a modulus of rupture (MOR) of 83 MPa is not consistent with a tensile strength σ_t of 240 MPa; as a general rule, the MOR is greater than the tensile strength. The discrepancy is too great to be correct; the data must either have come from two quite different materials or be just plain wrong.

All this is normal; one must expect it in materials which are still under development. It does not mean that zirconia is a bad choice for the valve. It means, rather, that we must identify suppliers and base the design on the properties they provide. Figure 14.4 shows what we get: supplier's data for the zirconia with the tradename AmZirOx. Odd mixture of units, but the conversion factors inside the covers of this book allow them to be restored to a consistent set. The supplier can give guidance on supply and cost (zirconia currently costs about three times more than alumina), and can be held responsible for errors in data. The design can proceed.

Postscript

The new ceramics offer design opportunities, but they can only be grasped if the designer has confidence that the material has a consistent quality, and properties with values that can be trusted. The handbooks and databases do their best, but they are, inevitably, describing average or 'typical' behaviour. The extremes can lie far from the average. Here is a case in which it is best, right from the start, to go to the supplier for help.

Related case studies

Case Study 6.21: Ceramic valves for taps
Case Study 12.5: Forming a ceramic tap valve

TECHNICAL DATA

AmZirOX (Astro Met Zirconium Oxide) is a yttria partially stabilized zirconia advanced ceramic material which features high strength and toughness making it a candidate material for use in severe structural applications which exhibit wear, corrosion abrasion and impact. AmZirOX has been developed with a unique microstructure utilizing transformation toughening which allows AmZirOX to absorb the energy of impacts that would cause most ceramics to shatter. AmZirOX components can be fabricated into a wide range of precision shapes and sizes utilizing conventional ceramic processing technology and finishing techniques.

PROPERTIES	UNITS	VALUE
Color	—	Ivory
Density	g/cm^3	6.01
Water Absorption	%	0
Gas Permeation	%	0
Hardness	Vickers	1250
Flexural Strength	MPa (KPSI)	1075 (156)
Modulus of Elasticity	GPa (10^6 psi)	207 (30)
Fracture Toughness	$MPa\,m^{1/2}$	9
Poisson's Ratio	—	100
Thermal Expansion (25°C–1000°C)	$10^{-6}/°C$ ($10^{-6}/°F$)	10.3 (5.8)
Thermal Conductivity	Btu in/ft^{2}h°F	15
Specific Heat	cal/°C gm	0.32
Maximum Temperature Use (no load)	°C (°F)	2400 (4350)

Fig. 14.4 A supplier's data sheet for a zirconia ceramic. The units can be converted to SI by using the conversion factors given inside the front and back covers of this book.

14.6 Data for a glass-filled polymer — nylon 30% glass

The main bronze rudder-bearings of large ships (Case Study 6.21) can be replaced by nylon, or, better, by a glass-filled nylon. The replacement requires redesign, and redesign requires data. Stiffness, strength and fatigue resistance are obviously involved; friction coefficient, wear rate and stability in sea water are needed too.

Start, as always, with the handbooks. Three yield information for 30% glass-filled Nylon 6/6. It is paraphrased in Table 14.5. The approach of the sources differs: two give a single 'typical' value for each property, and no information about friction, wear or corrosion. The third (Source C) gives a range of values, and encouragement, at least, that friction, wear and corrosion properties are adequate. The things to observe are, first, the consistency: the ranges of Source C contain the values of the other two. But — second — this range is so wide that it is not much help with detailed design. Something better is needed.

The database PLASCAMS could certainly help here, but we have already seen what PLASCAMS can do (Figure 14.2). We turn instead to dataPLAS and find what we want: 30% glass-filled Nylon 6/6. Figure 14.5 shows part of the output. It contains further helpful comments and addresses for

POLYAMIDE 6.6

FERRO

MECHANICAL PROPERTIES	Unit	Value
Tensile Yield Strength	psiE3	—
Ultimate Tensile Strength	psiE3	19.7
Elongation at Yield	%	—
Elongation at Break	%	2.8
Tensile Modulus	psiE3	942
Flexural Strength	psiE3	26.8
Flexural Modulus	psiE3	812
Compressive Strength	psiE3	23
Shear Strength	psiE3	11
Izod Impact Unnotched, 23 $\frac{1}{2}$ C	FLb/in	7
Izod Impact Unnotched, −40 $\frac{1}{2}$ C	FLb/in	6
Izod Impact Notched, 23 $\frac{1}{2}$ C	FLb/in	1.4
Izod Impact Notched, −40 $\frac{1}{2}$ C	FLb/in	0.7
Tensile Impact Unnotched, 23 $\frac{1}{2}$ C	FLP/i^2	—
Rockwell hardness M	—	90
Rockwell hardness R	—	115
Shore hardness D	—	85
Shore hardness A	—	—

THERMAL PROPERTIES	Unit	Value
DTUL @ 264 psi (1.80 MPa)	°F	401
DTUL @ 66 psi (0.45 MPa)	°F	428
Vicat B Temperature, 5 kg	°F	410
Vicat A Temperature, 1 kg	°F	—
Continuous Service Temperature	°F	284
Melting Temperature	°F	424
Glass Transition	°F	—
Thermal Conductivity	W/m K	0.35
Brittle Temperature	−°F	—
Linear Thermal Expansion Coeff.	E-5/F	1.67

Fig. 14.5 Part of the output of dataPLAS, a PC database for US engineering polymers, for 30% glass-filled Nylon 6/6. Details of this and other databases are given in the Appendix to Chapter 13, Section 13A.5.

suppliers (not shown), from whom data sheets and cost information, which we shall obviously need, can be obtained.

Postscript

Glass-filled polymers are classified as plastics, not as the composites they really are. Fillers are added to increase stiffness and abrasion resistance, and sometimes to reduce cost. Data for filled polymers can be found in all the handbooks and databases that include data for polymers.

Related case studies

Case Study 6.22: Bearings for ships' rudders

Table 14.5 Data for nylon 6/6, 30% glass filled

Property	Source A*	Source B*	Source C*
Density (Mg/m^3)	1.37	1.3	1.3–1.34
Melting point (°C)	265	—	120–250
Heat deflection temp. (°C)	260	260	—
Tensile modulus (GPa)	—	9	9
Tensile strength (MPa)	180	186	100–193
Compressive strength (MPa)	180	165	165–276
Elongation (%)	3	3–4	2.5–3.4
T-expansion ($10^{-6} K^{-1}$)	20	107	15–50
T-conductivity (W/m K)	—	0.49	0.21–0.48
Friction, wear, etc.	No comment	No comment	Uses include: unlubricated gears, bearings and anti-friction parts
Corrosion	No comment	No comment	Good in water

*Source A: *Reinforced Plastics: Properties and Applications* (1991); Source B: *Engineers Guide to Composite Materials* (1987); Source C: *ASM Engineered Materials Handbook, Vol. 2* (1989).

14.7 Data for a metal-matrix composite (MMC) — Ai/SiC$_p$

An astronomical telescope is a precision device; mechanical stability is of the essence. On earth, damaging distortions are caused by the earth's gravitational field — that was the subject of Case Study 6.2. If, like the Hubble telescope, it is to operate in space, gravity ceases to be a problem. Stability, though, is at an even greater premium; adjustments, in space, are difficult. The problem now is thermal and vibrational distortion. These were analysed in Case Study 6.19; they are minimized by high thermal conductivity λ and low expansion coefficient α, high modulus E and low density ρ.

One of the candidates for the precision device was aluminium. If aluminium is good, a metal-matrix composite made of aluminium reinforced with particles of silicon carbide (Al/SiC$_p$) is probably better; certainly, it is stiffer and it expands less. This composite is a new material, still under development, and for that reason it does not appear on the present generation of Materials Selection Charts. Its potential can be assessed by calculating the values of the two performance indices which appear in Case Study 6.19, and to do that we need data for the four properties listed above: λ, α, E and ρ. There are no accepted standards or specifications for metal matrix composites. Finding data for them could be a problem.

Handbooks published before 1986 will not help much here — most of the development has occurred since then. We turn to the *Engineers Guide to Composite Materials* (1987) and find limited data, part of it derived from a material of one producer, the rest from that of another (Table 14.6, Source A), leaving us uneasy about consistency.

This is a bit thin for something to be shot into space. Minor miscalculations here become major embarrassments, as the history of the Hubble demonstrates. Something better is needed. The resource to tap next is that of the producers' data sheets. BP International manufactures a range of aluminium–SiC composites and provides a standard booklet of properties to potential users. Data for 6061-20%SiC (the same alloy and reinforcement loading as before), abstracted from

Table 14.6 Data for 6061 aluminium with 20% particulate SiC

Properties	Source A*	Source B*	Source C*
Density (Mg/m^3)	2.91	2.9	2.9–2.95
Price ($/kg)	—	—	100–170
T-expansion (10^{-6} K^{-1})	14.4	13.5	12.4–13.5
T-conductivity (W/m K)	125	—	123–128
Specific heat (J/kg K)	800	—	800–840
Modulus (GPa)	121	125	121–125
Yield strength (MPa)	441	430	430–445
Ultimate strength (MPa)	593	610	590–610
Ductility (%)	4.5	5.0	4.0–6.0

*Source A: *Engineers Guide to Composite Materials* (1987) reporting data from DWA composites and Arco Chemical; Source B: BP Metal Composites Ltd, Technical Data Sheets for Metal Matrix Composites (1989); Source C: *CMS* database for metal matrix composites (1995)

the booklet, are listed under Source B in Table 14.6. The data from the two sources are remarkably consistent: density, modulus and strengths differ by less than 3%. But BP does not give a thermal conductivity; it will still be necessary to assume that it is the same as that of the Arco material.

Finally, a quick look at software. The CMS system contains records for a number of MMCs. That for an Al-20%SiC(p) material is listed under Source C. The ranges bracket the values of the other two sources, and there is an approximate price.

Postscript

Making this assumption, we can calculate values for the two 'precision instrument' performance indices of Case Study 6.19. As expected, they are both better than those for aluminium and its alloys, and in high-cost applications like a space telescope the temptation to exploit this improvement is strong.

And herein lies the difficulty in using 'new' materials: the documented properties, often, are very attractive, but others, not yet documented (corrosion behaviour; fracture toughness, fatigue strength) may catch you out. Risks exist. Accepting or rejecting them becomes an additional design decision.

Related case studies

Case Study 6.3: Mirrors for large telescopes
Case Study 6.20: Materials to minimize thermal distortion in precision devices

14.8 Data for a polymer-matrix composite — CFRP

If a design calls for a material which is light, stiff and strong (Case Studies 6.2, 6.3, 6.5 and 6.8), it is likely that carbon-fibre reinforced polymer (CFRP) will emerge as a candidate. Here we have a real problem: CFRP is made up of plies which can be laid-up in thousands of ways. It

Table 14.7 Data for 0/90/±45 carbon in epoxy

Properties	Source A*	Source B*	Source C*
Density (Mg/m^3)	1.54	1.55	1.55
Modulus (GPa)	65	72	60
Tensile strength (MPa)	503	550	700
Compressive strength (MPa)	503	400	—
T-expansion (10^{-6} K^{-1})	—	—	20
T-conductivity (W/m K)	—	8	5

*Source A: *ASM Engineers Guide to Composite Materials* (1987); Source B: *Engineered Materials Handbook, Vol. 1*, (1987); Source C: 'Reinforced Plastics, Properties and Applications' 1 (1991).

is not one material, but many. A report of data for CFRP which does not also report the lay-up is meaningless.

There are, though, some standard lay-ups, and for these, average properties can be measured. There is, in particular, the 'isotropic' lay-up, with equal number of plies with fibres in the 0, 90 and ±45 orientations. Let us suppose, by way of example, that this is what we want.

The best starting point for composite data is the *ASM Engineers Guide to Composite Materials* (Source A, Table 14.7). Comparing these data with those from Sources B and C (identified in the table) illustrates the problem. All are in the 'isotropic' lay up, but values differ by up to 50% — a more detailed analysis of this variability, documented in Source A, shows differences of up to 50% in modulus, 100% in strength.

Computer databases reveal the same problem. Here rescue, via material producers' data sheets, is not to hand: producers deliver epoxy and carbon fibres or prepreg — a premixed but uncured fibre-resin sheet; they do not supply finished laminates. We must accept the fact that published data are usually approximate.

There are two ways forward. The first is computational. Laminate theory allows the stiffness and strength of a given lay-up to be computed when the properties of fibres and matrix are known. Designers in large industries use laminate theory to decide on number and lay-up of plies, but few small industries have the resources to do this. The second is experimental: a trial lay-up is tested, measuring the responses which are critical to the design, and the lay-up is modified as necessary to bring these within acceptable limits.

Postscript

Conventional sources, this time, let us down. It is, perhaps, a mistake to think of CFRP as a 'material' with unique properties. It has 'properties' only when shaped to a component, and they depend on both the material and the shape.

The information for CFRP, GFRP and KFRP provided by the data sources is a starting point only; it should never be used, unchecked, in a critical design.

Related case studies

Case Study 6.3: Mirrors for large telescopes
Case Study 6.4: Materials for table legs
Case Study 6.6: Materials for flywheels
Case Study 6.9: Materials for springs

14.9 Data for a natural material — balsa wood

Woods are the oldest of structural materials. Surely, with their long history, they must be well characterized? They are. But the data are not so easy to find. Although woods are the world's principal material of building (even today), ordinary data books do not list their properties. One has to consult specialized sources.

Take a specific task: that of locating data for balsa, a possible material for the wind-spars of man-powered planes (Section 10.3). Of the data sources for woods listed in the Further reading section Chapter 13; one is particularly comprehensive. It is the massive compilation of the US Department of Agriculture Forest Services (Source A); it lists densities, moduli, strengths, and thermal properties for many different species, including balsa. Some of the others give some data too, but one quickly discovers that they got it from Source A. The scientific literature, some of it reviewed in Source B, gives a second, independent, set of data. The two are compared in Table 14.8. Considering that balsa is a natural material, subject to natural variability, the agreement is not bad.

Can databases help? Surprisingly, there are many, although they differ greatly. Print-out for balsa, from the *CMS*, is shown in Figure 14.6. Examining all this, we learn the following. First, woods are anisotropic: properties along the grain differ from those across it. Balsa is particularly anisotropic: the differences are as great as a factor of 40. Second, woods are variable: nature does not apply tight specifications. This initial variability is made worse by a dependence of the properties on humidity and on age, although these last two effects are documented and their effect can be estimated. Woods, generally, are used in low-performance applications (building, packaging) where safety margins are large; then a little uncertainty in properties does not matter. But there are other examples: balsa and spruce in aircraft; ash in automobile frames, vaulting poles, oars, yew in bows, hickory in skis, and so on. Then attention to these details is important.

Postscript

All natural materials have the difficulties encountered with balsa: anisotropy, variability, sensitivity to environment, and ageing. This is the main reason they are less-used now than in the past, despite

Table 14.8 Data for balsa wood

Properties	Source A*	Source B*
Density (Mg/m^3)	0.17	0.2
Modulus $\parallel$ (GPa)	3.8	6.3
$\perp$ (GPa)	0.1	0.1
Tensile strength $\parallel$ (MPa)	19.3	23
$\perp$ (MPa)	—	—
Compressive strength, $\parallel$ (MPa)	12	18
$\perp$ (MPa)	—	1
Fracture toughness $\parallel$ ($MPa^{1/2}$)	—	0.1
$\perp$ ($MPa^{1/2}$)	—	1.5

*Source A: *Wood Handbook, US Forest Service Handbook No. 72* (1974); Source B: Gibson and Ashby *Cellular Solids* (1997). The symbol $\parallel$ means parallel to the grain; $\perp$ means perpendicular.

Name	Ochroma spp. (MD), parallel to grain			
Common Name	Balsa (MD)L			

General Properties

Density	0.17	—	0.21	Mg/m^3
Diff. Shrinkage (Rad.)	0.05	—	0.06	% per % MC
Diff. Shrinkage (Tan.)	0.07	—	0.09	% per % MC
Rad. Shrinkage (green to oven-dry)	3.2	—	7	%
Tan. Shrinkage (green to oven-dry)	3.5	—	5.3	%
Vol. Shrinkage (green to oven-dry)	6	—	9	%

Mechanical Properties

Brinell Hardness	10.2	—	10.4	MPa
Bulk Modulus	0.08	—	0.1	GPa
Compressive Strength	8.5	—	12.5	MPa
Ductility	0.0103	—	0.0126	
Elastic Limit	11.4	—	14	MPa
Endurance Limit	5.4	—	6.6	MPa
Flexural Modulus	3.4	—	4.2	GPa
Fracture Toughness	0.5	—	0.6	$MPa.m^{1/2}$
Hardness	3.5	—	4.3	MPa
Impact Bending Strength	11.9	—	14.6	
Janka Hardness	0.35	—	0.43	kN
Loss Coefficient	0.0122	—	0.015	
Modulus of Rupture	18	—	22	MPa
Poisson's Ratio	0.35	—	0.4	
Shear Strength	3.2	—	3.9	MPa
Shear Modulus	0.31	—	0.38	GPa
Tensile Strength	16	—	25	MPa
Work to Maximum Strength	13	—	15.9	kJ/m^3
Young's Modulus	4.2	—	5.2	GPa

Thermal Properties

Glass Temperature	350	—	375	K
Maximum Service Temperature	390	—	410	K
Minimum Service Temperature	200	—	250	K
Specific Heat	1.66E+3	—	1.71E+3	J/kg.K
Thermal Conductivity	0.09	—	0.12	W/m.K
Thermal Expansion	2	—	11	$10^{-6}/K$

Electrical Properties

Breakdown Potential	4.85	—	4.9	10^6 V/m
Dielectric Constant	2.45	—	3	
Resistivity	6.00E+13	—	2.00E+14	10^{-8} ohm.m
Power Factor	0.021	—	0.026	

Typical uses   Cores for sandwich structures; model building; flotation; insulation; packaging.

References   Datasheets: Baltek SA
Gibson, L.J. and Ashby, M.F. 'Cellular Solids, Structure and Properties', CUP, Cambridge (1997) US Forestry Commission Handbook 72, (1974).

Supplier   Baltek SA, 61 rue de la Fontaine, 75016, Paris, FRANCE; Diab-Barracuda Inc., 1100 Avenue S., Grande Prairie, Texas 75050, USA; Flexicore UK Ltd, Earls Colne Industrial Park, Earls Colne, Colchester, Essex CO6 2NS, UK;

Fig. 14.6 Part of the record of the *CMS* database for the properties of balsa wood, parallel to the grain. A second record (not shown) gives the properties in the perpendicular direction.

their sometimes remarkable properties (think of bamboo, bone, antler and shell), their low cost and their environmental friendliness.

Related case studies

Case Study 8.2: Spars for man-powered planes

14.10 Summary and conclusions

One day there may be universal accepted standards and designations for all materials but it is a very long way off. If you want data today, you have to know your way around the sources, and the quirks and eccentricities of the ways in which they work.

Metals, because they have dominated engineering design for so long, are well specified, coded and documented in hard-copy and computerized databases. When data for metals are needed, they can be found; this chapter gave two examples. Organizations such as the American Society for Metals (ASM), the British Institute of Materials (IM), the French Societé de Metallurgie, and other similar organizations publish handbooks and guides which document properties, forming-processes and suppliers in easily accessed form.

Polymers are newer. Individual manufacturers tend to be jealous of their products: they give them strange names and withhold their precise compositions. This is beginning to change. Joint databases, listed at the end of the previous chapter, pool product information; and others, independently produced, document an enormous range of polymer types. But there remain difficulties: no two polyethylenes, for instance, are quite the same. And the data are not comprehensive: important bits are missing. Filled polymers, like the glass-filled nylon of this chapter, are in much the same state.

For ceramics it is worse. Ceramics of one sort have a very long history: pottery, sanitary ware, furnace linings, are all used to bear loads, but with large safety factors — design data can be badly wrong without compromising structural integrity. The newer aluminas, silicon carbides and nitrides, zirconias and sialons are used under much harsher conditions; here good design data are essential. They are coming, but it is a slow process. For the moment one must accept that handbook values are approximate; data from the materials supplier are better.

Metal-matrix composites are newer still. In their use they replace simple metals, for which well-tried testing and documentation procedures exist. Because they are metals, their properties are measured and recorded in well-accepted ways. Lack of standards, inevitable at this stage, creates problems. Further into the future lie ceramic-matrix composites. They exist, but cannot yet be thought of as engineering materials.

For fibre-reinforced polymers, the picture is different. The difficulty is not lack of experience; it is the enormous spread of properties which can be accessed by varying the lay-up. Approximate data for uniaxial and quasi-isotropic composite are documented; any other lay-up requires the use of laminate theory to calculate stiffness, and more approximate methods to predict strength. For critical applications, component tests are essential.

A lot is known about natural materials — wood, stone, bone — because they have been used for so long. Many of these uses are undemanding, with large safety margins, so much of the knowledge is undocumented. Their properties are variable, and depend also on environment and age, for which allowance must be made. Despite this, they remain attractive, not least because they are environmentally friendly (see Chapter 16).

So, in using data sources, it is sensible to be circumspect: the words in one context mean one thing, in another, another. Look for completeness, consistency, and documentation. Anticipate that newer materials cannot be subject to the standards which apply to the older ones. Turn to a supplier for data when you know what you want. And be prepared, if absolutely necessary, to test the stuff yourself.

14.11 Further reading

All the sources referenced in this chapter are detailed in the Appendix to Chapter 13, to which the reader is referred.

Chapter 15

Materials, aesthetics and industrial design

15.1 Introduction and synopsis

Good design works. Excellent design also gives pleasure.

Pleasure derives from form, colour, texture, feel, and the associations that these invoke. Pleasing design says something about itself; generally speaking, honest statements are more satisfying than deception, although eccentric or humorous designs can be appealing too.

Materials play a central role in this. A major reason for introducing new materials is the greater freedom of design that they allow. Metals, in the past century, allowed structures which could not have been built before: cast iron, the Crystal Palace; wrought iron, the Eiffel Tower; drawn steel, the Golden Gate Bridge, all undeniably beautiful. Polymers lend themselves to bright colours, satisfying textures and great freedom of form; they have opened new styles of design, of which some of the best examples are found in the household appliance sector: kitchen equipment, radio and CD-players, hair dryers, telephones and vacuum cleaners make extensive and imaginative use of materials to allow styling, weight, feel and form which give pleasure.

Those who concern themselves with this aesthetic dimension of engineering are known, rather confusingly, as 'industrial designers'. This chapter introduces some of the ideas of industrial design, emphasizing the role of materials. It ends with two illustrative case studies. But first a word of caution.

Previous chapters have dealt with systematic ways of choosing material and processes. 'Systematic' means that if *you* do it and *I* do it we will get the same result, and that the result, next year, will be the same as it is today. Industrial design is not, in this sense, systematic. Success, here, involves sensitivity to fashion, custom and educational background, and is influenced (manipulated, even) by advertising and association. The views of this chapter are partly those of writers who seem to me to say sensible things, and partly my own. You may not agree with them, but if they make you think about designing to give pleasure, the chapter has done what it should.

15.2 Aesthetics and industrial design

We have discussed the mechanical design of a product. But what of its appearance, its feel, its balance, its shape? Is it pleasing to look at? To handle? What associations does it suggest? In short, what of its *aesthetics*?

There are many books on the subject of Industrial Design (see Further reading at the end of this chapter). You will find — it may surprise you — that they hardly mention the issues of functionality and efficiency that have concerned us so far. They focus instead on qualities that cannot be measured: form, texture, proportion and style; and on subtler things: creative vision, historic perspective, honesty to the qualities of materials.

There is a view — one held by engineers as different as Brunel and Barnes-Wallis — that a design which is functional is automatically beautiful. When a thing is well made and well suited to its purpose, it is also pleasing to the eye. Its proponents cite the undeniable appeal of a beautiful bridge or of a modern aircraft. The craftsman Eric Gill (noted — among other things — for the elegant typefaces he designed) expresses it on a higher plane, saying: 'Look after goodness and truth in design and beauty will care of herself.' But there also exists a different and widely held view that design is an art, or if not that, then a craft with its basis in art, not in engineering. Its supporters — and they have included many distinguished designers — argue that the practice of fine arts and drawing must form the basis of the training of designers. Only this can give an appreciation of form, colour, line and quality, and the sensitivity to the possibilities of their right relationship.

Both views are extreme. The first argument is the one most likely to appeal to the engineer: that a functionally efficient machine is, of itself, aesthetically satisfying; it is the basis of what is called a 'machine aesthetic'. But something is obviously missing. It is part of the purpose of the machine to be *operated*, and the design is incomplete if the satisfaction of the operator is ignored. The missing elements include the ergonomics — the man–machine interface — and they include the idea of visual enjoyment and aesthetic pleasure for its own sake. It is as if eating had been reduced to the intake of measured quantities of carbohydrate and protein, depriving it of all gastronomic pleasure.

Empty decoration, on the other hand, is equally unsatisfying. Styling can give pleasure, but the pleasure is diminished if the appearance of the product bears no relationship to its function. The pleasure is transitory; you quickly grow tired of it; it is like living on a diet of chocolate and puff-pastry. The outside of a product should reflect the purpose and function of what is inside. Successful industrial design tells you what the product is and how to use it, *and* it gives pleasure.

So what is excellent design? It is the imaginative attempt to solve the problem in all its aspects: the use to which the article will be put, its proper working, the suitability of the materials of which it is made, its method of production, the quality of the workmanship, how it will be sold and packaged and serviced, and — by no means least important — the pleasure it will give the user. It seldom costs more to use a good shape than a bad one, good texture instead of bad.

But how are we to decide what is 'excellent'? That requires the development of an aesthetic sense. There are, in any country, exhibitions of industrial design (Table 15.1). Some are permanent, illustrating the way in which products have evolved; others are brought together to display current products. Visit these; examine the designs; ask yourself why they have survived or evolved or developed, and observe how the use of new materials has enabled their evolution. Browse through the books on industrial design listed at the end of this chapter. Don't expect them to explain how to design well — ideas of aesthetic design cannot be expressed as equations, or set down as procedures. Try, instead, to see how forms have evolved which are both functional and beautiful, that perform well and use materials in a way that exploits their natural texture and qualities, and that build, in a creative way, on the past. Case studies of the evolution of a product give a good way of developing an aesthetic sense (there are three later in this chapter). Examine, particularly, long-lived designs; things that are still pleasing long after they were made and have survived changes in taste and fashion: the Parthenon, St Paul's Cathedral, the Eiffel Tower, the Chippendale chair, the Victorian pillar-box, the XK 120 Jaguar, the 'tulip' telephone; the shape of a jug, a wine bottle, of candlesticks, certain cutlery — these have influence, and give satisfaction long after the designer has died.

Table 15.1 Design museums

Country	Design museum
Britain	• The Design Museum at Butlers Wharf, S. Thames Street, London SE1
	• The Victoria and Albert Museum (V & A) and the Science Museum, both in South Kensington, London
Czechoslovakia	• Musée National des Techniques, Prague
Denmark	• Musée des Arts Decoratifs, Copenhagen
France	• Musée Nationale des Techniques, CNAM, Paris
	• Musée des Arts Décoratifs, 107 Rue de Rivoli, Paris
	• Musée Natianal d'Art Modern, Centre George Pompidou, Paris
	• The Musée d'Orsay, Quai d'Orsay, Paris
	• Fondation National d'Art Contemporain, Ministère de Culture de la Francophonie
Germany	• Das Deutsche Museum, Munchen
	• Vitra Design Museum, Weil am Rhein
Holland	• The Stedljk Museeum, Amsterdam
	• The Booymans van Beumijen Museeum, Rotterdam
Switzerland	• Design Collection, Museum für Gestaltung, Zurich
USA	• The Smithsonian Museum, Washington DC
	• Industrial Design Collection of the Museum of Modern Art (MOMA), New York
	• The Museum of Fine Arts, Boston

One might consider the following approach to the design process. The quotation is from Misha Black, a Royal Designer for Industry (The Design Council, 1986):

We should approach each new problem on the basis of practicality — how can it most economically be made, how will it function most effectively, how can maintenance be simplified, how can the use of scarce materials be minimised? An absolute concern with practicalities will produce new formal solutions as technology constantly develops; when alternatives present themselves during the design process, the aesthetic sensitivity of the designer will determine his selective design.

So, when you look at an object (or, more important, when you design one) ask yourself the following questions. What is its NATURE — workaday or ornamental, useful or fun? What is its FUNCTION — does it achieve what it sets out to do? What is its STRUCTURE — well and appropriately made, too heavy or too light, made honestly or with unworthy tricks of concealment? How does it FEEL — are the weight and balance right? Its SCALE — is it the right size, and right for its use? What of its DECORATION and TEXTURE — is the colour attractive, the detailed design pleasing, harmonious, and giving delight? Is it GENUINE in its detail — is the decoration related to the structure and function, or is it coincidental or deceptive? Has the MATERIAL been used well, making the most of its properties and potential? And what ASSOCIATIONS does it suggest — speed? comfort? affluence? the past? the future? frugality? youth? culture and discernment? a bright awareness of latest trends? In short, does it have an appealing PERSONALITY? Why — if you do — do you like it? Why did you notice it at all?

We have seen, in the preceding chapters, that materials selection is intricately interwoven with the design process. A good design exploits the special properties of the materials used for each of its parts. Innovative design frequently does this in a new way, which either results in a cheaper product, or a product with better performance in some other sense (it is lighter, delivers more power, is easier to handle, more pleasing to look at and use). Much design is evolutionary, that is to say, the function and the basic scheme of achieving it does not change, but the details of shape, texture, and material do. Important markets are won in this way. The successful designer is, often, the one who exploits the potential of a new material more effectively than do his competitors.

Much can be learnt by examining the evolution of products in which the function has remained unchanged but the mode of achieving it has evolved with time. We now examine three such products: the telephone, the hand-held hair-dryer and the dinner fork. Lest they strike you as trivial, remember that all three are found in almost every household; their sales in Europe run to at least ten million units, or £200 m ($360 m), per year. You may disagree with the judgements I present here — aesthetics, as we have said, is a subjective and personal matter. But that means you have to decide what you like, or what your customer will like, and be able to express why. If you *do* disagree, see if you can formulate and express an alternative. But before that, something short and to the point: designing to please.

15.3 Why tolerate ugliness? The bar code

Few things are more functional, more information-intensive, than the bar code (Figure 15.1). And few are uglier. Their ugliness causes designers of book jackets, of wine labels, of food packages — of almost everything — to make them small and hide them at the bottom, round the back. And even there they are ugly.

Is that necessary? Could they not give, in some small degree, pleasure? Bar codes are read by a horizontal sweep; no information is contained vertically. Those in Figure 15.1 come from a pharmaceutical product and from the end of a bobbin of thread. Why not, at least, acknow-ledge this?

One response is shown in Figure 15.2. These are designs from the Ecole Supèrieure des Arts Graphiques in Paris, commissioned by the US firm Intermec which markets the most widely used coding system. They succeed at two levels. They are novel — other bar codes are not like this — and because they are novel, they entertain, they turn dullness into interest, they please. And because they are to-be-seen, not to-be-hidden, the designer can make them bigger and display them prominently where they can be scanned easily.

And making this change has cost nothing at all*. It is no more expensive to print a bar code which appeals as an abstract design, or as a caricature, or has humour, or conveys visual information (the examples of Figure 15.2 do all these things) than it is to print an ugly one. So why not? Designing

Fig. 15.1 Bar codes. The first is from a pharmaceutical product, the second from a bobbin of thread.

---

* A disingenuous statement. It cost the design time.

Fig. 15.2 The same bar codes, redesigned.

for pleasure as well as functionality is a worthy goal. The case studies which follow illustrate successes and failures in this, and the way in which materials have contributed.

15.4 The evolution of the telephone

The function of the telephone and the manner of achieving it has hardly changed since the days of Alexander Graham Bell (1847–1916). It consists of a device for turning electrical signals into sound, one for turning sound into electrical signals, and a system for sending digital information to the exchange.

Figure 15.3 shows how telephones have evolved. Note, first, materials; they follow the evolutionary pattern of Figure 1.1. The telephone of 1900, shown at (a), was largely made of wood; only the parts that had to conduct electricity or respond to a magnetic field are metallic. In the tulip phone (b), standard from 1901 to 1925, metal has replaced wood: a cast iron base supports a pressed steel cover from which rises a column of iron or brass, supporting the mouthpiece. The receiver, made of turned brass, is long and slender in order to accommodate the soft iron magnet. The whole thing is metal except for the bakelite mouthpiece and the rim of the ear-piece, but even these are turned and threaded, an inheritance from metal technology.

From here on the transition to polymers begins, although it takes 50 years to complete. Phone (c) of 1928–1970 (an Ericsson design of extraordinarily long life) has, technically speaking, only two significant changes. First, it uses magnets with a higher remanence and coercive force, allowing the ear-piece to be made smaller. Second, the body is moulded from bakelite — a polymer — but still with a metal base screwed to it. The 'metal design' mentality persists. Screw threads are cut or moulded into the mouthpiece and ear-piece, screw fasteners are widely used, and the only other major change — the shape and structure of the body — is designed much as one would design a metal die casting. There is a reduction in weight and, presumably, a saving in cost of manufacture, but the unique properties of the new material have not been exploited.

The later 'phones of 1970–1975 (Figure 15.3(d)) show some advances in the way the materials are used. Instead of the numerous fasteners, the case (made of acrylic) is held to the base (still metal) by a smaller number of screws and by moulded protrusions which locate in slots in the base. The full exploitation of the potential of polymers is found only in the 'phones of 1982 and later, like

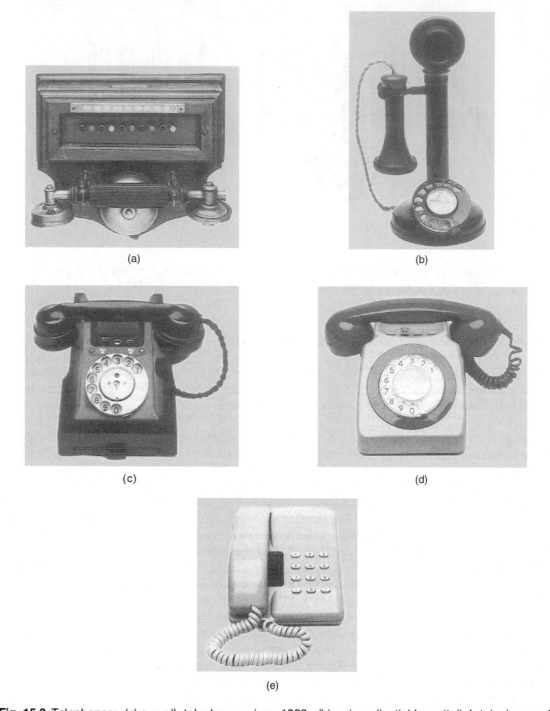

Fig. 15.3 Telephones: (a) a wall telephone, circa 1900; (b) a 'candlestick' or 'tulip' telephone of 1920–1928; (c) the standard Ericsson telephone of 1928–1970; (d) a telephone of the period 1970–1980; (e) the telephone of 1982 to 1992, making good use of polymers, but unappealing in its form, weight and proportion.

that shown at (e). Here snap fasteners and moulded clips are used throughout and there are very few fasteners. Polymer properties are exploited in elastic hinges (replacing pivots) and in the bi-stable 'touch-sensitive' supports for the keys. Both the cover and the base are injection mouldings of ABS shaped to give good stiffness despite the low modulus of the polymer itself. The design has at last escaped from the 'metal technology' mind-set.

But at a different level, that of the aesthetics, it might be argued that the design has not improved. The early telephones (a) and (b) express their function well. The tulip 'phone, particularly, is pleasing to look at; there is no confusion about which bit you speak into and which you listen to; and the dial is well positioned and displayed; its one drawback is that two hands are needed to work it. Its successor, (c), overcomes this by combining mouth- and ear-piece in one, and it does so in a bold, sculpted design: it sits solidly on the desk, has pleasing angular lines and suggests — or did in its day — the power and success of technology. Both it and its 'tulip' predecessor had long lives; they influenced the designs which followed; and they are both sought after and reproduced today, some 60 years later. Those are the characteristics of a 'classic' design — one which successfully combines functionality with consumer satisfaction.

The subsequent period might be called the decadent era of telephone design. The model shown in (d) uses polymers more effectively, but lacks vigour; it has historical perspective, but dilutes rather than innovates. The rounded edges and pastel colours must have appealed to the consumer of the 1970s, but it lacks the lasting quality of (c).

Still, it is much better than the last phone of all, (e). This design has none of the directness and elegance of the earlier ones. It ignores its past; historical perspective, visible in each of the earlier phones, is absent here. The keys are too small for ordinary fingers and it is so light that it slides away from you when they are pressed. Its gooey, lava-like shape in no way suggests its function; nor does it suggest any other satisfying image. The phone works, but it provides very little of the further pleasure that is inherent in really good design.

15.5 The design of hair dryers

Electric hair dryers first became available about 1925. As with telephones, there has been very little change in the way their function is achieved: an electric motor drives a fan which propels air through heating elements whence it is directed by a nozzle onto the hair (Figure 15.4). But the materials of which hair dryers are made, and the consumer appeal which has been created by these materials, has evolved steadily.

Early hair dryers (Figure 15.4(a)) had a power of barely 100 watts. They made from pressed steel or zinc die castings, and they were bulky and heavy. Their engineering was dominated by the 'metal mentality': parts which could be easily cast or machined were held together by numerous fasteners. Metals conduct both electricity and heat, so internal insulation was necessary to prevent the dryee getting electrocuted or roasted. This, together with inefficient motors and fans, made for a bulky product, the casing of which, typically, was made up of five or more parts held together by numerous fasteners (Table 15.2).

The emergence of polymers led to hair dryers which at first used bakelite, then other polymeric materials, for the casing and handle (Figure 15.4(b), (c), (d) and (e)). The early versions are plastic imitations of their metal counterparts; the bakelite model shown at (b) has the same shape, the same number of parts, and even more fasteners than the metal one shown at (a). Polymers were at first attractive because of the freedom of decorative moulding they allowed. Dryers (c) and (d) have lost some of the machine-tool look of (a); they were aimed at a fashion-conscious public;

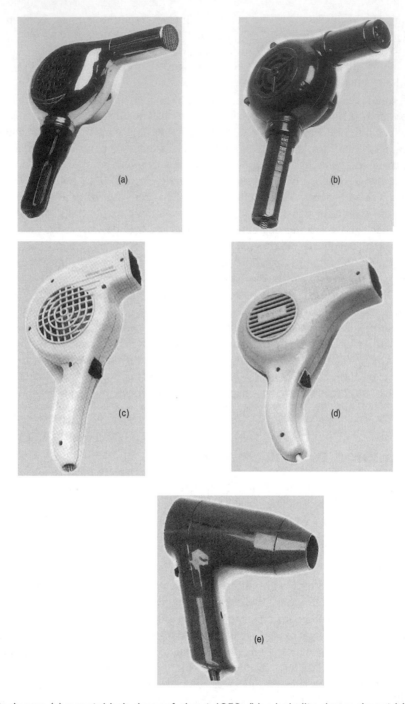

Fig. 15.4 Hair dryers: (a) a metal hair dryer of about 1950; (b) a bakelite dryer, almost identical in form to (a); (c) a plastic dryer of 1960, still influenced by 'metal' thinking, but with attractive moulding; (d) a dryer of 1965 — it has fewer fasteners than (c), but is undistinguished in design; (e) a hair dryer of 1986, exploiting fully and effectively the properties of polymers, and with a racy, youthful look. Their characteristics are given in Table 15.2.

Table 15.2 Characteristics of hair dryers and their casings

Model and date	Power (W)	Weight (kg)	Parts	Fasteners
Schott, 1940	300	1.0	5	7
Ormond, 1950	500	0.85	5	7
Morphy–Richards, 1960	400	0.82	3	6
Pifco, 1965	300	0.80	3	4
Braun, 1986	1200	0.27	3	1

they are boudoir-compatible. But their designers did not appreciate fully the advantages which could be gained from the use of the polymer: brighter colours, more complex mouldings which interlock, snap-fasteners for easy assembly and the like. There were some gains: the unit was a little lighter, and (because the thermal conductivity of polymers is low) it didn't get quite so hot. But if the fan stalled, the softening point of the polymer was quickly exceeded; most old hair dryers that survive today from this era are badly distorted by heat. Nonetheless, more efficient motors and better thermal design slowly pushed the power up and the weight down (Table 15.2 again).

The pace of change has been faster in the most recent decade than ever before. The modern hair dryer (Figure 15.4(e)), cheaper than any of its predecessors, sells for around £10 ($18) and delivers up to 1500 watts of heat. This is an enormous increase over the earlier designs, and from a unit which is smaller and lighter. This has been achieved in a number of ways, most of them relating to new materials. The fan is axial, not centrifugal. The motor is much smaller, and uses ceramic magnets and a ceramic frame to give a high power density. Sensors detect overheating and shut the unit down when necessary. The higher velocity of air-flow allows a heater of higher power density and reduces the need for insulation between the heating element and the casing. This casing is now designed in a way that exploits fully the attributes of poly-mers: it is moulded in two parts, with only one fastener. An adjustable nozzle can be removed by twisting it off; it is a snap-fit, exploiting the high strength/modulus ratio of plastics. The whole thing is youthfully attractive in appearance, light and extremely efficient. Any company left producing pressed-metal hair dryers when a unit like this becomes available finds that its market has disappeared.

15.6 The design of forks

The term 'cutlery' derives from the Latin cutelus, a knife; the word 'fork' from furka, a hay-fork. The cutlery industry, with a recorded history which dates back to the 12th century, was originally concerned only with the making of knives. Forks as eating-irons came later: they first appear in the 14th century, a gesture to improved table manners. Table manners in the middle ages, it must be said, lacked finesse. Jean Suplice, writing in 1480, councils that it is unseemly to grab your food with both hands at once, and that one should not scratch oneself at meals and then put ones fingers into the communal bowl. Erasmus, in Britain a little later (1530), remarks that it is not good manners to wipe your hands on your jacket after eating. But almost anything else went. Sophistication in eating, as in so many other things, seems to be an Italian import. Thomas Coryat, returning to Britain from a visit to Rome in 1611 reported that it had become customary for the Italian 'to use both a knife and a little fork, because he could by no means endure to have his dish touched with fingers, seeing that all men's fingers are not alike clean'. The British saw the point.

The forks of the 15th century had two long, straight prongs (Figure 15.5 top); it looks like a dagger, and was probably used like one. It evolved slowly towards the decorous, elegant and yet functional object it is today (Figure 15.5, centre). Or perhaps one should say: was, yesterday. Not all contemporary forks function well, nor are they all elegant — but we are getting ahead of ourselves.

The function of a fork is to transfix bite-sized morsels and transport them to the mouth. In use it is loaded in bending, and must be designed to stand this without flexing too much or collapsing completely — everyone knows the cheap cafe fork which succumbs, bending if metal, breaking if plastic, when used. And there are other design requirements. Food should not slip off the prongs and down the shirt-front on the way to the mouth; long slender prongs are better here than short wedge-shaped ones. The business-end of the instrument should enter and leave the mouth without causing injury; gentle curvature helps here. The tail should be shaped in such a way that it does not hurt the palm when the fork is used. The balance should be right; a fork which tips backwards when picked up is poorly designed. And its form, finish, and decoration should be such as to give pleasure.

With these criteria in mind, re-examine the forks of Figure 15.5. The one in the middle is a classic design known as Old English. The four prongs are long, slender, rounded on the shaft and well finished at the root. The tip of the handle is smoothly rounded with a gentle upward curve which fits well in the palm of the hand. It balances when picked up at the natural point — the high point of the neck. The form is exceptionally pleasing; it flows, and looks every inch what it is: a

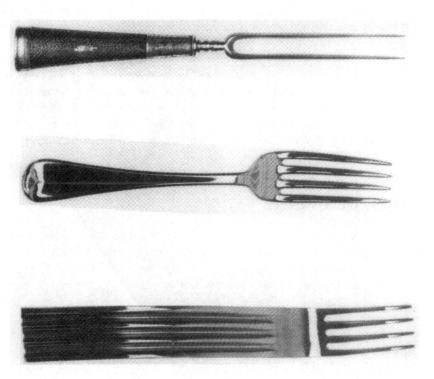

Fig. 15.5 Top: 15th century fork, more a weapon than a domestic object, but the progenitor of the elegant forks of later centuries. Middle: Old English, a classic design. Bottom: a fork of post-modernist design.

shape which has evolved to meet a human need effectively, gracefully and without pretence. Such a shape has little need of decoration, and there is little here: only the discreet double crescent, or 'rattail' as it is called, at the end of the handle.

Contrast this with the fork at the bottom, a product of the post-modernist movement. The shape is certainly striking: it has the shear linearity loved by modernists (think of modern office blocks). The longitudinal channels suggest a machine element, and by implication, mechanical efficiency. But where modernist design emphasized function, post-modernism takes modernist (and other) forms and uses them in ways which, sometimes, do not function well. This is an example. The prongs are adequate enough, but the handle is too long and too wide and because of this, it balances at completely the wrong point. The end of the handle has sharp corners which dig into the palm. The harsh form and linear decoration are better suited to the office than the dinner table, though poor balance and awkward corners are a drawback there too. Here is an example of function sacrificed to style.

15.7 Summary and conclusions

Competitive design requires the innovative use of new materials and the intelligent exploitation of their special properties. The case studies illustrate how one generation of materials replaces another, with the most successful designs exploiting the special properties of new materials. We live in an age in which polymers are replacing metals in many applications. The case studies illustrate how this can allow an enormous saving in the number of components, the use of elastic design in place of kinematic design for hinges and pivots and the use of moulded snap-fasteners to replace older screws and rivets, simplifying assembly. The successful designer has escaped from the mentality associated with the previous generation of materials, and has exploited the special properties and design freedom of the new ones. It will not end there. Novel composites now drive change in the way that polymers did in the 1980s. Ceramics, functionally graded materials, and novel manufacturing routes which allow greater freedom of shape and assembly are all just round the corner.

But today this is not nearly enough. Consumers look for more than functionality in the products they purchase. In the sophisticated market places of developed nations, the 'consumer durable' is a thing of the past. The challenge for the designer no longer lies in meeting the functional requirements alone, but in doing so in a way that also satisfies the aesthetic and emotional needs. The product must carry the image and convey the meaning that the consumer seeks: timeless elegance, perhaps; or racy newness. One Japanese manufacturer goes so far as to say: '*Desire* replaces *need* as the engine of design'.

Not everyone, perhaps, would wish to accept that. So we end with simpler words — the same ones with which we started. Good design works. Excellent design also gives pleasure.

15.8 Further reading

Bayley, S. (1979) *In Good Shape: Style in Industrial Products 1900–1960*, The Design Council, Haymarket, London.
Bayley, S. (1987) *The Conran Directory of Design*, Conran-Octopus, London.
Design Council (1966) *On Design*, published by The Design Council, Haymarket, London.
Flurscheim, C.H. (1983) *Industrial Design in Engineering*, The Design Council, Haymarket, London.
Lucie-Smith, E. (1983) (ed.) *The History of Industrial Design*, Oxford University Press, Oxford.

On telephones

Emmerson, A. (1986) *Old Telephones*, Shire Publications, Shire Album 161.
Myerson, J. and Katz, S., Conran Design Guide: *Home and Office*.

On forks

The book by Major Bailey combines knowledge with anecdote and superb illustration.
Bailey, C.T.P. (1927) *Knives and Forks*, The Medici Society, London and Boston.
Himsworth, J.B. (1953) *The Story of Cutlery*, Ernest Benn Ltd, London.

Chapter 16

Forces for change

16.1 Introduction and synopsis

Materials are evolving faster now than at any previous time in history. The speed of change was suggested by Figure 1.2: new polymers, elastomers, ceramics and composites are under development; and new processing routes offer cheaper, more reproducible production of conventional materials. These changes are driven by a number of forces. First, there is the *market-pull*: the demand from industry for materials which are lighter, stiffer, stronger, tougher, cheaper and more tolerant of extremes of temperature and environment. Then there is the *science-push*: the curiosity-driven researches of materials experts in the laboratories of universities, industries and government. Beyond this, there are *global issues*: the desire of society to minimize environmental damage, to save energy, and to reuse rather than discard. Finally, there is the driving force of what might be called *mega-projects*: historically, the Manhattan Project, the space-race and various defence programmes; today, one might think of alternative energy technology, the problems of maintaining an ageing infrastructure of drainage, roads, bridges and aircraft, and environmental problems associated with industrialization.

This chapter examines these forces for change and the directions in which they push materials and their deployment.

16.2 The market pull: economy versus performance

The end-users of materials are the manufacturing industries. They decide which material they will purchase, and adapt their designs to make best use of them. Their decisions are based on the nature of their products. Materials for large civil structures (which might weigh 10 000 tonnes or more) must be cheap; economy is the overriding consideration. By contrast, the cost of the materials for biomedical applications (an artificial heart valve, for instance) is almost irrelevant; performance, not economy, dictates the choice.

The market price of a product has several contributions. One is the cost of the materials of which the product is made, but there is also the cost of the research and development which went into its design, the cost of manufacture and marketing and the perceived value associated with fashion, scarcity, lack of competition and such like. When the material costs are a large part of the market value (50%, say) — that is, when the value added to the material is small — the manufacturer seeks to economize on materials to increase profit or market share. When, by contrast, material costs are a tiny fraction of the market value (1%, say), the manufacturer seeks the materials which will most improve the performance of the produce with little concern for their cost.

With this background, examine Figures 16.1 and 16.2. The vertical axis is the price per unit weight (£/kg or $/kg), applied to both materials and products: it gives a common measure by which

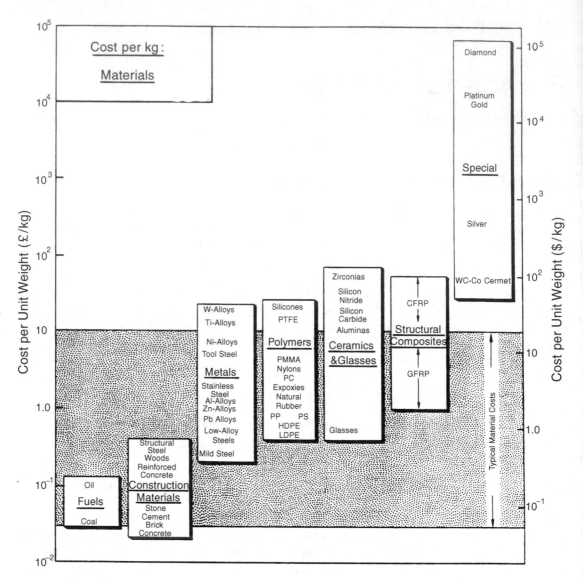

Fig. 16.1 The cost-per-unit-weight diagrams for materials. The shaded band spans the range in which lie the widely used commodity materials of manufacture and construction.

materials and products can be compared. The measure is a crude one but has the great merit that it is unambiguous, easily determined, and bears some relationship to value-added. A product with a price/kg which is twice that of its materials is material-intensive and is sensitive to material costs; one with a price/kg which is 100 times that of its materials is insensitive to material costs, and is probably performance-driven rather than cost-driven. On this scale the cost per kg of a contact lens differs from that of a glass bottle by a factor of 10^5, even though both are made of almost the same glass; the cost per kg of a heart valve differs from that of a plastic bottle by a similar factor, even though both are made of polyethylene. There is obviously something to be learned here.

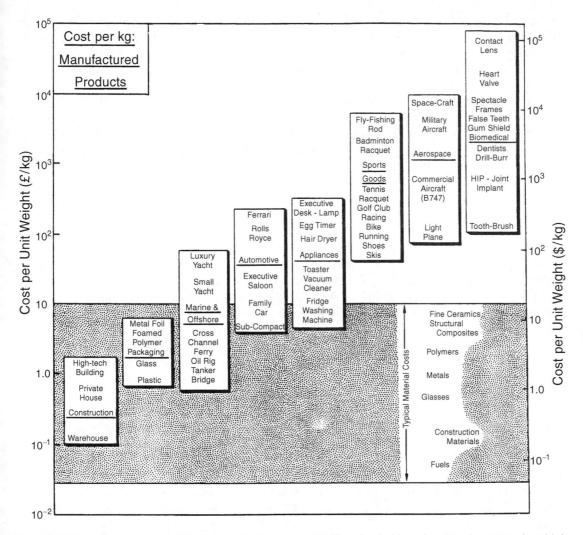

Fig. 16.2 The cost-per-unit-weight diagram for products. The shaded band spans the range in which lie most of the materials of which they are made. Products in the shaded band are material-intensive; those above it are not.

Look first at the price per unit weight of materials (Figure 16.1). The bulk, 'commodity' materials of construction and manufacture lie in the shaded band; they all cost between £0.05 and £10/kg, or $0.7 and $16/kg. Construction materials like brick, concrete, wood and structural steel, lie at the lower end; high-tech materials, like titanium alloys, lie at the upper. Polymers span a similar range: polyethylene at the bottom, polytetrafluorethylene (PTFE) near the top. Composites lie higher, with GFRP at the bottom and CFRP at the top of the range. Engineering ceramics, at present, lie higher still, though this will change as production increases. Only the low-volume 'exotic' materials lie much above the shaded band.

The price per kg of products (Figure 16.2) shows a different distribution. Eight market sectors are shown, covering much of the manufacturing industry. The shaded band on this figure spans the cost of commodity materials, exactly as in the previous figure. Sectors and their products within the shaded band shave the characteristic that material cost is a major fraction of product price: about 50% in civil construction, large marine structures and some consumer packaging, falling to perhaps 20% as the top of the band is approached (family car — around 25%). The value added in converting material to product in these sectors is relatively low, but the market volume is large. These constraints condition the choice of materials: they must meet modest performance requirements at the lowest possible cost. The associated market sectors generate a driving force for improved processing of conventional materials in order to reduce cost without loss of performance, or to increase reliability at no increase in cost. For these sectors, incremental improvements in well-tried materials are far more important than revolutionary research-findings. Slight improvements in steels, in precision manufacturing methods, or in lubrication technology are quickly assimilated and used.

The products in the upper half of the diagram are technically more sophisticated. The materials of which they are made account for less than 10% — sometimes less than 1% — of the price of the product. The value added to the material during manufacture is high. Product competitiveness is closely linked to material performance. Designers in these sectors have greater freedom in their choice of material and there is a readier acceptance of new materials with attractive property-profiles. The market-pull here is for performance, with cost as a secondary consideration. These smaller volume, higher value-added sectors drive the development of new or improved materials with enhanced performance: materials which are lighter, or stiffer, or stronger, or tougher, or expand less, or conduct better — or all of these at once.

The sectors have been ordered to form an ascending sequence, prompting the question: what does the horizontal axis measure? Many factors are involved here, one of which can be identified as 'information content'. The accumulated knowledge involved in the production of a contact lens or a heart valve is clearly greater than that in a beer-glass or a plastic bottle. The sectors on the left make few demands on the materials they employ; those on the right push materials to their limits, and at the same time demand the highest reliability. These features make them information-intensive. But there are also other factors: market size, competition (or lack of it), perceived value, fashion and taste, and so on. For this reason the diagram should not be over-interpreted: it is a help in structuring information, but it is not a quantitative tool.

The manufacturing industry, even in times of recession, has substantial resources; and it is in the interests of government to support their needs. The market pull is, ultimately, the strongest force for change.

16.3 The science-push: curiosity-driven research

Curiosity may kill cats, but it is the life-blood of innovative engineering. Technically advanced countries sustain the flow of new ideas by supporting research in three kinds of organization: universities, government laboratories and industrial research laboratories. Some of the scientists and engineers working in these institutions are encouraged to pursue ideas which may have no immediate economic objective, but which can evolve into the materials and manufacturing methods of future decades. Numerous now-commercial materials started in this way. Aluminium, in the time of Napoleon III, was a scientific wonder — he commissioned a set of aluminium spoons for which he paid more than those of solid silver. Aluminium was not, at that time, a commercial success; now it is. Titanium, more recently, has had a similar history. Amorphous (= non-crystalline) metals, now important in transformer technology and in recording-heads of tape decks, were, for years, of only academic interest. It seems improbable that superconductors or semiconductors would have been discovered in response to market forces alone; it took long-term curiosity-driven research to carry them to the point that they became commercially attractive. Polyethylene was discovered by chemists studying the effect of pressure on chemical reactions, not by the sales or marketing departments of multinational corporations. History is dotted with examples of materials and processes which have developed from the inquisitiveness of individuals.

What new ideas are churning in the minds of the materials scientists of today? There are many, some already on the verge of commercialization, others for which the potential is not yet clear. Some, at least, will provide opportunities for innovation; the best may create new markets.

Monolithic ceramics, now produced in commercial quantities, offer high hardness, chemical stability, wear resistance and resistance to extreme temperatures. Their use as substrates for microcircuits is established; their use in wear-resistant applications is growing, and their use in heat engines is being explored. The emphasis in the development of *composite materials* is shifting towards those which can support loads at higher temperatures. Metal-matrix composites (example: the aluminium containing particles or fibres of silicon-carbide of Section 14.7) and intermetallic-matrix composites (titanium-aluminide or molybdenum-disilicide containing silicon-carbide, for instance) can do this. So, potentially, can ceramic-matrix composites (alumina with silicon carbide fibres) though the extreme brittleness of these materials requires new design techniques. Metallic foams, up to 90% less dense than the parent metal, promise light, stiff sandwich structures competing with composites.

A number of new techniques of *surface engineering* allows the alloying, coating or heat treating of a thin surface layer of a component, modifying its properties to enhance its performance. They include: laser hardening, coatings of well-adhering polymers and ceramics, ion implantation, and even the deposition of ultra-hard carbon films with a structure and properties like those of diamond. New *bio-materials*, designed to be implanted in the human body, have structures onto which growing tissue will bond without rejection. New *polymers* which can be used at temperatures up to 350°C allow plastics to replace metals in even more applications — the inlet manifold of the automobile engine, for example. New *elastomers* are flexible but strong and tough; they allow better seals, elastic hinges, and resilient coatings. Techniques for producing *functionally-graded materials* can give tailored gradients of composition and structure through a component so that it could be corrosion resistant on the outer surface, tough in the middle and hard on the inner surface. *'Intelligent' materials* which can sense and report their condition (via embedded sensors) allow safety margins to be reduced. New *adhesives* could displace rivets and spot-welds; the glue-bonded automobile is a real possibility. And new techniques of *mathematical modelling* and *process control* allow much tighter control of composition and structure in manufacture, reducing cost and increasing reliability and safety.

All these and many more are in the pipeline. They have the potential to enable new design, or, more often, potential for the redesign of a product which already has a market, increasing its market share. Some are already commercial or near commercial; others may not become commercially viable for two decades. The designer must stay alert.

16.4 Materials and the environment: green design

Technical progress and environmental stewardship are not incompatible goals. History contains many examples of civilizations that have adopted environmentally conscious life-styles while making technological and sociological progress. But since the start of the industrial revolution, the acceleration of industrial development has overwhelmed the environment, with local and global consequences which cannot be ignored.

There is a growing pressure to reduce and reverse this environmental impact. It requires processes which are less toxic and products which are lighter, less energy-intensive and easier to recycle; and this must be achieved without compromising product quality. New technologies must (and can) be developed which allow an increase in production with diminished impact on the environment. Concern for the environment must be injected into the design process — brought 'behind the drawing-board', so to speak — taking a life-cycle view of the product which includes manufacture, distribution, use and final disposal.

Energy-content as a measure of environmental impact

All materials contain energy (Table 16.1). Energy is used to mine, refine, and shape metals; it is consumed in the firing of ceramics and cements; and it is intrinsic to oil-based polymers and elastomers. When you use a material, you are using energy, and energy carries with it an environmental penalty: CO_2, oxides of nitrogen, sulphur compounds, dust, waste heat. Energy is only one of the eco-influences of material production and use, but it is one which is easier to quantify than most others. We take it as an example.

Performance indices which include energy content are derived in the same way as those for weight or cost (Chapter 5). An example: the selection of a material for a beam which must meet a stiffness constraint, at minimum energy content. If the energy content per kilogram of a material is q (data in Table 16.1), that per unit volume is ρq where ρ is the density of the material. Repeating the derivations of Chapter 5 but with the objective of minimizing the energy content of the beam rather than its mass leads to performance equations and material indices which are simply those of Chapter 5 with ρ replaced by ρq. Thus the best materials to minimize energy content of a beam of specified stiffness and length are those with large values of the index

$$M = \frac{E^{1/2}}{\rho q} \qquad (16.1)$$

where E is the modulus of the material of the beam. The stiff tie of minimum energy content is best made of a material of high $E/\rho q$; the stiff plate, of a material with high $E^{1/3}/\rho q$.

Strength works the same way. The best choice of material for a beam of specified bending strength and minimum energy content is that with the highest value of

$$M = \frac{\sigma_f^{2/3}}{\rho q} \qquad (16.2)$$

where σ_f is the failure strength of the beam-material. The equivalent calculation for the tie gives the index $\sigma_f/\rho q$; that for a plate gives $\sigma_f^{1/2}/\rho q$. The calculation is easily adapted to include shape; then the indices of Table 8.1 apply, with ρ replaced by ρq.

Figures 16.3 and 16.4 are a pair of Materials Selection Charts for minimizing energy content per unit of function. The first show modulus, E, plotted against energy content, ρq; the design guidelines give the slopes for three of the commonest performance indices. The second shows strength σ_f (defined as in Chapter 4) against ρq; again, design guide-lines give the slopes.

The charts are used in exactly the same way as before. Energy consumption, and the potential for saving, are significant when large quantities of material are used, as they are in civil construction. The reader can quickly establish that the most energy-efficient beam, whether the design is based on stiffness or on strength, is that made of wood; steel, even with a large shape factor, consumes far more. Columns of brick or stone are more energy-efficient than concrete, though more labour intensive.

Table 16.1 Energy content and eco-indicator values for materials

Class	Material	Energy/wt q (MJ/kg)	Energy/vol $\rho q (GJ/m^3)$	Eco-indicator (millipoints/kg)
Metals	Titanium and alloys	555–565	2400–2880	80–100 (est.)
	Magnesium and alloys	410–420	717–756	20–30 (est.)
	Cast irons	60–260	468–1500	3–10
	Aluminium and alloys	290–305	754–884	10–18
	Stainless steels	110–120	825–972	16–18
	Copper and alloys	95–115	712–1035	60–85
	Zinc and alloys	67–73	348–525	60–85 (est.)
	Carbon steels	50–60	390–468	4.0–4.3
	Lead and alloys	28–32	300–360	60–85 (est.)
Polymers	Nylon 66	170–180	187–216	12–14
	Polypropylene	108–113	95–102	3.2–3.4
	H.D. polyethylene	103–120	97–116	2.8–3.0
	L.D. polyethylene	80–104	73–94	3.7–3.9
	Polystyrene	96–140	96–154	8.0–8.5
	PVC	67–92	87–147	4.2–4.3
	Synthetic rubber	120–140	108–126	13–15
	Natural rubber	5.5–6.5	5–6	14–16
Ceramics and glasses	Glasses	13–23	32–57	2.0–2.2
	Glass fibres	38–64	95–160	2.1–2.3
	Bone china	270	540–580	1.0–1.5 (est.)
	Bricks	3.4–6.0	6.8–12	0.5–1.0
	Refractories	1–50	3–100	10–20 (est.)
	Pottery	6–15	12–30	0.5–1.5
	Cement	4.5–8.0	9–18	1.0–2.0 (est.)
	Concrete	3–6	7–15	0.6–1.0 (est.)
	Stone	1.8–4.0	4–8.8	0.5–1.0
	Gravel	0.1	0.2–0.4	0.2–0.5
Composites (estimates)	GFRP	90–120	160–220	12–12 (est.)
	CFRP	130–300	230–540	20–25 (est.)
Other	Hard and soft woods	1.8–4.0	1.2–3.6	0.6–0.8
	Reinforced concrete	8–20	20–50	1.5–2.5 (est.)
	Crude oil	44	38–40	—
	Coal	29	27–30	—
	Natural gas		0.033–0.039	—

(1 MJ = 0.278 kWh = 9.48×10^2 Btu)

Most polymers are derived from oil. This leads to statements that they are energy-intensive, with implications for their future. The two charts show that, per unit of function in bending (the commonest mode of loading), most polymers are less energy-intensive than primary aluminium, magnesium or titanium, and that several are competitive with steel. Most of the energy consumed in the production of light alloys such as aluminium and magnesium is used to reduce the ore to the elemental metal, so that these materials, when recycled, are much less energy intensive. Efficient collection and recycling makes important contributions to energy saving.

Eco-indicators

Energy content, as said earlier, is only one measure of the environmental impact of material usage. In many circumstances it is not the important one; the emission of a toxic by-product, the difficulty

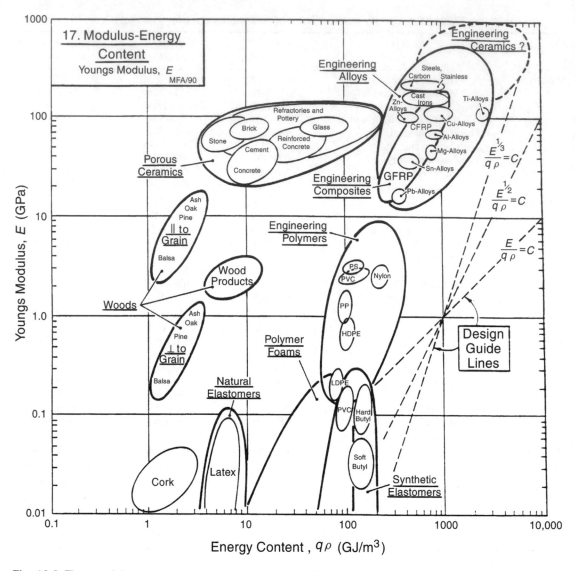

Fig. 16.3 The modulus versus energy-content chart, with guide-lines for selecting materials for stiff structures at minimum energy-content. It is an example of charts which allow selection to minimize environmental impact.

of recycling, or the resistance to biodegrading can be the real environmental threat. Table 16.2, as an example, lists the eco-profile associated with the production of 1 kg of aluminium from bauxite. Energy is a contribution, and linked to energy is the consumption of certain resources and the production of greenhouse gasses such as CO_2, CH_4 and CO. But there are other contributions too: acidification (SO_X, NO_X) particulates, solid waste etc. They are measured in strange units and the numbers in these units vary by enormous factors. Which are important and which are not? What is the designer to do with information like this? In this form it is not helpful.

What does the designer need? Ideally, a single number characterizing, in a properly balanced way, the eco-impact associated with the production of 1 kg of each material. We deal with cost in exactly

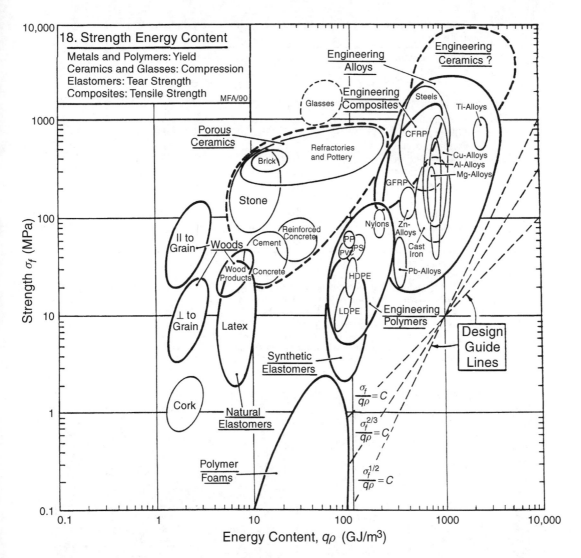

Fig. 16.4 The strength versus energy-content chart, with guide-lines for selecting materials for strong structures at minimum energy-content.

this way — it is, after all, an aggregated measure of the costs of resources, labour, capital and energy required to make 1 kg of material. Can a similar aggregate be constructed for eco-burden?

Efforts are underway in Europe to devise such a lumped measure, called the *eco-indicator* value, associated with the manufacture or processing of 1 kg of each material. Evaluating it involves three steps (Figure 16.5). First, values for the individual contributions of Table 16.2 are normalized to remove the strange units. To do this, the contribution is divided by the average contribution per (European) person per year. Thus the energy is normalized by the energy consumption per person per year (the total European energy consumption per year divided by the population). Second, the normalized contributions are weighted to take account of the severity of the problems they cause. Thus if acidification is a serious problem it is weighted heavily, and if summer smog is not a problem

Table 16.2 Eco-profile: production of 1 kg of aluminium from bauxite

Environmental load	Value	Units* (all per kg)
Energy	220	MJ
Resources	2.0	kg
Greenhouse	10.6	GWP
Ozone	0	ODP
Acidification	0.11	AP
Eutrophication	0.002	NP
Heavy metals	0	Pb equiv.
Carciogenicity	0	PAH equiv
Wintersmog	0.13	SO$_2$ equiv
Summersmog	0.003	POCP
Pesticides	0	kg
Solid	0.083	kg

*Units (all per kg):
   MJ = megajoules of energy
   GWP = global warming potential relative to 1 kg of CO_2
   ODP = ozone depletion potential relative to 1 kg of CFC-111
   AP = acidification potential relative to 1 kg of SO_2
   NP = nutrification potential relative to 1 kg of PO_4
   Pb equiv. = heavy metal toxicity relative to 1 kg of Pb ion
   POCP = photochemical oxidant formation relative to 1 kg of ethylene
   SO$_2$ equiv. = equivalent smog-potential relative to 1 kg of SO_2

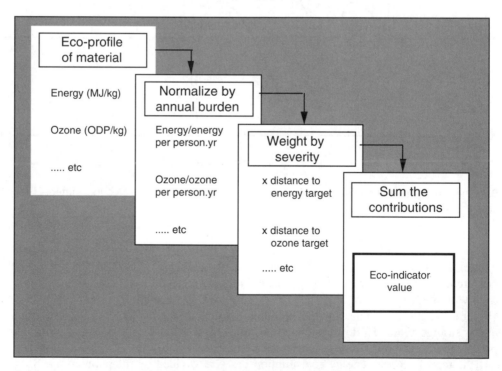

Fig. 16.5 The steps in deriving an eco-indicator value for a material or process. The raw data are first normalized by the average output per European person per year, then weighted by the severity of their effect, then summed. For details, see Goedkoop *et al.* (1995).

it is given a light weight. Finally, the weighted, normalized contributes are *summed* to give the eco-indicator value. There is a lot more to it than that, but this outline gives the essentials. The last column of Table 16.1 lists values based on weight-factors appropriate to a European nation. A high value means that the use of 1 kg of the material carries a high eco-burden; a low value, a low one.

These eco-indicators (symbol: I_e) are only an approximate measure of the eco-burden, but they are a useful one because they allow the initial election of material to minimize overall eco-impact per unit of function. The reasoning, as with energy in the last section, follows the method of Chapter 5. This leads to a set of indices which are simply those given above with ρq replaced by ρI_e.

Despite often-expressed reservations about the low resolution of eco-indicators, several large industries now use them to guide the selection of materials and processes. As the documentation of the eco profile of materials improves and broader agreement is reached on procedures for normalizing and weighting, it can be expected that their use will grow. The right way to exploit them is that described here, seeking materials which minimize the eco-impact, not per unit of weight, but per unit of function.

16.5 The pressure to recycle and reuse

There are many good reasons for not throwing things away. Discarded materials damage the environment; they are a form of pollution. Materials removed from the manufacturing cycle must be replaced by drawing on a natural resource. And materials contain energy, lost when they are dumped. Recycling is obviously desirable. But in a market economy it will happen only if there is profit to be made. What is needed to allow this?

Look, first, at where recycling works well and where it does not. *Primary scrap* — the turnings, trimmings and tailings which are a by-product of manufacture — has high value: it is virtually all recycled. That is because it is uncontaminated and because it is not dispersed. *Secondary scrap* has been through a consumption cycle — the paper of newsprint, the aluminium of a drink-can, the steel of an automobile — all are contaminated by other materials to which they are joined; by corrosion products; by ink and paint. And they are dispersed, some, like the tungsten in the filaments of lamp bulbs, very widely dispersed. In this form they are worth nothing or less-than-nothing, meaning that the cost of collection is greater than the value of the scrap itself. Yet this is by far the largest component of the material cycle. Newsprint and bottles are present examples: in a free market it is not economic to recycle either of these. Recycling *does* take place, but it relies on social conscience and good will, local subsidies and publicity. It is precarious for just those reasons.

Two things can change all that. Legislation (a departure from a true free market economy) is the obvious one. A deposit or 'dispersal cost', built into the price of each product, profoundly changes the economics and effectiveness of recycling; numerous societies have tried it, and it works. The other is design. The great obstacles in recycling are recognition, separation and decontamination; all are problems the designer can address. Finger-printing materials by colour or emblem or bar code allows recognition. Design for disassembly and the avoidance of mutually contaminating combinations allow economic separation. Clever chemistry (strippable paints; soluble glues) help with decontamination. And finally: design to by-pass the need to recycle: longer primary life; and more thought, at the initial design stage, of secondary usage.

16.6 Summary and conclusions

Powerful forces drive the development of new and improved materials, encourage substitution, and modify the way in which materials are produced and used. Market forces, historically the most influential, remain the strongest. The ingenuity of research scientists, too, drives change by revealing a remarkable spectrum of new materials with exciting possibilities, though the time it takes to develop and commercialize them is long: typically 15 years from laboratory to market.

Until recently, these were the evolutionary forces of materials technology. But man's damaging impact on the environment can no longer be ignored. Materials contribute to this damage at three points: in their production, in the use of products made from them, and in the disposal of these products. Concern about this, backed by legislation, already drives the development of new processing routes, the elimination of particularly damaging materials, and requirements for more effective recycling. The need, today, is to inject concern for environmental friendliness into the design process. Only the designer can do that.

16.7 Further reading

Boustead, I. and Hancock, G.F. (1979) *Handbook of Industrial Energy Analysis*, Wiley, New York.

Chapman, P.F. and Roberts, F. (1983) *Model Resources and Energy*. Butterworths, London.

Goedkoop, M.J., Demmers, M. and Collignon, M.X. (1995) *Eco-Indicator '95, Manual*, Pré Consultants, and the Netherland Agency for Energy and the Environment, Amersfort, ISBN 90-72130-80-4.

Kreigger, P.C. (1981) Energy analysis of materials and structures in the building industry, IEEE.

Navichandra, D. (1991) Design for environmentability, ASME Design Theory and Methodology Conference, American Society of Mechanical Engineers, Miami, Florida.

van Griethuysen, A.J. (1987) (ed.) *New Applications of Materials*. Scientific and Technical Publications, The Hague.

Appendix A

Useful solutions to standard problems

Introduction and synopsis

Modelling is a key part of design. In the early stage, approximate modelling establishes whether the concept will work at all, and identifies the combination of material properties which maximize performance. At the embodiment stage, more accurate modelling brackets values for the forces, the displacements, the velocities, the heat fluxes and the dimensions of the components. And in the final stage, modelling gives precise values for stresses, strains and failure probability in key components; power, speed, efficiency and so forth.

Many components with simple geometries and loads have been modelled already. Many more complex components can be modelled approximately by idealizing them as one of these. There is no need to reinvent the beam or the column or the pressure vessel; their behaviour under all common types of loading has already been analysed. The important thing is to know that the results exist and where to find them.

This appendix summarizes the results of modelling a number of standard problems. Their usefulness cannot be overstated. Many problems of conceptual design can be treated, with adequate precision, by patching together the solutions given here; and even the detailed analysis of non-critical components can often be tackled in the same way. Even when this approximate approach is not sufficiently accurate, the insight it gives is valuable.

The appendix contains 15 double page sections which list, with a short commentary, results for constitutive equations; for the loading of beams, columns and torsion bars; for contact stresses, cracks and other stress concentrations; for pressure vessels, vibrating beams and plates; and for the flow of heat and matter. They are drawn from numerous sources, listed under Further reading in Section A.16.

A.1 Constitutive equations for mechanical response

The behaviour of a component when it is loaded depends on the *mechanism* by which it deforms. A beam loaded in bending may deflect elastically; it may yield plastically; it may deform by creep; and it may fracture in a brittle or in a ductile way. The equation which describes the material response is known as a *constitutive equation*. Each mechanism is characterized by a different constitutive equation. The constitutive equation contains one or more than one *material property*: Young's modulus, E, and Poisson's ratio, ν, are the material properties which enter the constitutive equation for linear-elastic deformation; the yield strength, σ_y, is the material property which enters the constitutive equation for plastic flow; creep constants, $\dot{\varepsilon}_0$, σ_0 and n enter the equation for creep; the fracture toughness, K_{IC}, enters that for brittle fracture.

The common constitutive equations for mechanical deformation are listed on the facing page. In each case the equation for uniaxial loading by a tensile stress σ is given first; below it is the equation for multiaxial loading by principal stresses σ_1, σ_2 and σ_3, always chosen so that σ_1 is the most tensile and σ_3 the most compressive (or least tensile) stress. They are the basic equations which determine mechanical response.

Constitutive equations for mechanical response

Elastic deformation

Uniaxial	$\varepsilon_1 = \dfrac{\sigma_1}{E}$
General	$\varepsilon_1 = \dfrac{\sigma_1}{E} - \dfrac{\nu}{E}(\sigma_2 + \sigma_3)$

Plastic deformation

Uniaxial	$\sigma_1 \geq \sigma_y$
General $(\sigma_1 > \sigma_2 > \sigma_3)$	$\sigma_1 - \sigma_3 = \sigma_y$ (Tresca) $\sigma_e \geq \sigma_y$ (Von Mises) with $\sigma_e^2 = \dfrac{1}{2}\left[(\sigma_1 - \sigma_2)^2 + (\sigma_2 - \sigma_3)^2 + (\sigma_3 - \sigma_1)^2\right]$

Creep deformation

Uniaxial	$\dot{\varepsilon}_1 = \dot{\varepsilon}_0 \left(\dfrac{\sigma}{\sigma_0}\right)^n$
General	$\dot{\varepsilon}_1 = \dot{\varepsilon}_0 \left(\dfrac{\sigma^{n-1}}{\sigma_0^n}\right)\left(\sigma_1 - \dfrac{1}{2}(\sigma_2 + \sigma_3)\right)$

Fracture

Uniaxial	$\sigma_1 = \dfrac{CK_{Ic}}{\sqrt{\pi a}}$
General	$\sigma_1 = \dfrac{CK_{Ic}}{\sqrt{\pi a}}(\sigma_1 > \sigma_2 > \sigma_3)$

Material properties

E, ν Elastic constants	K_{Ic} Fracture toughness
σ_y Plastic yield strength	a Crack length
$\dot{\varepsilon}_0, \sigma_0, n$ Creep constants	$C \approx 1$ tension $C \approx 15$ compression

A.2 Moments of sections

A beam of uniform section, loaded in simple tension by a force F, carries a stress $\sigma = F/A$ where A is the area of the section. Its response is calculated from the appropriate constitutive equation. Here the important characteristic of the section is its area, A. For other modes of loading, higher moments of the area are involved. Those for various common sections are given on the facing page. They are defined as follows.

The second moment I measures the resistance of the section to bending about a horizontal axis (shown as a broken line). It is

$$I = \int_{\text{section}} y^2 \, b(y) \, dy$$

where y is measured vertically and $b(y)$ is the width of the section at y. The moment K measures the resistance of the section to twisting. It is equal to the polar moment J for circular sections, where

$$J = \int_{\text{section}} 2\pi r^3 \, dr$$

where r is measured radially from the centre of the circular section. For non-circular sections K is less than J.

The section modulus $Z = I/y_m$ (where y_m is the normal distance from the neutral axis of bending to the outer surface of the beam) measures the surface stress generated by a given bending moment, M:

$$\sigma = \frac{M \, y_m}{I} = \frac{M}{Z}$$

Finally, the moment H, defined by

$$H = \int_{\text{section}} y \, b(y) \, dy$$

measures the resistance of the beam to fully-plastic bending. The fully plastic moment for a beam in bending is

$$M_p = H\sigma_y$$

Thin or slender shapes may buckle before they yield or fracture. It is this which sets a practical limit to the thinness of tube walls and webs.

Moments of sections

Section Shape	$A(m^2)$	$I_{xx}(m^4)$	$K(m^4)$	$Z(m^3)$	$Q(m^3)$
	πr^2	$\dfrac{\pi}{4}r^4$	$\dfrac{\pi}{2}r^4$	$\dfrac{\pi}{4}r^3$	$\dfrac{\pi}{2}r^3$
	b^2	$\dfrac{b^4}{12}$	$0.14b^4$	$\dfrac{b^3}{6}$	$0.21b^3$
	πab	$\dfrac{\pi}{4}a^3b$	$\dfrac{\pi a^3 b^3}{(a^2+b^2)}$	$\dfrac{\pi}{4}a^2b$	$\dfrac{\pi a^2 b}{2}$ $(a<b)$
	bh	$\dfrac{bh^3}{12}$	$\dfrac{b^3h}{3}\left(1-0.58\dfrac{b}{h}\right)$ $(h>b)$	$\dfrac{bh^2}{6}$	$\dfrac{b^2h^2}{3h+1.8b}$ $(h>b)$
	$\dfrac{\sqrt{3}}{4}a^2$	$\dfrac{a^4}{32\sqrt{3}}$	$\dfrac{a^4\sqrt{3}}{80}$	$\dfrac{a^3}{32}$	$\dfrac{a^3}{20}$
	$\pi(r_o^2-r_i^2)$ $\approx 2\pi rt$	$\dfrac{\pi}{4}(r_o^4-r_i^4)$ $\approx \pi r^3 t$	$\dfrac{\pi}{2}(r_o^4-r_i^4)$ $\approx 2\pi r^3 t$	$\dfrac{\pi}{4r_o}(r_o^4-r_i^4)$ $\approx \pi r^2 t$	$\dfrac{\pi}{2r_o}(r_o^4-r_i^4)$ $\approx 2\pi r^2 t$
	$4bt$	$\dfrac{2}{3}b^3t$	$b^3t\left(1-\dfrac{t}{b}\right)^4$	$\dfrac{4}{3}b^2t$	$2b^2t\left(1-\dfrac{t}{b}\right)^2$
	$\pi(a+b)t$	$\dfrac{\pi}{4}a^3t\left(1+\dfrac{3b}{a}\right)$	$\dfrac{4\pi(ab)^{5/2}t}{(a^2+b^2)}$	$\dfrac{\pi a^2 t}{4}\left(1+\dfrac{3b}{a}\right)$	$2\pi t(a^3b)^{1/2}$ $(b>a)$
	$b(h_o-h_i)$ $\approx 2bt$	$\dfrac{b}{12}(h_o^3-h_i^3)$ $\approx \dfrac{1}{2}bth_o^2$	—	$\dfrac{b}{6h_o}(h_o^3-h_i^3)$ $\approx bth_o$	—
	$2t(h+b)$	$\dfrac{1}{6}h^3t\left(1+\dfrac{3b}{h}\right)$	$\approx \dfrac{2tb^2h^2}{h+b}$ [I] $\dfrac{2}{3}bt^3\left(1+\dfrac{4h}{b}\right)$ [□]	$\dfrac{h^2t}{3}\left(1+\dfrac{3b}{h}\right)$ [I] $\dfrac{2}{3}bt^2\left(1+\dfrac{4h}{b}\right)$ [□]	$2tbh$ [I] $\dfrac{2}{3}bt^2\left(1+\dfrac{4h}{b}\right)$ [□]
	$2t(h+b)$	$\dfrac{t}{6}(h^3+4bt^2)$	$\dfrac{t^3}{3}(8b+h)$ [H] $\dfrac{2}{3}ht^3\left(1+\dfrac{4b}{h}\right)$ [⊢]	$\dfrac{t}{3h}(h^3+4bt^2)$ [H] $\dfrac{2}{3}ht^2\left(1+\dfrac{4b}{h}\right)$ [⊢]	$\dfrac{t^2}{3}(8b+h)$ [H] $\dfrac{2}{3}ht^2\left(1+\dfrac{4b}{h}\right)$ [⊢]
	$t\lambda\left(1+\dfrac{\pi^2d^2}{4\lambda^2}\right)$	$\dfrac{t\lambda d^2}{8}$	—	$\dfrac{t\lambda d}{4}$	—

A.3 Elastic bending of beams

When a beam is loaded by a force F or moments M, the initially straight axis is deformed into a curve. If the beam is uniform in section and properties, long in relation to its depth and nowhere stressed beyond the elastic limit, the deflection δ, and the angle of rotation, θ, can be calculated using elastic beam theory (see Further reading in Section A.16). The basic differential equation describing the curvature of the beam at a point x along its length is

$$EI\frac{\mathrm{d}y^2}{\mathrm{d}x^2} = M$$

where y is the lateral deflection, and M is the bending moment at the point x on the beam. E is Young's modulus and I is the second moment of area (Section A.2). When M is constant this becomes

$$\frac{M}{I} = E\left(\frac{1}{R} - \frac{1}{R_0}\right)$$

where R_0 is the radius of curvature before applying the moment and R the radius after it is applied. Deflections δ and rotations θ are found by integrating these equations along the beam. Equations for the deflection, δ, and end slope, θ, of beams, for various common modes of loading are shown on the facing page.

The stiffness of the beam is defined by

$$S = \frac{F}{\delta} = \frac{C_1 EI}{\ell^3}$$

It depends on Young's modulus, E, for the material of the beam, on its length, ℓ, and on the second moment of its section, I. The end-slope of the beam, θ, is given by

$$\theta = \frac{F\ell^2}{C_2 EI}$$

Values of C_1 and C_2 are listed opposite.

Elastic bending of beams

	C_1	C_2
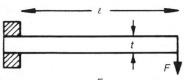	3	2
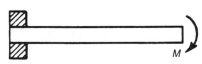	8	6
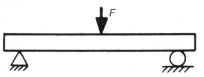	2	1
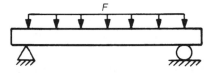	48	16
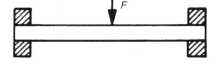	$\frac{384}{5}$	24
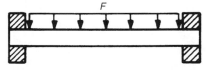	192	–
	384	–
	6	–
	–	4
	–	3

$$\delta = \frac{F\ell^3}{C_1 EI} = \frac{M\ell^2}{C_1 EI}$$

$$\theta = \frac{F\ell^2}{C_2 EI} = \frac{M\ell}{C_2 EI}$$

$E =$ Young's modulus (N/m^2)

$\delta =$ deflection (m)

$F =$ force (N)

$M =$ moment (Nm)

$\ell =$ length (m)

$b =$ width (m)

$t =$ depth (m)

$\theta =$ end slope $(-)$

$I =$ see Table 2 (m^4)

$y =$ distance from N.A. (m)

$R =$ radius of curvature (m)

$$\frac{\sigma}{y} = \frac{M}{I} = \frac{E}{R}$$

A.4 Failure of beams and panels

The longitudinal (or 'fibre') stress σ at a point y from the neutral axis of a uniform beam loaded elastically in bending by a moment M is

$$\frac{\sigma}{y} = \frac{M}{I} = E \left(\frac{1}{R} - \frac{1}{R_0} \right)$$

where I is the second moment of area (Section A.2), E is Young's modulus, R_0 is the radius of curvature before applying the moment and R is the radius after it is applied. The tensile stress in the outer fibre of such a beam is

$$\sigma = \frac{M\,y_m}{I} = \frac{M}{Z}$$

where y_m is the perpendicular distance from the neutral axis to the outer surface of the beam. If this stress reaches the yield strength σ_y of the material of the beam, small zones of plasticity appear at the surface (top diagram, facing page). The beam is no longer elastic, and, in this sense, has failed. If, instead, the maximum fibre stress reaches the brittle fracture strength, σ_f (the 'modulus of rupture', often shortened to MOR) of the material of the beam, a crack nucleates at the surface and propagates inwards (second diagram); in this case, the beam has certainly failed. A third criterion for failure is often important: that the plastic zones penetrate through the section of the beam, linking to form a plastic hinge (third diagram).

The failure moments and failure loads, for each of these three types of failure, and for each of several geometries of loading, are given on the diagram. The formulae labelled 'Onset' refer to the first two failure modes; those labelled 'Full plasticity' refer to the third. Two new functions of section shape are involved. Onset of failure involves the quantity Z; full plasticity involves the quantity H. Both are listed in the table of Section A.2, and defined in the text which accompanies it.

Failure of beams and panels

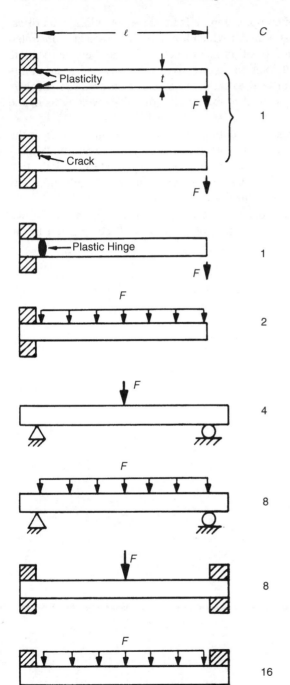

$$M_f = \left(\frac{I}{y_m}\right)\sigma^* \quad \text{(Onset)}$$

$$M_f = H\sigma_y \quad \text{(Full plasticity)}$$

$$F_f = C\left(\frac{I}{y_m}\right)\frac{\sigma^*}{\ell} \quad \text{(Onset)}$$

$$F_f = \frac{CH\sigma_y}{\ell} \quad \text{(Full plasticity)}$$

M_f = failure moment (Nm)

F_f = force at failure (N)

ℓ = length (m)

t = depth (m)

b = width (m)

I = see Table 2 (m^4)

$\dfrac{I}{y_m}$ = see Table 2 (m^3)

H = see Table 2 (m^3)

σ_y = yield strength (N/m^2)

σ_f = modulus of rupture (N/m^2)

$\sigma^* = \sigma_y$ (plastic material)

$\quad = \sigma_f$ (brittle material)

A.5 Buckling of columns and plates

If sufficiently slender, an elastic column, loaded in compression, fails by elastic buckling at a critical load, F_{crit}. This load is determined by the end constraints, of which four extreme cases are illustrated on the facing page: an end may be constrained in a position and direction; it may be free to rotate but not translate (or 'sway'); it may sway without rotation; and it may both sway and rotate. Pairs of these constraints applied to the ends of column lead to the five cases shown opposite. Each is characterized by a value of the constant n which is equal to the number of half-wavelengths of the buckled shape.

The addition of the bending moment M reduces the buckling load by the amount shown in the second box. A negative value of F_{crit} means that a tensile force is necessary to prevent buckling.

An elastic foundation is one that exerts a lateral restoring pressure, p, proportional to the deflection ($p = ky$ where k is the foundation stiffness per unit depth and y the local lateral deflection). Its effect is to increase F_{crit}, by the amount shown in the third box.

A thin-walled elastic tube will buckle inwards under an external pressure p', given in the last box. Here I refers to the second moment of area of a section of the tube wall cut parallel to the tube axis.

Buckling of columns and plates

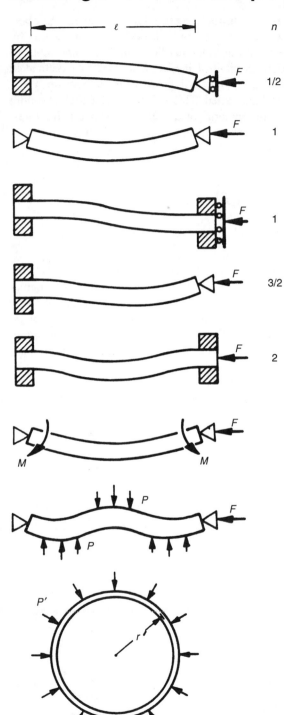

n

$1/2$

1

1

$3/2$

2

$$F_{CRIT} = \frac{n^2 \pi^2 EI}{\ell^2}$$

or

$$\frac{F_{CRIT}}{A} = \frac{n^2 \pi^2 E}{(\ell/r)^2}$$

F = force (N)

M = moment (Nm)

E = Young's modulus (N/m^2)

ℓ = length (m)

A = section area (m^2)

I = see Table 2 (m^4)

r = gyration rad. $\left(\dfrac{I}{A}\right)^{1/2}$ (m)

k = foundation stiffness (N/m^2)

n = half-wavelengths in buckled shape

p' = pressure (N/m^2)

$$F_{CRIT} = \frac{\pi^2 EI}{\ell^2} - \frac{M^2}{4EI}$$

$$F_{CRIT} = \frac{n^2 \pi^2 EI}{\ell^2} + \frac{k\ell^2}{n^2}$$

$$p'_{CRIT} = \frac{3EI}{(r')^3}$$

A.6 Torsion of shafts

A torque, T, applied to the ends of an isotropic bar of uniform section, and acting in the plane normal to the axis of the bar, produces an angle of twist θ. The twist is related to the torque by the first equation on the facing page, in which G is the shear modulus. For round bars and tubes of circular section, the factor K is equal to J, the polar moment of inertia of the section, defined in Section A.2. For any other section shape K is less than J. Values of K are given in Section A.2.

If the bar ceases to deform elastically, it is said to have failed. This will happen if the maximum surface stress exceeds either the yield strength σ_y of the material or the stress at which it fractures. For circular sections, the shear stress at any point a distance r from the axis of rotation is

$$\tau = \frac{Tr}{K} = \frac{G\theta r}{\ell}$$

The maximum shear stress, τ_{max}, and the maximum tensile stress, σ_{max}, are at the surface and have the values

$$\tau_{max} = \sigma_{max} = \frac{Td_0}{2K} = \frac{G\theta d_0}{2\ell}$$

If τ_{max} exceeds $\sigma_y/2$ (using a Tresca yield criterion), or if σ_{max} exceeds the MOR, the bar fails, as shown on the figure. The maximum surface stress for the solid ellipsoidal, square, rectangular and triangular sections is at the points on the surface closest to the centroid of the section (the mid-points of the longer sides). It can be estimated approximately by inscribing the largest circle which can be contained within the section and calculating the surface stress for a circular bar of that diameter. More complex section-shapes require special consideration, and, if thin, may additionally fail by buckling.

Helical springs are a special case of torsional deformation. The extension of a helical spring of n turns of radius R, under a force F, and the failure force F_{crit}, are given on the facing page.

Torsion of shafts

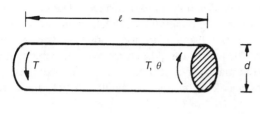

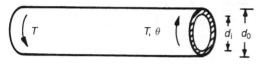

ETC

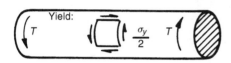

Elastic deflection

$$\theta = \frac{\ell T}{KG}$$

Failure

$$T_f = \frac{K\sigma_y}{d_0} \text{ (Onset of yield)}$$

$$T_f = \frac{2K\sigma_f}{d_0} \text{ (Brittle fracture)}$$

T = torque (Nm)

θ = angle of twist

G = shear modulus (N/m^2)

ℓ = length (m)

d = diameter (m)

K = see Table 1 (m^4)

σ_y = yield strength (N/m^2)

σ_f = modulus of rupture (N/m^2)

Spring deflection and failure

$$u = \frac{64FR^3 n}{Gd^4}$$

$$F_f = \frac{\pi}{32} \frac{d^3 \sigma_y}{R}$$

F = force (N)

u = deflection (m)

R = coil radius (m)

n = number of turns

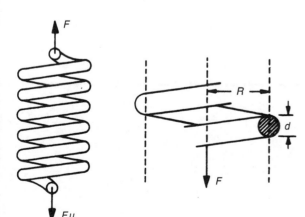

A.7 Static and spinning discs

A thin disc deflects when a pressure difference Δp is applied across its two surfaces. The deflection causes stresses to appear in the disc. The first box on the facing page gives deflection and maximum stress (important in predicting failure) when the edges of the disc are simply supported. The second gives the same quantities when the edges are clamped. The results for a thin horizontal disc deflecting under its own weight are found by replacing Dp by the mass-per-unit-area, ρgt, of the disc (here ρ is the density of the material of the disc and g is the acceleration due to gravity). Thick discs are more complicated; for those, see Further reading.

Spinning discs, rings and cylinders store kinetic energy. Centrifugal forces generate stresses in the disc. The two boxes list the kinetic energy and the maximum stress $\sigma_{\max}$ in discs and rings rotating at an angular velocity ω (radians/sec). The maximum rotation rate and energy are limited by the burst-strength of the disc. They are found by equating the maximum stress in the disc to the strength of the material.

Static and spinning discs

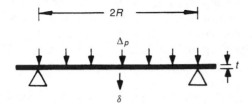

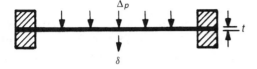

Simple

$$\delta = \frac{3}{4}(1 - \nu^2)\frac{\Delta pR^4}{Et^3}$$

$$\sigma_{max} = \frac{3}{8}(3 + \nu)\frac{\Delta pR^2}{t^2}$$

Clamped

$$\delta = \frac{3}{16}(1 - \nu^2)\frac{\Delta pR^4}{Et^3}$$

$$\sigma_{max} = \frac{3}{8}(1 + \nu)\frac{\Delta pR^2}{t^2}$$

δ = deflection (m)

E = Young's modulus (N/m)

Δp = pressure diff. (N/m)

ν = Poisson's ratio

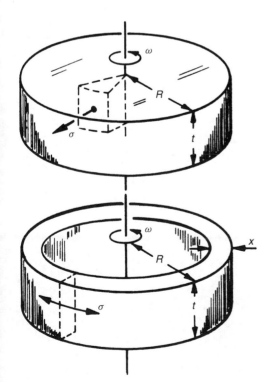

Disc

$$u = \frac{\pi}{4}\rho t\omega^2 R^4$$

$$\sigma_{max} = \frac{1}{8}(3 + \nu)\rho\omega^2 R^2$$

Ring

$$u = \pi\rho t\omega^2 R^3 x$$

$$\sigma_{max} = \rho R^2\omega^2$$

u = energy (J)

ω = angular vel. (rad/s)

ρ = density kg/m^3

A.8 Contact stresses

When surfaces are placed in contact they touch at one or a few discrete points. If the surfaces are loaded, the contacts flatten elastically and the contact areas grow until failure of some sort occurs: failure by crushing (caused by the compressive stress, σ_c), tensile fracture (caused by the tensile stress, σ_t) or yielding (caused by the shear stress σ_s). The boxes on the facing page summarize the important results for the radius, a, of the contact zone, the centre-to-centre displacement u and the peak values of σ_c, σ_t and σ_s.

The first box shows results for a sphere on a flat, when both have the same moduli and Poisson's ratio has the value 1/3. Results for the more general problem (the 'Hertzian Indentation' problem) are shown in the second box: two elastic spheres (radii R_1 and R_2, moduli and Poisson's ratios E_1, v_1 and E_2, v_2) are pressed together by a force F.

If the shear stress σ_s exceeds the shear yield strength $\sigma_y/2$, a plastic zone appears beneath the centre of the contact at a depth of about $a/2$ and spreads to form the fully-plastic field shown in the two lower figures. When this state is reached, the contact pressure is approximately 3 times the yield stress, as shown in the bottom box.

Contact stresses

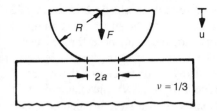

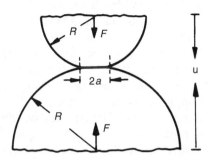

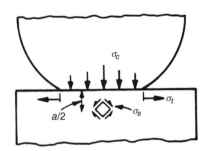

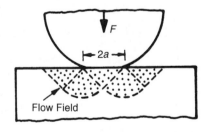

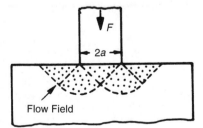

$$a = 0.7 \left(\frac{FR}{E} \right)^{1/3}$$

$$u = 1.0 \left(\frac{F^2}{E^2 R} \right)^{1/3}$$

$$\left.\right\} \quad v = \frac{1}{3}$$

$$a = \left(\frac{3}{4} \frac{F}{E^*} \frac{R_1 R_2}{(R_1 + R_2)} \right)^{1/3}$$

$$u = \left(\frac{9}{16} \frac{F^2}{(E^*)^2} \frac{(R_1 + R_2)}{R_1 R_2} \right)^{1/3}$$

$$(\sigma_c)_{\max} = \frac{3F}{2\pi a^2}$$

$$(\sigma_s)_{\max} = \frac{F}{2\pi a^2}$$

$$(\sigma_t)_{\max} = \frac{F}{6\pi a^2}$$

$R_1 R_2$ radii of spheres (m)

$E_1 E_2$ modulii of spheres (N/m²)

$v_1 v_2$ Poisson's ratios

F load (N)

a radius of contact (m)

u displacement (m)

σ stresses (N/m²)

σ_y yield stress (N/m²)

$$E^* \quad \left(\frac{1 - v_1^2}{E_1} + \frac{1 - v_2^2}{E_2} \right)^{-1}$$

$$\frac{F}{\pi a^2} = 3\sigma_y$$

A.9 Estimates for stress concentrations

Stresses and strains are concentrated at holes, slots or changes of section in elastic bodies. Plastic flow, fracture and fatigue cracking start at these places. The local stresses at the stress concentrations can be computed numerically, but this is often unnecessary. Instead, they can be estimated using the equation shown on the facing page.

The stress concentration caused by a change in section dies away at distances of the order of the characteristic dimension of the section-change (defined more fully below), an example of St Venant's principle at work. This means that the maximum local stresses in a structure can be found by determining the nominal stress distribution, neglecting local discontinuities (such as holes or grooves), and then multiplying the nominal stress by a stress concentration factor. Elastic stress concentration factors are given approximately by the equation. In it, σ_{nom} is defined as the load divided by the minimum cross-section of the part, r is the minimum radius of curvature of the stress-concentrating groove or hole, and c is the characteristic dimension: either the half-thickness of the remaining ligament, the half-length of a contained crack, the length of an edge-crack or the height of a shoulder, whichever is *least*. The drawings show examples of each such situation. The factor α is roughly 2 for tension, but is nearer 1/2 for torsion and bending. Though inexact, the equation is an adequate working approximation for many design problems.

The maximum stress is limited by plastic flow or fracture. When plastic flow starts, the strain concentration grows rapidly while the stress concentration remains constant. The strain concentration becomes the more important quantity, and may not die out rapidly with distance (St Venant's principle no longer applies).

Estimates for stress concentrations

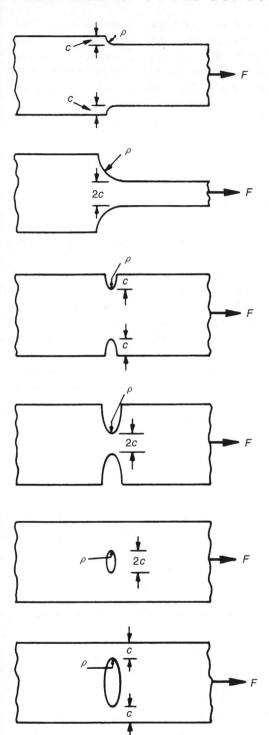

$$\frac{\sigma_{max}}{\sigma_{nom}} = 1 + \alpha \left(\frac{c}{\rho}\right)^{1/2}$$

F = force (N)

A_{min} = minimum section (m^2)

$\sigma_{nom} = F/A_{min}$ (N/m^2)

ρ = radius of curvature (m)

c = characteristic length (m)

$\alpha \approx 0.5$ (tension)

$\alpha \approx 2.0$ (torsion)

A.10 Sharp cracks

Sharp cracks (that is, stress concentrations with a tip radius of curvature of atomic dimensions) concentrate stress in an elastic body more acutely than rounded stress concentrations do. To a first approximation, the local stress falls off as $1/r^{1/2}$ with radial distance r from the crack tip. A tensile stress σ, applied normal to the plane of a crack of length $2a$ contained in an infinite plate (as in the top figure on the facing page) gives rise to a local stress field σ_ℓ which is tensile in the plane containing the crack and given by

$$\sigma_\ell = \frac{C\sigma\sqrt{\pi a}}{\sqrt{2\pi r}}$$

where r is measured from the crack tip in the plane $\theta = 0$, and C is a constant. The mode 1 *stress intensity factor* K_I, is defined as

$$K_I = C\sigma\sqrt{\pi a}$$

Values of the constant C for various modes of loading are given on the figure. (The stress σ for point loads and moments is given by the equations at the bottom.) The crack propagates when $K_I > K_{IC}$, the *fracture toughness*.

When the crack length is very small compared with all specimen dimensions and compared with the distance over which the applied stress varies, C is equal to 1 for a contained crack and 1.1 for an edge crack. As the crack extends in a uniformly loaded component, it interacts with the free surfaces, giving the correction factors shown opposite. If, in addition, the stress field is non-uniform (as it is in an elastically bent beam), C differs from 1; two examples are given on the figure. The factors, C, given here, are approximate only, good when the crack is short but not when the crack tips are very close to the boundaries of the sample. They are adequate for most design calculations. More accurate approximations, and other less common loading geometries can be found in the references listed in Further reading.

Sharp cracks

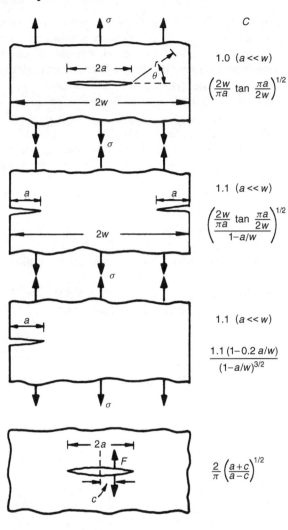

C

1.0 $(a \ll w)$

$$\left(\frac{2w}{\pi a} \tan \frac{\pi a}{2w}\right)^{1/2}$$

1.1 $(a \ll w)$

$$\left(\frac{\frac{2w}{\pi a} \tan \frac{\pi a}{2w}}{1 - a/w}\right)^{1/2}$$

1.1 $(a \ll w)$

$$\frac{1.1 \left(1 - 0.2\, a/w\right)}{(1 - a/w)^{3/2}}$$

$$\frac{2}{\pi} \left(\frac{a+c}{a-c}\right)^{1/2}$$

$$\frac{1.1 \left(1 - \frac{3}{2} \frac{a}{t}\right)}{(1 - a/t)^{3/2}}$$

$$\frac{1.1 \left(1 - \frac{3}{2} \frac{a}{t}\right)}{(1 - a/t)^{3/2}}$$

$$K_1 = C\sigma\sqrt{\pi a}$$

failure when

$$K_1 \geq K_{IC}$$

K_1 = stress intensity (N/m$^{3/2}$)

σ = remote stress (N/m^2)

F = load (N)

M = moment (Nm)

a = crack half-length

 = surface crack length (m)

w = half-width (centre) (m)

 = width (edge crack) (m)

b = sample depth (m)

t = beam thickness (m)

point load on crack face:

$$\sigma = \frac{F}{2ab}$$

moment on beam:

$$\sigma = \frac{6M}{bt}$$

3-point bending:

$$\sigma = \frac{3F\ell}{2bt}$$

A.11 Pressure vessels

Thin-walled pressure vessels are treated as membranes. The approximation is reasonable when $t < b/4$. The stresses in the wall are given on the facing page; they do not vary significantly with radial distance, r. Those in the plane tangent to the skin, σ_θ and σ_z for the cylinder and σ_θ and σ_ϕ for the sphere, are just equal to the internal pressure amplified by the ratio b/t or $b/2t$, depending on geometry. The radial stress σ_r is equal to the mean of the internal and external stress, $p/2$ in this case. The equations describe the stresses when an external pressure p_e is superimposed if p is replaced by $(p - p_e)$.

In thick-walled vessels, the stresses vary with radial distance r from the inner to the outer surfaces, and are greatest at the inner surface. The equations can be adapted for the case of both internal and external pressures by noting that when the internal and external pressures are equal, the state of stress in the wall is

$$\sigma_\theta = \sigma_r = -p \quad \text{(cylinder)}$$

or

$$\sigma_\theta = \sigma_\phi = \sigma_r = -p \quad \text{(sphere)}$$

allowing the term involving the external pressure to be evaluated. It is not valid to just replace p by $(p - p_e)$.

Pressure vessels fail by yielding when the Von Mises equivalent stress first exceed the yield strength, σ_y. They fail by fracture if the largest tensile stress exceeds the fracture stress σ_f, where

$$\sigma_f = \frac{CK_{IC}}{\sqrt{\pi a}}$$

and K_{IC} is the fracture toughness, a the half-crack length and C a constant given in Section A.10.

Pressure vessels

Thin Walled

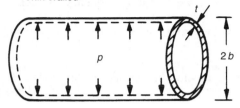

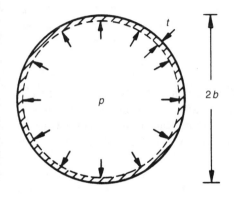

Thick Walled

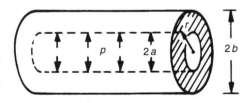

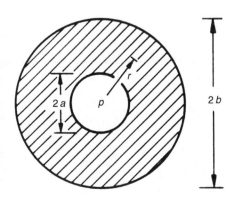

Cylinder

$$\sigma_\theta = \frac{pb}{t}$$

$$\sigma_r = -p/2$$

$$\sigma_z = \frac{pb}{2t} \quad \text{(closed ends)}$$

Sphere

$$\sigma_\theta = \sigma_\phi = \frac{pb}{2t}$$

$$\sigma_r = -p/2$$

$p =$ pressure (N/m^2)

$t =$ wall thickness (m)

$a =$ inner radius (m)

$b =$ outer radius (m)

$r =$ radial coordinate (m)

$$\sigma_\theta = \frac{pa^2}{r^2}\left(\frac{b^2 - r^2}{b^2 - a^2}\right)$$

$$\sigma_r = -\frac{pa^2}{r^2}\left(\frac{b^2 + r^2}{b^2 - a^2}\right)$$

$$\sigma_\theta = \sigma_\phi = \frac{pa^3}{2r^3}\left(\frac{b^3 + 2r^3}{b^3 - a^3}\right)$$

$$\sigma_r = -\frac{pa^3}{r^3}\left(\frac{b^3 - r^3}{b^3 - a^3}\right)$$

A.12 Vibrating beams, tubes and discs

Any undamped system vibrating at one of its natural frequencies can be reduced to the simple problem of a mass m attached to a spring of stiffness K. The lowest natural frequency of such a system is

$$f = \frac{1}{2\pi}\sqrt{\frac{K}{m}}$$

Specific cases require specific values for m and K. They can often be estimated with sufficient accuracy to be useful in approximate modelling. Higher natural frequencies are simple multiples of the lowest.

The first box on the facing page gives the lowest natural frequencies of the flexural modes of uniform beams with various end-constraints. As an example, the first can be estimated by assuming that the effective mass of the beam is one quarter of its real mass, so that

$$m = \frac{m_0 \ell}{4}$$

where m_0 is the mass per unit length of the beam and that K is the bending stiffness (given by F/δ from Section A.3); the estimate differs from the exact value by 2%. Vibrations of a tube have a similar form, using I and m_0 for the tube. Circumferential vibrations can be found approximately by 'unwrapping' the tube and treating it as a vibrating plate, simply supported at two of its four edges.

The second box gives the lowest natural frequencies for flat circular discs with simply-supported and clamped edges. Discs with doubly-curved faces are stiffer and have higher natural frequencies.

Vibrating beams, tubes and discs

Beams, tubes

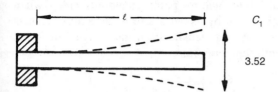

C_1

3.52

$$F_1 = \frac{C_1}{2\pi}\sqrt{\frac{EI}{m_0\ell^4}}$$

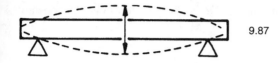

9.87

f = natural frequency (s^{-1})

$m_0 = \rho A$ = mass/length (kg/m)

ρ = density (kg/m^3)

A = section area (m^2)

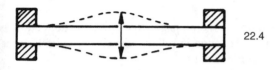

22.4

I = see Table A1

$$\begin{cases} \text{with } A = 2\pi R \\ \qquad I = \pi R^3 t \end{cases}$$

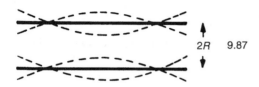

2R   9.87

$$\begin{cases} \text{with } A = \dfrac{\ell t^3}{12} \end{cases}$$

Discs

$$f_1 = \frac{C_2}{2\pi}\sqrt{\frac{Et^3}{m_1 R^4(1-v^2)}}$$

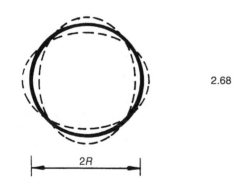

2.68

$m_1 = \rho t$ = mass/area (kg/m^2)

t = thickness (m)

R = radius (m)

v = Poisson's ratio

2R

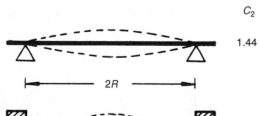

C_2

1.44

2R

2.94

A.13 Creep and creep fracture

At temperatures above $1/3\ T_m$ (where T_m is the absolute melting point), materials creep when loaded. It is convenient to characterize the creep of a material by its behaviour under a tensile stress σ, at a temperature T_m. Under these conditions the steady-state tensile strain rate $\dot\varepsilon$ is often found to vary as a power of the stress and exponentially with temperature:

$$\dot\varepsilon_{SS} = A\left(\frac{\sigma}{\sigma_0}\right)^n \exp-\frac{Q}{RT}$$

where Q is an activation energy, A is a kinetic constant and R is the gas constant. At constant temperature this becomes

$$\dot\varepsilon_{SS} = \dot\varepsilon_0\left(\frac{\sigma}{\sigma_0}\right)^n$$

where $\dot\varepsilon_0(s^{-1})$, $\sigma_0(N/m^2)$ and n are creep constants.

The behaviour of creeping components is summarized on the facing page which give the deflection rate of a beam, the displacement rate of an indenter and the change in relative density of cylindrical and spherical pressure vessels in terms of the tensile creep constants.

Prolonged creep causes the accumulation of creep damage which ultimately leads, after a time t_f, to fracture. To a useful approximation

$$t_f\dot\varepsilon_{SS} = C$$

where C is a constant characteristic of the material. Creep-ductile material have values of C between 0.1 and 0.5; creep-brittle materials have values of C as low as 0.01.

Creep and creep fracture

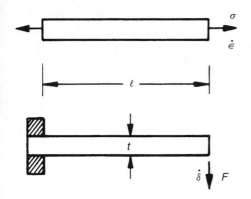

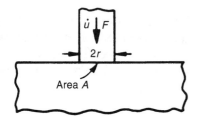

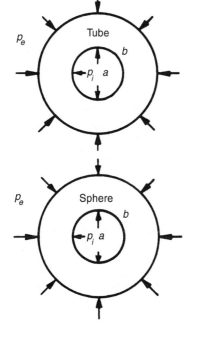

$$\dot{\varepsilon} = \dot{\varepsilon}_0 \left(\frac{\sigma}{\sigma_0} \right)^n$$

$$\dot{\delta} = \frac{2\dot{\varepsilon}_0}{n+2} \left(\frac{(2n+1)}{n\sigma_0} \frac{E}{bt} \right)^n \left(\frac{\ell}{t} \right)^{n+1} \ell$$

$$\dot{u} = C_1 \dot{\varepsilon}_0 \sqrt{A} \left(\frac{C_2 F}{\sigma_0 A} \right)^n$$

$$\dot{\rho} = 2\dot{\varepsilon}_0 \frac{\rho(1-\rho)}{\left(1 - (1-\rho)^{1/n}\right)^n}$$
$$\times \left(\frac{2}{n} \frac{(p_e - p_i)}{\sigma_0} \right)^n$$

$$\dot{\rho} = \frac{3}{2} \dot{\varepsilon}_0 \frac{\rho(1-\rho)}{\left(1 - (1-\rho)^{1/n}\right)^n}$$
$$\times \left(\frac{3}{2n} \frac{(p_e - p_i)}{\sigma_0} \right)^n$$

$\sigma =$ stress N/m^2

$F =$ force (N)

$\dot{\delta}, \dot{u} =$ displacement rates (m/s)

$n, \dot{\varepsilon}_0, \sigma_0 =$ creep constants

$\ell, b, t =$ beam dimensions (m)

$a, b =$ radii of pressure vessels (m)

$\rho =$ relative density, $\dfrac{b^3 - a^3}{b^3}$

$C_1, C_2 =$ constants

A.14 Flow of heat and matter

Heat flow can be limited by conduction, convection or radiation. The constitutive equations for each are listed on the facing page. The first equation is Fourier's first law, describing steady-state heat flow; it contains the thermal conductivity, λ. The second is Fourier's second law, which treats transient heat-flow problems; it contains the thermal diffusivity, a, defined by

$$a = \frac{\lambda}{\rho C}$$

where r is the density and C the specific heat at constant pressure. Solutions to these two differential equations are given in Section A.15.

The third equation describes convective heat transfer. It, rather than conduction, limits heat flow when the Biot number

$$B_i = \frac{hs}{\lambda} < 1$$

where h is the heat-transfer coefficient and s is a characteristic dimension of the sample. When, instead, $B_i > 1$, heat flow is limited by conduction. The final equation is the Stefan-Boltzmann law for radiative heat transfer. The emissivity, ε, is unity for black bodies; less for all other surfaces.

Diffusion of matter follows a pair of differential equations with the same form as Fourier's two laws, and with similar solutions. They are commonly written

$$J = -D\nabla C = -D\frac{dC}{dx} \quad \text{(steady state)}$$

and

$$\frac{\partial C}{\partial t} = D\nabla C^2 = D\frac{\partial^2 C}{\partial x^2} \quad \text{(time-dependent flow)}$$

where J is the flux, C is the concentration, x is the distance and t is time. Solutions are given in the next section.

Flow of heat and matter

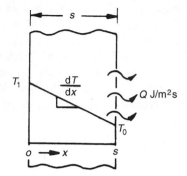

$$Q = -\lambda \nabla T = -\lambda \frac{dT}{dx}$$

Q = heat flux (J/m^{2}s)

T = temperature (K)

x = distance (m)

λ = thermal conductivity (W/mK)

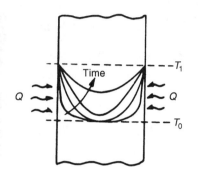

$$\frac{\partial T}{\partial t} = a\nabla^2 T = a\frac{\partial^2 T}{\partial x^2}$$

t = time (s)

ρ = density (kg/m^3)

C = specific heat (J/m^{3}K)

a = thermal diffusivity, $\dfrac{\lambda}{\rho c}$ (m^2/s)

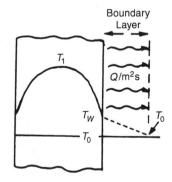

$$Q = h(T_w - T_0)$$

T_w = surface temperature (K)

T = fluid temperature (K)

h = heat transfer coeff. (W/m^{2}K)

   = 5–50 W/m^{2}K in air

   = 1000–5000 W/m^{2}K in water

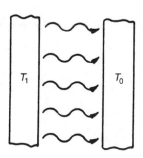

$$Q = \varepsilon\sigma(T_1^4 - T_0^4)$$

ε = emissivity (1 for black body)

σ = Stefan constant

   = 5.67 × 10^{-8} W/m^{2}K^4

A.15 Solutions for diffusion equations

Solutions exist for the diffusion equations for a number of standard geometries. They are worth knowing because many real problems can be approximated by one of these.

At steady-state the temperature or concentration profile does not change with time. This is expressed by equations in the box within the first box at the top of the facing page. Solutions for these are given below for uniaxial flow, radial flow in a cylinder and radial flow in a sphere. The solutions are fitted to individual cases by matching the constants A and B to the boundary conditions. Solutions for matter flow are found by replacing temperature, T, by concentration, C, and conductivity, λ, by diffusion coefficient, D.

The box within the second large box summarizes the governing equations for time-dependent flow, assuming that the diffusivity (a or D) is not a function of position. Solutions for the temperature or concentration profiles, $T(x, t)$ or $C(x, t)$, are given below. The first equation gives the 'thin-film' solution: a thin slab at temperature T_1, or concentration C_1 is sandwiched between two semi-infinite blocks at T_0 or C_0, at $t = 0$, and flow allowed. The second result is for two semi-infinite blocks, initially at T_1 and T_0, (or C_1 or C_0) brought together at $t = 0$. The last is for a T or C profile which is sinusoidal, of amplitude A at $t = 0$.

Note that all transient problems end up with a characteristic time constant t^* with

$$t^* = \frac{x^2}{\beta a} \quad \text{or} \quad \frac{x^2}{\beta D}$$

where x is a dimension of the specimen; or a characteristic length x^* with

$$x^* = \sqrt{\beta a t} \quad \text{or} \quad \sqrt{\beta D t}$$

where t is the timescale of observation, with $1 < \beta < 4$, depending on geometry.

Solutions for diffusion equations

Steady state

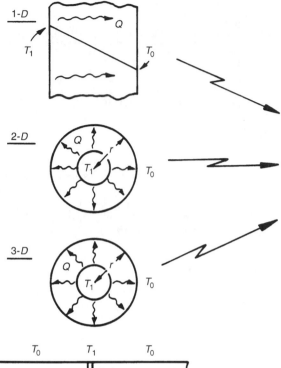

1-D

2-D

3-D

$$\lambda \nabla^2 T = 0$$

$$D \nabla^2 C = 0$$

$$\lambda T(x) = Ax + B(1 - D)$$

$$\lambda T(r) = A \ln r + B(2 - D)$$

$$\lambda T(r) = A/r + B(3 - D) \text{ etc.}$$

$A, B = $ constants of integration

$x = $ linear distance (m)

$r = $ radial distance (m)

Transient

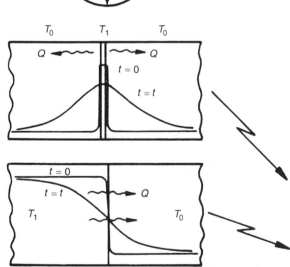

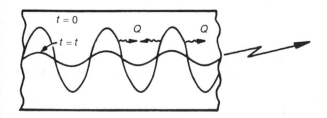

$$\frac{\partial T}{\partial t} = a \nabla^2 T$$

$$\frac{\partial C}{\partial t} = D \nabla^2 C$$

$$T(x, t) = \frac{A}{\sqrt{at}} \exp - \left(\frac{x^2}{4at} \right) + B$$

$$T(x, t) = A \left(1 + erf \left(\frac{x}{2\sqrt{at}} \right) \right) + B$$

$$T(x, t) = A \sin \lambda x, \exp -\lambda^2 at \text{ etc.}$$

A.16 Further reading

Constitutive laws

Cottrell, A.H. *Mechanical Properties of Matter*, Wiley NY (1964).
Gere, J.M. and Timoshenko, S.P. *Mechanics of Materials*, 2nd SI edition, Wadsworth International, California (1985).

Moments of area

Young, W.C. *Roark's Formulas for Stress and Strain*, 6th edition, McGraw-Hill (1989).

Beams, shafts, columns and shells

Calladine, C.R. *Theory of Shell Structures*, Cambridge University Press, Cambridge (1983).
Gere, J.M. and Timoshenko, S.P. *Mechanics of Materials*, 2nd edition, Wadsworth International, California USA (1985).
Timoshenko, S.P. and Goodier, J.N. *Theory of Elasticity*, 3rd edition, McGraw-Hill, (1970).
Timoshenko, S.P. and Gere, J.M. *Theory of Elastic Stability*, 2nd edition, McGraw-Hill (1961).
Young, W.C. *Roark's Formulas for Stress and Strain*, 6th edition, McGraw-Hill (1989).

Contact stresses and stress concentration

Timoshenko, S.P. and Goodier, J.N. *Theory of Elasticity*, 3rd edition, McGraw-Hill, (1970).
Hill, R. *Plasticity*, Oxford University Press, Oxford (1950).
Johnson, K.L. *Contact Mechanics*, Oxford University Press, Oxford (1985).

Sharp cracks

Hertzberg, R.W. *Deformation and Fracture of Engineering Materials*, 3rd edition, Wiley, New York, 1989.
Tada, H., Paris, P.C. and Irwin, G.R. *The Stress Analysis of Cracks Handbook*, 2nd edition, Paris Productions and Del Research Group, Missouri.

Pressure vessels

Timoshenko, S.P. and Goodier, J.N. *Theory of Elasticity*, 3rd edition, McGraw-Hill, (1970).
Hill, R. *Plasticity*, Oxford University Press, Oxford (1950).
Young, W.C. *Roark's Formulas for Stress and Strain*, 6th edition, McGraw-Hill (1989).

Vibration

Young, W.C. *Roark's Formulas for Stress and Strain*, 6th edition, McGraw-Hill (1989).

Creep

Finnie, I. and Heller, W.R. *Creep of Engineering Materials*, McGraw-Hill, New York. (1976).

Heat and matter flow

Hollman, J.P. *Heat Transfer*, 5th edition, McGraw-Hill, New York (1981).
Carslaw, H.S. and Jaeger, J.C. *Conduction of Heat in Solids*, 2nd edition, Oxford University Press, Oxford (1959).
Shewmon, P.G. *Diffusion in Solids*, 2nd edition, TMS Warrendale, PA (1989).

Appendix B

Material indices

Introduction

The performance, p, of a component is measured by a performance equation. The performance equation contains groups of material properties. These groups are the material indices. Sometimes the 'group' is a single property; thus if the performance of a beam is measured by its stiffness, the performance equation contains only one property, the elastic modulus E. It is the material index for this problem. More commonly the performance equation contains a group of two or more properties. Familiar examples are the specific stiffness, E/ρ, and the specific strength, σ_y/ρ, (where E is Young's modulus, σ_y is the yield strength or elastic limit, and ρ is the density), but there are many others. They are a key to the optimal selection of materials. Details of the method, with numerous examples are given in references [1] and [2]. PC-based software systems [3] which implement the method are available. This Appendix compiles indices for a range of common applications.

Uses of material indices

Material selection

Components have functions: to carry loads safely, to transmit heat, to store energy, to insulate, and so forth. Each function has an associated material index. Materials with high values of the appropriate index maximize that aspect of the performance of the component. For reasons given in reference [1], the material index is generally independent of the details of the design. Thus the indices for beams in the tables which follow are independent of the detailed shape of the beam; that for minimizing thermal distortion of precision instruments is independent of the configuration of the instrument. This gives them great generality.

Material deployment or substitution

A new material will have potential application in functions for which its indices have unusually high values. Fruitful applications for a new material can be identified by evaluating its indices and comparing these with those of incumbent materials. Similar reasoning points the way to identifying viable substitutes for an incumbent material in an established application.

Table B1 Stiffness-limited design at minimum mass (cost, energy, environmental impact*)

Function and constraints*	Maximize[†]
Tie (tensile strut)	
stiffness, length specified; section area free	E/ρ
Shaft (loaded in torsion)	
stiffness, length, shape specified, section area free	$G^{1/2}/\rho$
stiffness, length, outer radius specified; wall thickness free	G/ρ
stiffness, length, wall-thickness specified; outer radius free	$G^{1/3}/\rho$
Beam (loaded in bending)	
stiffness, length, shape specified; section area free	$E^{1/2}/\rho$
stiffness, length, height specified; width free	E/ρ
stiffness, length, width specified; height free	$E^{1/3}/\rho$
Column (compression strut, failure by elastic buckling)	
buckling load, length, shape specified; section area free	$E^{1/2}/\rho$
Panel (flat plate, loaded in bending)	
stiffness, length, width specified, thickness free	$E^{1/3}/\rho$
Plate (flat plate, compressed in-plane, buckling failure)	
collapse load, length and width specified, thickness free	$E^{1/3}/\rho$
Cylinder with internal pressure	
elastic distortion, pressure and radius specified; wall thickness free	E/ρ
Spherical shell with internal pressure	
elastic distortion, pressure and radius specified, wall thickness free	$E/(1-\nu)\rho$

*To minimize cost, use the above criteria for minimum weight, replacing density ρ by $C_m\rho$, where C_m is the material cost per kg. To minimize energy content, use the above criteria for minimum weight replacing density ρ by $q\rho$ where q is the energy content per kg. To minimize environmental impact, replace density ρ by $I_e\rho$ instead, where I_e is the eco-indicator value for the material (references [1] and [4]).

[†]E = Young's modulus for tension, the flexural modulus for bending or buckling; G = shear modulus; ρ = density, q = energy content/kg; I_e = eco-indicator value/kg.

Table B2 Strength-limited design at minimum mass (cost, energy, environmental impact*)

Function and constraints*‡	Maximize†
Tie (tensile strut)	
stiffness, length specified; section area free	σ_f/ρ
Shaft (loaded in torsion)	
load, length, shape specified, section area free	$\sigma_f^{2/3}/\rho$
load, length, outer radius specified; wall thickness free	σ_f/ρ
load, length, wall-thickness specified; outer radius free	$\sigma_f^{1/2}/\rho$
Beam (loaded in bending)	
load, length, shape specified; section area free	$\sigma_f^{2/3}/\rho$
load length, height specified; width free	σ_f/ρ
load, length, width specified; height free	$\sigma_f^{1/2}/\rho$
Column (compression strut)	
load, length, shape specified; section area free	σ_f/ρ
Panel (flat plate, loaded in bending)	
stiffness, length, width specified, thickness free	$\sigma_f^{1/2}/\rho$
Plate (flat plate, compressed in-plane, buckling failure)	
collapse load, length and width specified, thickness free	$\sigma_f^{1/2}/\rho$
Cylinder with internal pressure	
elastic distortion, pressure and radius specified; wall thickness free	σ_f/ρ
Spherical shell with internal pressure	
elastic distortion, pressure and radius specified, wall thickness free	σ_f/ρ
Flywheels, rotating discs	
maximum energy storage per unit volume; given velocity	ρ
maximum energy storage per unit mass; no failure	σ_f/ρ

*To minimize cost, use the above criteria for minimum weight, replacing density ρ by $C_m\rho$, where C_m is the material cost per kg. To minimize energy content, use the above criteria for minimum weight replacing density ρ by $q\rho$ where q is the energy content per kg. To minimize environmental impact, replace density ρ by $I_e\rho$ instead, where I_e is the eco-indicator value for the material (references [1] and [4]).

†σ_f = failure strength (the yield strength for metals and ductile polymers, the tensile strength for ceramics, glasses and brittle polymers loaded in tension; the flexural strength or modulus of rupture for materials loaded in bending); ρ = density.

‡For design for infinite fatigue life, replace σ_f by the endurance limit σ_e.

Table B3 Strength-limited design: springs, hinges etc. for maximum performance*

Function and constraints*‡	Maximize†
Springs	
maximum stored elastic energy per unit volume; no failure	σ_f^2/E
maximum stored elastic energy per unit mass; no failure	$\sigma_f^2/E\rho$
Elastic hinges	
radius of bend to be minimized (max flexibility without failure)	σ_f/E
Knife edges, pivots	σ_f^3/E^2 and
minimum contact area, maximum bearing load	H
Compression seals and gaskets	$\sigma_f^{3/2}/E$ and
maximum conformability; limit on contact pressure	$1/E$
Diaphragms	
maximum deflection under specified pressure or force	$\sigma_f^{3/2}/E$
Rotating drums and centrifuges	
maximum angular velocity; radius fixed; wall thickness free	σ_f/ρ

*To minimize cost, use the above criteria for minimum weight, replacing density ρ by $C_m\rho$, where C_m is the material cost per kg. To minimize energy content, use the above criteria for minimum weight replacing density ρ by $q\rho$ where q is the energy content per kg. To minimize environmental impact, replace density ρ by $I_e\rho$ instead, where I_e is the eco-indicator value for the material (references [1] and [4]).

†σ_f = failure strength (the yield strength for metals and ductile polymers, the tensile strength for ceramics, glasses and brittle polymers loaded in tension; the flexural strength or modulus of rupture for materials loaded in bending); ρ = density; H = hardness.

‡For design for infinite fatigue life, replace σ_f by the endurance limit σ_e.

Table B4 Vibration-limited design

Function and constraints	Maximize*
Ties, columns	
maximum longitudinal vibration frequencies	E/ρ
Beams, all dimensions prescribed	
maximum flexural vibration frequencies	E/ρ
Beams, length and stiffness prescribed	
maximum flexural vibration frequencies	$E^{1/2}/\rho$
Panels, all dimensions prescribed	
maximum flexural vibration frequencies	E/ρ
Panels, length, width and stiffness prescribed	
maximum flexural vibration frequencies	$E^{1/3}/\rho$
Ties, columns, beams, panels, stiffness prescribed	
minimum longitudinal excitation from external drivers, ties	$\eta E/\rho$
minimum flexural excitation from external drivers, beams	$\eta E^{1/2}/\rho$
minimum flexural excitation from external drivers, panels	$\eta E^{1/3}/\rho$

*E = Young's modulus for tension, the flexural modulus for bending; G = shear modulus; ρ = density; η = damping coefficient (loss coefficient).

Table B5 Damage-tolerant design

Function and constraints	Maximize*
Ties (tensile member)	
Maximize flaw tolerance and strength, load-controlled design	K_{Ic} and σ_f
Maximize flaw tolerance and strength, displacement-control	K_{Ic}/E and σ_f
Maximize flaw tolerance and strength, energy-control	K_{Ic}^2/E and σ_f
Shafts (loaded in torsion)	
Maximize flaw tolerance and strength, load-controlled design	K_{Ic} and σ_f
Maximize flaw tolerance and strength, displacement-control	K_{Ic}/E and σ_f
Maximize flaw tolerance and strength, energy-control	K_{Ic}^2/E and σ_f
Beams (loaded in bending)	
Maximize flaw tolerance and strength, load-controlled design	K_{Ic} and σ_f
Maximize flaw tolerance and strength, displacement-control	K_{Ic}/E and σ_f
Maximize flaw tolerance and strength, energy-control	K_{Ic}^2/E and σ_f
Pressure vessel	
Yield-before-break	K_{Ic}/σ_f
Leak-before-break	K_{Ic}^2/σ_f

*K_{Ic} = fracture toughness; E = Young's modulus; σ_f = failure strength (the yield strength for metals and ductile polymers, the tensile strength for ceramics, glasses and brittle polymers loaded in tension; the flexural strength or modulus of rupture for materials loaded in bending).

Table B6 Thermal and thermo-mechanical design

Function and constraints	Maximize*
Thermal insulation materials	
minimum heat flux at steady state; thickness specified	$1/\lambda$
minimum temp rise in specified time; thickness specified	$1/a = \rho C_p/\lambda$
minimize total energy consumed in thermal cycle (kilns, etc)	$\sqrt{a}/\lambda = \sqrt{1/\lambda\rho C_p}$
Thermal storage materials	
maximum energy stored/unit material cost (storage heaters)	C_p/C_m
maximize energy stored for given temperature rise and time	$\lambda/\sqrt{a} = \sqrt{\lambda\rho C_p}$
Precision devices	
minimize thermal distortion for given heat flux	λ/α
Thermal shock resistance	$\sigma_f/E\alpha$
maximum change in surface temperature; no failure	
Heat sinks	
maximum heat flux per unit volume; expansion limited	$\lambda/\Delta\alpha$
maximum heat flux per unit mass; expansion limited	$\lambda/\rho\Delta\alpha$
Heat exchangers (pressure-limited)	
maximum heat flux per unit area; no failure under Δp	$\lambda\sigma_f$
maximum heat flux per unit mass; no failure under Δp	$\lambda\sigma_f/\rho$

*λ = thermal conductivity; a = thermal diffusivity; C_p = specific heat capacity; C_m = material cost/kg; T_{max} = maximum service temperature; α = thermal expansion coeff.; E = Young's modulus; ρ = density; σ_f = failure strength (the yield strength for metals and ductile polymers, the tensile strength for ceramics, glasses and brittle polymers).

Table B7 Electro-mechanical design

Function and constraints	Maximize*
Bus bars	
minimum life-cost; high current conductor	$1/\rho_e \rho C_m$
Electro-magnet windings	
maximum short-pulse field; no mechanical failure	σ_y
maximize field and pulse-length, limit on temperature rise	$C_p \rho / \rho_e$
Windings, high-speed electric motors	
maximum rotational speed; no fatigue failure	σ_e / ρ_e
minimum ohmic losses; no fatigue failure	$1/\rho_e$
Relay arms	
minimum response time; no fatigue failure	$\sigma_e / E \rho_e$
minimum ohmic losses; no fatigue failure	$\sigma_e^2 / E \rho_e$

*C_m = material cost/kg; E = Young's modulus; ρ = density; ρ_e = electrical resistivity; σ_y = yield strength; σ_e = endurance limit.

References

Derivations for almost all the indices listed in this Appendix can be found in references [1] and [2].

[1] Ashby, M.F. (1999) *Materials Selection in Mechanical Design*, 2nd edn, Chapter 6, Butterworth-Heinemann, Oxford.
[2] Ashby, M.F. and Cebon, D. (1995) *Case Studies in Materials Selection*, Granta Design, Trumpington.
[3] CMS 2.1(1995) and CMS 3.0 (1999) Granta Design, Trumpington.
[4] Goedkoop, M.J., Demmers, M. and Collignon, M.X. 'Eco-Indicator '95, Manual', Pré Consultants, and the Netherlands Agency for Energy and the Environment, Amersfort, Holland (1995).

Appendix C

Material and process selection charts

C.1 Introduction

The charts in this booklet summarize *material properties* and *process attributes*. Each chart appears on a single page with a brief commentary about its use on the facing page. Background and data sources can be found in the appendix to Chapter 13, pp. 313–333.

The material charts map the areas of property space occupied by each material class. They can be used in two ways:

(a) to retrieve approximate values for material properties
(b) to select materials which have prescribed property profiles

The collection of process charts, similarly, can be used as a data source or as a selection tool. Sequential application of several charts allows several design goals to be met simultaneously. More advanced methods are described in the book cited above.

The best way to tackle selection problems is to work directly on the appropriate charts. Permission is given to copy charts for this purpose. Normal copyright restrictions apply to reproduction for other purposes.

It is not possible to give charts which plot all the possible combinations: there are too many. Those presented here are the most commonly useful. Any other can be created easily using the CMS2 (1995) or CES (1999) software.

1.1 Cautions

The data on the charts and in the tables are approximate: they typify each class of material (stainless steels, or polyethylenes, for instance) or processes (sand casting, or injection moulding, for example), but within each class there is considerable variation. They are adequate for the broad comparisons required for conceptual design, and, often, for the rough calculations of embodiment design. **They are not appropriate for detailed design calculations**. For these, it is essential to seek accurate data from handbooks and the data sheets provided by material suppliers. The charts help in narrowing the choice of candidate materials to a sensible short list, but not in providing numbers for final accurate analysis.

Every effort has been made to ensure the accuracy of the data shown on the charts. No guarantee can, however, be given that the data are error-free, or that new data may not supersede those given here. The charts are an aid to creative thinking, not a source of numerical data for precise analysis.

1.2 Material classes, class members and properties

The materials of mechanical and structural engineering fall into nine broad classes listed in Table 1.1.

Within each class, the Materials Selection Charts show data for a representative set of materials, chosen both to span the full range of behaviour for that class, and to include the most widely used members of it. In this way the envelope for a class (heavy lines) encloses data not only for the materials listed on Table 1.2 (next two pages) but for virtually all other members of the class as well.

As far as possible, the same materials appear on all the charts. There are exceptions. Invar is only interesting because of its low thermal expansion: it appears on the thermal expansion charts (10 and 11) but on no others. Mn−Cu alloys have high internal damping: they are shown on the loss-coefficient chart (8) but not elsewhere. And there are others. But, broadly, the material and classes which appear on one chart appear on them all.

Table 1.1 Material classes

Engineering alloys	(metals and their alloys)
Engineering polymers	(thermoplastics and thermosets)
Engineering ceramics	('fine' ceramics)
Engineering composites	(GFRP, KFRP and CFRP)
Porous ceramics	(brick, cement, concrete, stone)
Glasses	(silicate glasses)
Woods	(common structural timbers)
Elastomers	(natural and artificial rubbers)
Foams	(foamed polymers)

Table 1.2 Members of each material class

Class	Members	Short name
Engineering alloys (The metals and alloys of engineering)	Aluminium alloys	Al Alloys
	Beryllium alloys	Be Alloys
	Cast irons	Cast iron
	Copper alloys	Cu Alloys
	Lead alloys	Lead Alloys
	Magnesium alloys	Mg Alloys
	Molybdenum alloys	Mo Alloys
	Nickel alloys	Ni Alloys
	Steels	Steels
	Tin alloys	Tin Alloys
	Titanium alloys	Ti Alloys
	Tungsten alloys	W Alloys
	Zinc alloys	Zn Alloys
Engineering polymers (The thermoplastics and thermosets of engineering)	Epoxies	EP
	Melamines	MEL
	Polycarbonate	PC
	Polyesters	PEST
	Polyethylene, high density	HDPE
	Polyethylene, low density	LDPE
	Polyformaldehyde	PF
	Polymethylmethacrylate	PMMA
	Polypropylene	PP
	Polytetrafluorethylene	PTFE
	Polyvinylchloride	PVC

Table 1.2 (*continued*)

Engineering ceramics (Fine ceramics capable of load-bearing applications)	Alumina	Al_2O_3
	Beryllia	BeO
	Diamond	Diamond
	Germanium	Ge
	Magnesium	MgO
	Silicon	Si
	Sialons	Sialons
	Silicon carbide	SiC
	Silicon nitride	Si_3N_4
	Zirconia	ZrO_2
Engineering composites (The composites of engineering practice A distinction is drawn between the properties of a ply – 'uniply' – and of a laminate 'laminates')	Carbon fibre reinforced polymer	CFRP
	Glass fibre reinforced polymer	GFRP
	Kevlar fibre reinforced polymer	KFRP
Porous ceramics (Traditional ceramics cements, rocks and minerals)	Brick	Brick
	Cement	Cement
	Common rocks	Rocks
	Concrete	Concrete
	Porcelain	Pcln
	Pottery	Pot
Glasses (Silicate glass and silica itself)	Borosilicate glass	B-glass
	Soda glass	Na-glass
	Silica	SiO_2
Woods (Separate envelopes describe properties parallel to the grain and normal to it, and wood products)	Ash	Ash
	Balsa	Balsa
	Fir	Fir
	Oak	Oak
	Pine	Pine
	Wood products (ply, etc)	Wood products
Elastomers (Natural and artificial rubbers)	Natural rubber	Rubber
	Hard butyl rubber	Hard butyl
	Polyurethane	PU
	Silicone rubber	Silicone
	Soft Butyl rubber	Soft butyl
Polymer foams (Foamed polymers of engineering)	These include:	
	Cork	Cork
	Polyester	PEST
	Polystyrene	PS
	Polyurethane	PU
Special materials (Materials included on one or a few charts only, because of their special characteristics)	Beryllium–copper alloys	BeCu
	Invar	Invar
	WC–Co Cermets	WC–Co
	Mn–Cu alloys	Mn–Cu Alloys

You will not find specific materials listed on the charts. The aluminium alloy 7075 in the T6 condition (for instance) is contained in the property envelopes for *Al-alloys*; the Nylon 66 in those for *nylons*. The charts are designed for the broad, early stages of materials selection, not for retrieving the precise values of material properties needed in the later, detailed design, stage.

The Material Selection Charts which follow display, for the nine classes of materials, the properties listed in Table 1.3.

The charts let you pick off the subset of materials with a property within a specified range: materials with modulus E between 100 and 200 GPa for instance; or materials with a thermal conductivity above 100 W/mK.

More usually, performance is maximized by selecting the subset of materials with the greatest value of a grouping of material properties. A light, stiff beam is best made of a material with a high value of $E^{1/2}/\rho$; safe pressure vessels are best made of a material with a high value of $K_{Ic}^{1/2}/\sigma_f$, and so on. Table 1.4 lists some of these performance-maximizing groups or 'material indices'. The charts are designed to display these, and to allow you to pick off the subset of materials which maximize them. Details of the method, with worked examples, are given in Chapters 5 and 6.

Multiple criteria can be used. You can pick off the subset of materials with both high $E^{1/2}/\rho$ and high E (good for light, stiff beams) from Chart 1; that with high σ_f^3/E^2 and high E (good materials for pivots) from Chart 4. Throughout, the goal is to identify from the charts a *subset* of materials, not a single material. Finding the best material for a given application involves many considerations, many of them (like availability, appearance and feel) not easily quantifiable. The charts do not give you the final choice — that requires the use of your judgement and experience. Their power is that they guide you quickly and efficiently to a subset of materials worth considering; and they make sure that you do not overlook a promising candidate.

1.4 Process classes and class members

A *process* is a method of shaping, finishing or joining a material. *Sand casting*, *injection moulding*, *polishing* and *fusion welding* are all processes. The choice, for a given component, depends on the material of which it is to be made, on its size, shape and precision, and on how many are required.

The manufacturing processes of engineering fall into nine broad classes:

Table 1.3 Material properties shown on the charts

Property	Symbol	Units
Relative cost	C_m	(−)
Density	ρ	(Mg/m^3)
Young's modulus	E	(GPa)
Strength	σ_f	(MPa)
Fracture toughness	K_{Ic}	(MPam$^{1/2}$)
Toughness	G_{Ic}	(J/m^2)
Damping coefficient	η	(−)
Thermal conductivity	λ	(W/mK)
Thermal diffusivity	a	(m^2/s)
Volume specific heat	$C_{P\rho}$	(J/m^{3}K)
Thermal expansion coefficient	a	(1/K)
Thermal shock resistance	ΔT	(K)
Strength at temperature	$\sigma(T)$	(MPa)
Specific wear rate	W/AP	(1/MPa)

Table 1.4 Examples of material-indices

Function	Index
Tie, minimum weight, stiffness prescribed	$\dfrac{E}{\rho}$
Beam, minimum weight, stiffness prescribed	$\dfrac{E^{1/2}}{\rho}$
Beam, minimum weight, strength prescribed	$\dfrac{\sigma_y^{2/3}}{\rho}$
Beam, minimum cost, stiffness prescribed	$\dfrac{E^{1/2}}{C_m\rho}$
Beam, minimum cost, strength prescribed	$\dfrac{\sigma_y^{2/3}}{C_m\rho}$
Column, minimum cost, buckling load prescribed	$\dfrac{E^{1/2}}{C_m\rho}$
Spring, minimum weight for given energy storage	$\dfrac{\sigma_y^2}{E\rho}$
Thermal insulation, minimum cost, heat flux prescribed	$\dfrac{1}{\lambda C_m\rho}$

(ρ = density; E = Young's modulus; σ_y = elastic limit; C_m = cost/kg; λ = thermal conductivity; κ = electrical conductivity; C_p = specific heat)

Table 1.5 Process classes

Casting	(sand, gravity, pressure, die, etc.)
Pressure moulding	(direct, transfer, injection, etc.)
Deformation processes	(rolling, forging, drawing, etc.)
Powder methods	(slip cast, sinter, hot press, hip)
Special methods	(CVD, electroform, lay up, etc.)
Machining	(cut, turn, drill, mill, grind, etc.)
Heat treatment	(quench, temper, solution treat, age, etc.)
Joining	(bolt, rivet, weld, braze, adhesives)
Finish	(polish, plate, anodize, paint)

Each process is characterized by a set of *attributes*: the materials it can handle, the shapes it can make and their precision, complexity and size. Process Selection Charts map the attributes, showing the ranges of size, shape, material, precision and surface finish of which each class of process is capable. The procedure does not lead to a final choice of process. Instead, it identifies a subset of processes which have the potential to meet the design requirements. More specialized sources must then be consulted to determine which of these is the most economical.

C.2 THE MATERIALS SELECTION CHARTS

Chart 1: Young's modulus, *E* against density, ρ

This chart guides selection of materials for light, stiff, components. The contours show the longitudinal wave speed in m/s; natural vibration frequencies are proportional to this quantity. The guide lines show the loci of points for which

(a) $E/\rho = C$ (minimum weight design of stiff ties; minimum deflection in centrifugal loading, etc.)

(b) $E^{1/2}/\rho = C$ (minimum weight design of stiff beams, shafts and columns)

(c) $E^{1/3}/\rho = C$ (minimum weight design of stiff plates)

The value of the constant C increases as the lines are displaced upwards and to the left. Materials offering the greatest stiffness-to-weight ratio lie towards the upper left-hand corner.

Other moduli are obtained approximately from E using

   (a) $\nu = 1/3$; $G = 3/8E$; $K \approx E$ (metals, ceramics, glasses and glassy polymers)

or (b) $\nu \approx 1/2$; $G \approx 1/3E$; $K \approx 10E$ (elastomers, rubbery polymers)

where ν is Poisson's ratio, G the shear modulus and K the bulk modulus.

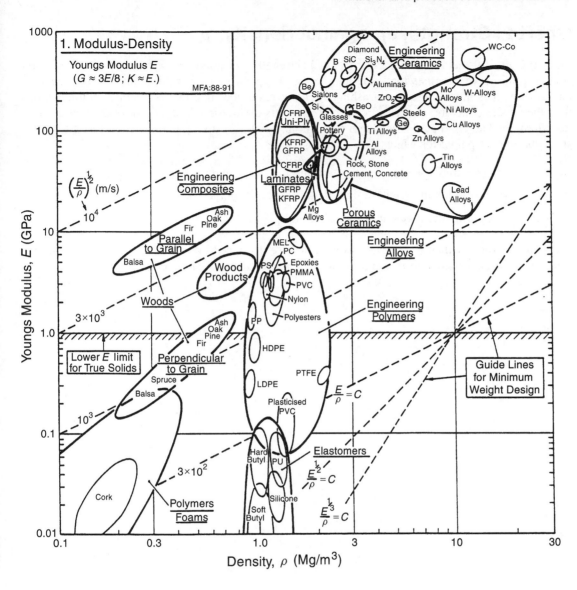

1. Modulus-Density

Youngs Modulus E
($G \approx 3E/8$; $K \approx E.$)

MFA:88-91

Chart 2: Strength, σ_f, against density, ρ

The 'strength' for metals is the 0.2% offset yield strength. For polymers, it is the stress at which the stress–strain curve becomes markedly non-linear — typically, a strain of about 1%. For ceramics and glasses, it is the compressive crushing strength; remember that this is roughly 15 times larger than the tensile (fracture) strength. For composites it is the tensile strength. For elastomers it is the tear-strength. The chart guides selection of materials for light, strong, components. The guide lines show the loci of points for which:

(a) $\sigma_f/\rho = C$ (minimum weight design of strong ties; maximum rotational velocity of discs)

(b) $\sigma_f^{2/3}/\rho = C$ (minimum weight design of strong beams and shafts)

(c) $\sigma_f^{1/2}/\rho = C$ (minimum weight design of strong plates)

The value of the constant C increases as the lines are displaced upwards and to the left. Materials offering the greatest strength-to-weight ratio lie towards the upper left corner.

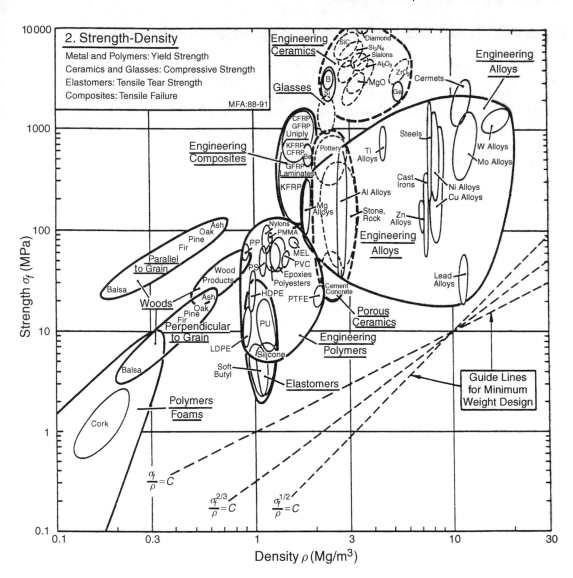

2. Strength-Density

Metal and Polymers: Yield Strength
Ceramics and Glasses: Compressive Strength
Elastomers: Tensile Tear Strength
Composites: Tensile Failure

MFA:88-91

Chart 3: Fracture toughness, K_{Ic}, against density, ρ

Linear-elastic fracture mechanics describes the behaviour of cracked, brittle solids. It breaks down when the fracture toughness is large and the section is small; then J-integral methods should be used. The data shown here are adequate for the rough calculations of conceptual design and as a way of ranking materials. The chart guides selection of materials for light, fracture-resistant components. The guide lines show the loci of points for which:

(a) $K_{Ic}^{4/3}/\rho = C$ (minimum weight design of brittle ties, maximum rotational velocity of brittle discs, etc.)

$K_{Ic}/\rho = C$

(b) $K_{Ic}^{4/5}/\rho = C$ (minimum weight design of brittle beams and shafts)

$K_{Ic}^{2/3}/\rho = C$

(c) $K_{Ic}^{2/3}/\rho = C$ (minimum weight design of brittle plates)

$K_{Ic}^{1/2}/\rho = C$

The value of the constant C increases as the lines are displaced upwards and to the left. Materials offering the greatest toughness-to-weight ratio lie towards the upper left corner.

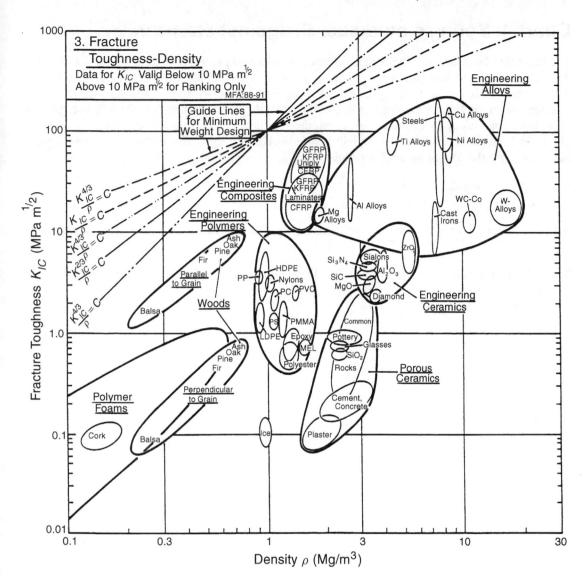

3. Fracture Toughness-Density
Data for K_{IC} Valid Below 10 MPa m$^{1/2}$
Above 10 MPa m$^{1/2}$ for Ranking Only
MFA:88-91

Chart 4: Young's modulus, E, against strength, σ_f

The chart for elastic design. The contours show the failure strain, σ_f/E. The 'strength' for metals is the 0.2% offset yield strength. For polymers, it is the 1% yield strength. For ceramics and glasses, it is the compressive crushing strength; remember that this is roughly 15 times larger than the tensile (fracture) strength. For composites it is the tensile strength. For elastomers it is the tear-strength. The chart has numerous applications among them: the selection of materials for springs, elastic hinges, pivots and elastic bearings, and for yield-before-buckling design. The guide lines show three of these; they are the loci of points for which:

(a) $\sigma_f/E = C$ (elastic hinges)

(b) $\sigma_f^2/E = C$ (springs, elastic energy storage per unit volume)

(c) $\sigma_f^{3/2}/E = C$ (selection for elastic constants such as knife edges; elastic diaphragms, compression seals)

The value of the constant C increases as the lines are displaced downward and to the right.

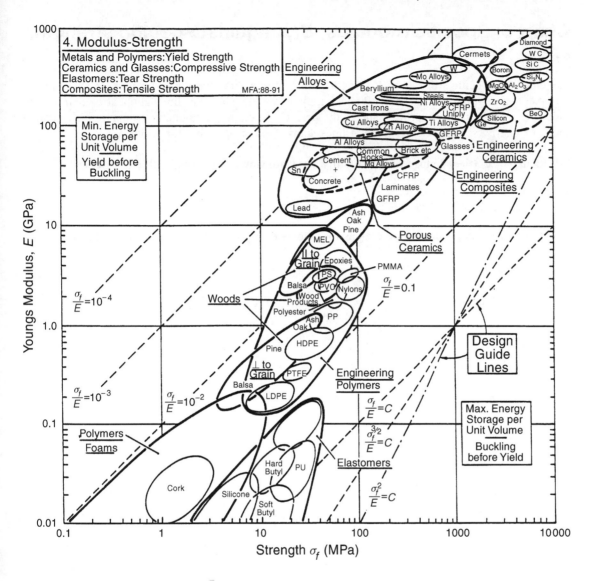

4. Modulus-Strength

Metals and Polymers:Yield Strength
Ceramics and Glasses:Compressive Strength
Elastomers:Tear Strength
Composites:Tensile Strength

MFA:88-91

Chart 5: Specific modulus, E/ρ, against specific strength, σ_f/ρ

The chart for specific stiffness and strength. The contours show the yield strain, σ_f/E. The qualifications on strength given for Charts 2 and 4 apply here also. The chart finds application in minimum weight design of ties and springs, and in the design of rotating components to maximize rotational speed or energy storage, etc. The guide lines show the loci of points for which

(a) $\sigma_f^2/E\rho = C$ (ties, springs of minimum weight; maximum rotational velocity of discs)

(b) $\sigma_f^{3/2}/E_\rho^{1/2} = C$

(c) $\sigma_f/E = C$ (elastic hinge design)

The value of the constant C increases as the lines are displaced downwards and to the right.

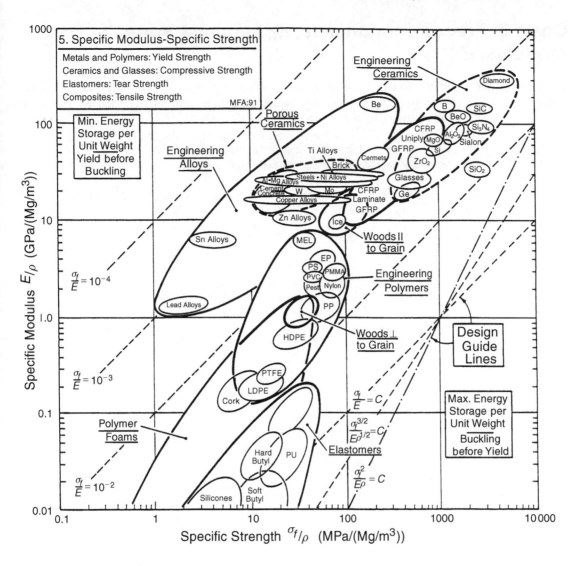

5. Specific Modulus-Specific Strength

Metals and Polymers: Yield Strength
Ceramics and Glasses: Compressive Strength
Elastomers: Tear Strength
Composites: Tensile Strength

MFA:91

Chart 6: Fracture toughness, K_{Ic}, against Young's modulus, E

The chart displays both the fracture toughness, K_{Ic}, and (as contours) the toughness, $G_{Ic} \approx K_{Ic}^2/E$; and it allows criteria for stress and displacement-limited failure criteria (K_{Ic} and K_{Ic}/E) to be compared. The guide lines show the loci of points for which

(a) $K_{Ic}^2/E = C$ (lines of constant toughness, G_c; energy-limited failure)
(b) $K_{Ic}/E = C$ (guideline for displacement-limited brittle failure)

The values of the constant C increases as the lines are displaced upwards and to the left. Tough materials lie towards the upper left corner, brittle materials towards the bottom right.

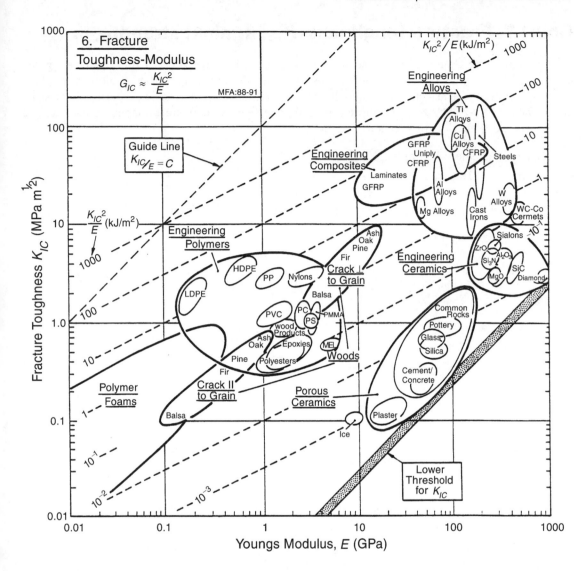

Chart 7: Fracture toughness, K_{Ic}, against strength, σ_f

The chart for safe design against fracture. The contours show the process-zone diameter, given approximately by $K_{Ic}^2/\pi\sigma_f^2$. The qualifications on 'strength' given for Charts 2 and 4 apply here also. The chart guides selection of materials to meet yield-before-break design criteria, in assessing plastic or process-zone sizes, and in designing samples for valid fracture toughness testing. The guide lines show the loci of points for which

(a) $K_{Ic}/\sigma_f = C$ (yield-before-break)

(b) $K_{Ic}^2/\sigma_f = C$ (leak-before-break)

The value of the constant C increases as the lines are displaced upward and to the left.

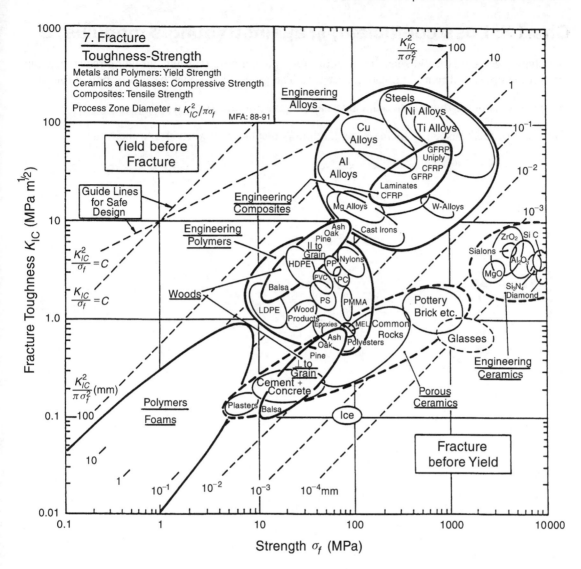

Chart 8: Loss coefficient, η, against Young's modulus, E

The chart gives guidance in selecting material for low damping (springs, vibrating reeds, etc.) and for high damping (vibration-mitigating systems). The guide line shows the loci of points for which

(a) $\eta E = C$ (rule-of-thumb for estimating damping in polymers)

The value of the constant C increases as the line is displaced upward and to the right.

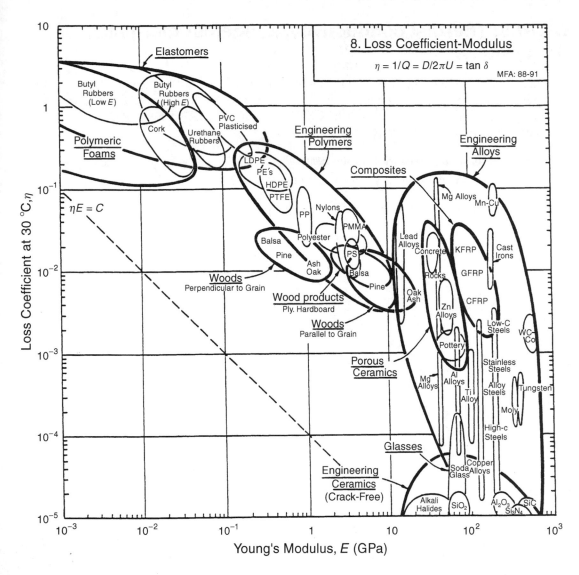

8. Loss Coefficient-Modulus

$$\eta = 1/Q = D/2\pi U = \tan \delta$$

MFA: 88-91

Chart 9: Thermal conductivity, λ, against thermal diffusivity, a

The chart guides in selecting materials for thermal insulation, for use as heat sinks and so forth, both when heat flow is steady, (λ) and when it is transient ($a = \lambda/\rho C_p$ where ρ is the density and C_p the specific heat). Contours show values of the volumetric specific heat, $\rho C_p = \lambda/a$ (J/m^{3}K). The guide lines show the loci of points for which

(a) $\lambda/a = C$ (constant volumetric specific heat)

(b) $\lambda/a^{1/2} = C$ (efficient insulation; thermal energy storage)

The value of constant C increases towards the upper left.

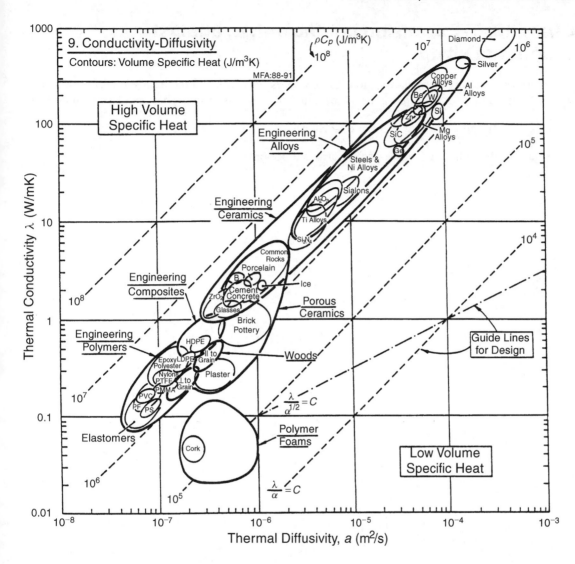

9. Conductivity-Diffusivity

Contours: Volume Specific Heat (J/m³K)

MFA:88-91

Chart 10: T-Expansion coefficient, α, against T-conductivity, λ

The chart for assessing thermal distortion. The contours show value of the ratio λ/α (W/m). Materials with a large value of this design index show small thermal distortion. They define the guide line

(a) $\lambda/\alpha = C$ (minimization of thermal distortion)

The value of the constant C increases towards the bottom right.

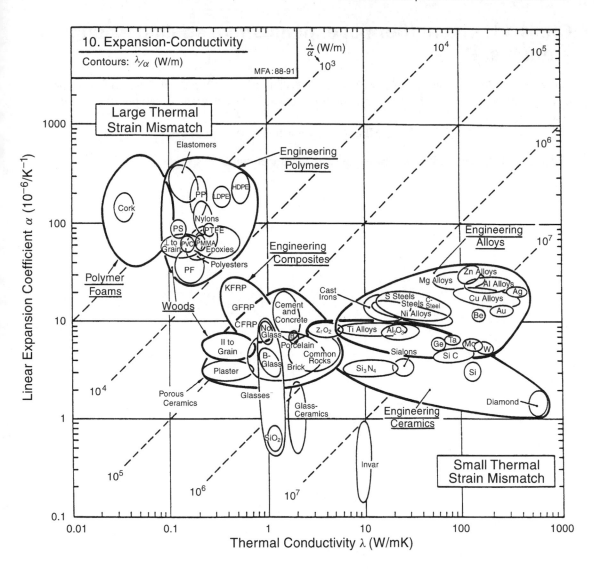

10. Expansion-Conductivity

Contours: λ/α (W/m)

MFA:88-91

$\frac{\lambda}{\alpha}$ (W/m)

Chart 11: Linear thermal expansion, α, against Young's modulus, E

The chart guides in selecting materials when thermal stress is important. The contours show the thermal stress generated, per °C temperature change, in a constrained sample. They define the guide line

$$\alpha E = C \text{ MPa/K (constant thermal stress per K)}$$

The value of the constant C increases towards the upper right.

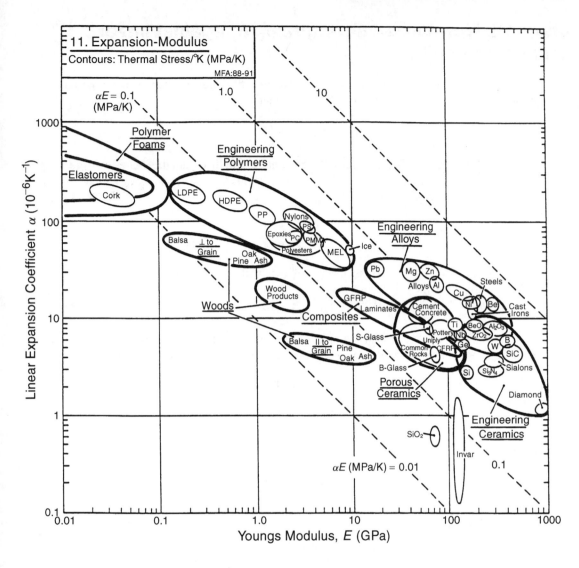

Chart 12: Normalized strength, σ_t/E, against linear expansion coeff., α

The chart guides in selecting materials to resist damage in a sudden change of temperature ΔT. The contours show values of the thermal shock parameter

$$B\Delta T = \frac{\sigma_t}{\alpha E}$$

in $^\circ$C. Here σ_t is the tensile failure strength (the yield strength of ductile materials, the fracture strength of those which are brittle), E is Young's modulus and B is a factor which allows for constraint and for heat-transfer considerations:

$$B = 1/A \quad \text{(axial constraint)}$$

$$B = (1 - \nu)/A \quad \text{(biaxial constraint)}$$

$$B = (1 - 2\nu)/A \quad \text{(triaxial constraint)}$$

with
$$A = \frac{th/\lambda}{1 + th/\lambda}$$

Here ν is Poisson's ratio, t a typical sample dimension, h is the heat-transfer coefficient at the sample surface and λ is its thermal conductivity. The contours define the guide line

$$B\Delta T = C \quad \text{(thermal shock resistance)}$$

The value of the constant C increases towards the top left.

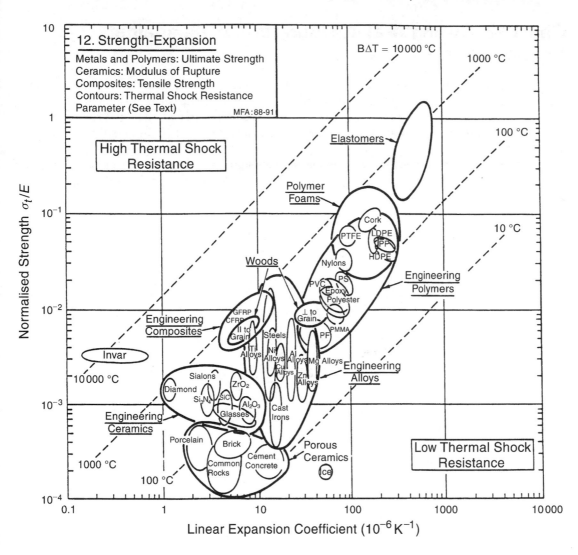

12. Strength-Expansion

Metals and Polymers: Ultimate Strength
Ceramics: Modulus of Rupture
Composites: Tensile Strength
Contours: Thermal Shock Resistance
Parameter (See Text)

MFA:88-91

Chart 13: Strength-at-temperature, $\sigma(T)$, against temperature, T

Materials tend to show a strength which is almost independent of temperature up to a given temperature (the 'onset-of-creep' temperature); above this temperature the strength falls, often steeply. The lozenges show this behaviour (see inset at the bottom right). The 'strength' here is a short-term yield strength, corresponding to 1 hour of loading. For long loading times (10 000 hours for instance), the strengths are lower.

This chart gives an overview of high temperature strength, giving guidance in making an initial choice. Design against creep and creep-fracture requires further information and techniques.

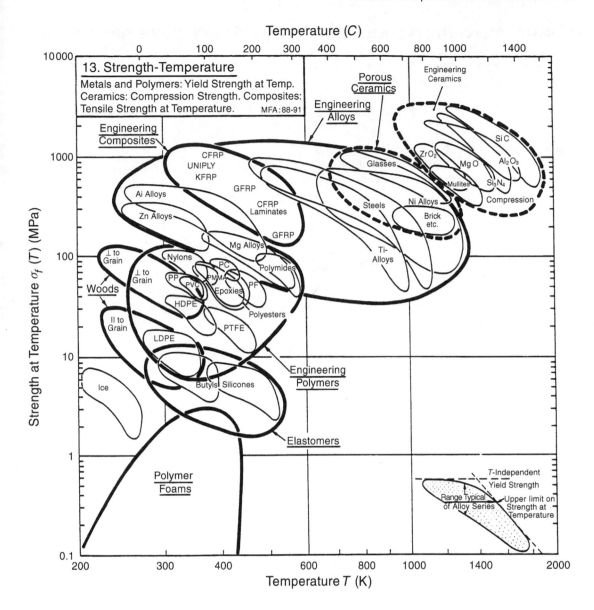

Temperature (*C*)

13. Strength-Temperature

Metals and Polymers: Yield Strength at Temp.
Ceramics: Compression Strength. Composites:
Tensile Strength at Temperature.   MFA:88-91

Engineering
Composites

CFRP
UNIPLY
KFRP
GFRP
Ai Alloys
Zn Alloys
CFRP
Laminates
GFRP
Mg Alloys
Nylons
⊥ to
Grain
PC
PP
PMMA
PVC
Epoxies
Polymides
PF
⊥ to
Grain
Woods
HDPE
Polyesters
II to
Grain
PTFE
LDPE
Ice
Butyls Silicones

Polymer
Foams

Engineering
Polymers

Elastomers

Porous
Ceramics

Engineering
Alloys

Engineering
Ceramics

Glasses
Steels
Ni Alloys
Brick
etc.
Ti-
Alloys

ZrO₂
MgO
Mullites
Si₃N₄
SiC
Al₂O₃

Compression

T-Independent
Yield Strength
Range Typical
of Alloy Series
Upper limit on
Strength at
Temperature

Strength at Temperature σ_f (*T*) (MPa)

Temperature *T* (K)

Chart 14: Young's modulus, E, against relative cost, $C_R\rho$

The chart guides selection of materials for cheap, stiff, components (material cost only). The relative cost C_R is calculated by taking that for mild steel reinforcing-rods as unity; thus

$$C_R = \frac{\text{Cost per unit weight of material}}{\text{Cost per unit weight of mild steel}}$$

The guide lines show the loci of points for which

(a) $E/C_R\rho = C$ (minimum cost design of stiff ties, etc.)
(b) $E^{1/2}/C_R\rho = C$ (minimum cost design of stiff beams and columns)
(c) $E^{1/3}/C_R\rho = C$ (minimum cost design of stiff plates)

The value of the constant C increases as the lines are displayed upwards and to the left. Materials offering the greatest stiffness per unit cost lie towards the upper left corner. Other moduli are obtained approximately from E by

(a) $\nu = 1/3$; $G = 3/8E$; $K \approx E$ (metals, ceramics, glasses and glassy polymers)
or (b) $\nu \approx 1/2$; $G \approx 1/3E$; $K \approx 10E$ (elastomers, rubbery polymers)

where ν is Poisson's ratio, G the shear modulus and K the bulk modulus.

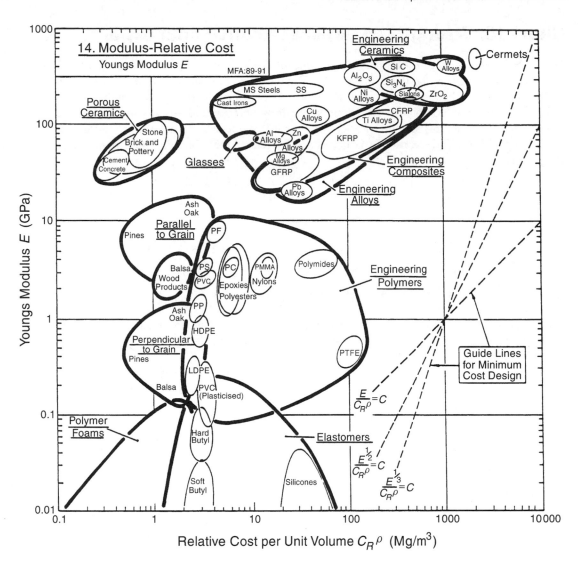

14. Modulus-Relative Cost
Youngs Modulus E

MFA:89-91

Chart 15: Strength, σ_f, against relative cost, $C_R\rho$

The chart guides selection of materials for cheap strong, components (material cost only). The 'strength' for metals is the 0.2% offset yield strength. For polymers, it is the stress at which the stress–strain curve becomes markedly non-linear — typically, a strain of about 1%. For ceramics and glasses, it is the compressive crushing strength; remember that this is roughly 15 times larger than the tensile (fracture) strength. For composites it is the tensile strength. For elastomers it is the tear-strength. The relative cost C_R is calculated by taking that for mild steel reinforcing-rods as unity; thus

$$C_R = \frac{\text{cost per unit weight of material}}{\text{cost per unit weight of mild steel}}$$

The guide lines show the loci of points for which

(a) $\sigma_f/C_R\rho = C$ (minimum cost design of strong ties, rotating discs, etc.)

(b) $\sigma_f^{2/3}/C_R\rho = C$ (minimum cost design of strong beams and shafts)

(c) $\sigma_f^{1/2}/C_R\rho = C$ (minimum cost design of strong plates)

The value of the constants C increase as the lines are displaced upwards and to the left. Materials offering the greatest strength per unit cost lie towards the upper left corner.

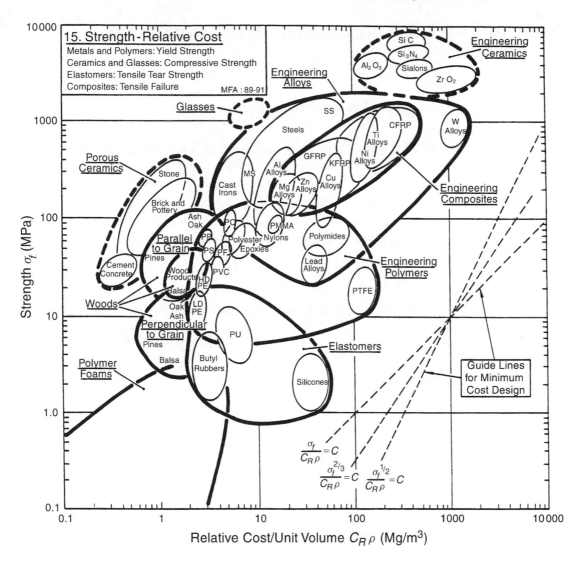

15. Strength-Relative Cost

Metals and Polymers: Yield Strength
Ceramics and Glasses: Compressive Strength
Elastomers: Tensile Tear Strength
Composites: Tensile Failure

MFA : 89-91

$$\frac{\sigma_f}{C_R \rho} = C$$

$$\frac{\sigma_f^{2/3}}{C_R \rho} = C$$

$$\frac{\sigma_f^{1/2}}{C_R \rho} = C$$

Guide Lines for Minimum Cost Design

Chart 16: Dry wear rate against maximum bearing pressure, P_{max}

The wear rate is defined as

$$W = \frac{\text{volume removed from contact surface}}{\text{distance slid}}$$

Archard's law, broadly describing wear rates at sliding velocities below 1 m/s, states that

$$W = k_a A_n P$$

where A_n is the nominal contact area, P the bearing pressure (force per unit area) at the sliding surfaces and k_a is Archard's wear-rate constant. At low bearing pressures k_a is a true constant, but as the maximum bearing pressure is approached it rises steeply. The chart shows Archard's constant,

$$k_a = \frac{W}{A_n P}$$

plotted against the hardness H of the material. In any one class of materials, high hardness correlates with low k_a.

Materials which have low k_a have low wear rates at a given bearing pressure, P. Efficient bearings, in terms of size or weight, will be loaded to a safe fraction of their maximum bearing pressure, which is proportional to hardness. For these, materials with low values of the product $k_a H$ are best. The diagonal lines show values of $k_a H$.

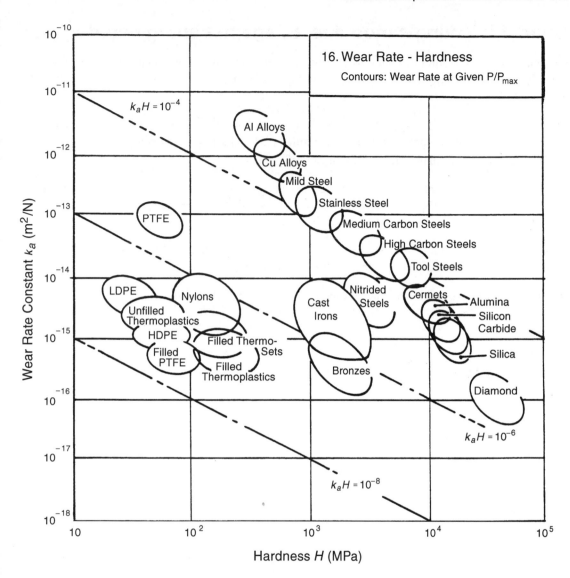

Chart 17: Young's modulus, E, against energy content, $q\rho$

The chart guides selection of materials for stiff, energy-economic components. The energy content per m^3, $q\rho$, is the energy content per kg, q, multiplied by the density ρ. The guide-lines show the loci of points for which

(a) $E/q\rho = C$ (minimum energy design of stiff ties; minimum deflection in centrifugal loading etc.)
(b) $E^{1/2}/q\rho = C$ (minimum energy design of stiff beams, shafts and columns)
(c) $E^{1/3}/q\rho = C$ (minimum energy design of stiff plates)

The value of the constant C increases as the lines are displaced upwards and to the left. Materials offering the greatest stiffness per energy content lie towards the upper left corner.
  Other moduli are obtained approximately from E using

  (a) $\nu = 1/3$; $G = 3/8E$; $K \approx E$ (metals, ceramics, glasses and glassy polymers)
or (b) $\nu \approx 1/2$; $G \approx 1/3E$; $K \approx 10E$ (elastomers, rubbery polymers)

where ν is Poisson's ratio, G the shear modulus and K the bulk modulus.

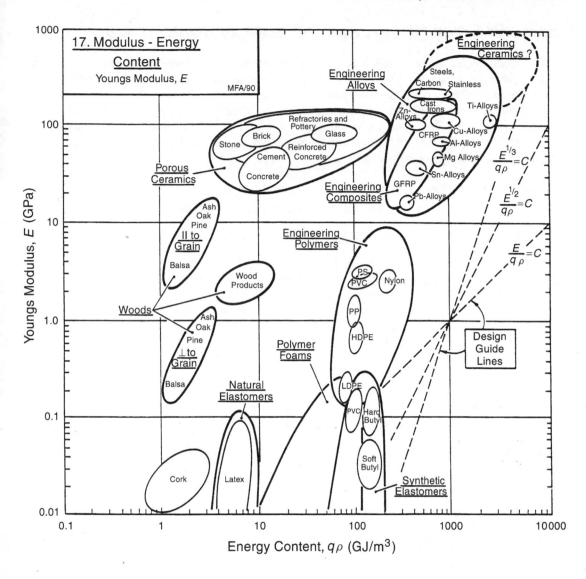

17. Modulus – Energy Content

Youngs Modulus, E

MFA/90

Chart 18: Strength, σ_f, against energy content, $q\rho$

The chart guides selection of materials for strong, energy-economic components. The 'strength' for metals is the 0.2% offset yield strength. For polymers, it is the stress at which the stress–strain curve becomes markedly non-linear — typically, a strain of about 1%. For ceramics and glasses, it is the compressive crushing strength; remember that this is roughly 15 times larger than the tensile (fracture) strength. For composites it is the tensile strength. For elastomers it is the tear-strength. The energy content per m^3, $q\rho$, is the energy content per kg, q, multiplied by the density, ρ. The guide lines show the loci of points for which

(a) $\sigma_f/q\rho = C$ (minimum energy design of strong ties; maximum rotational velocity of disks)

(b) $\sigma_f^{2/3}/q\rho = C$ (minimum energy design of strong beams and shafts)

(c) $\sigma_f^{1/2}/q\rho = C$ (minimum energy design of strong plates)

The value of the constant C increases as the lines are displaced upwards and to the left. Materials offering the greatest strength per unit energy content lie towards the upper left corner.

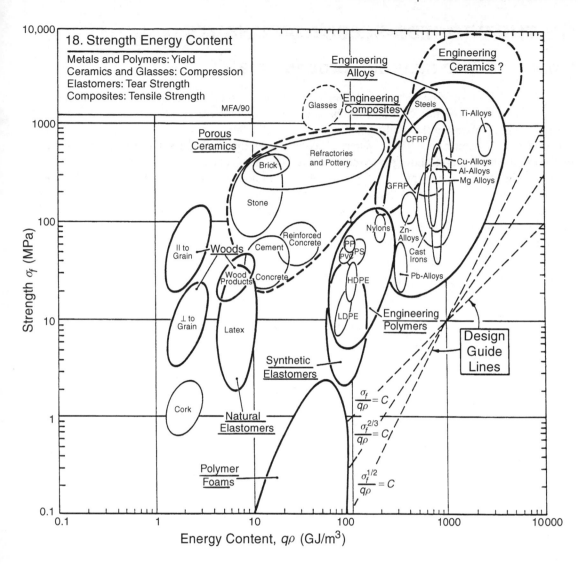

18. Strength Energy Content

Metals and Polymers: Yield
Ceramics and Glasses: Compression
Elastomers: Tear Strength
Composites: Tensile Strength

MFA/90

$$\frac{\sigma_f}{q\rho} = C$$

$$\frac{\sigma_f^{2/3}}{q\rho} = C$$

$$\frac{\sigma_f^{1/2}}{q\rho} = C$$

C.3 THE PROCESS-SELECTION CHARTS

Chart P1: The material–process matrix

The great number of processes used in manufacture can be classified under the broad headings on the vertical axis of this chart, which is a matrix relating material class to process class. The material classes, listed horizontally, are the usual ones: metals, ceramics and glasses, polymers and elastomers, and composites. These generic classes are subdivided: ferrous and non-ferrous metals, thermoplastic and thermosetting polymers, and so on. The number at a row-column intersection indicates the viability of a process for a material: 2 indicates that it is viable; 1 that it could be under special circumstances; 0 that it is not viable. Because the materials and processes are listed as subclasses (not individuals) some generalizations are inevitable. For a given material-subclass the table yields two short-lists: one of viable processes, the other of those which are possible or potentially viable.

Process Class		Ferrous	Refractory	Precious	Heavy	Light	Cementitous	Vitreous	Fine	Glasses	Thermosets	Thermoplastics	Elastomers	PMCs	MMCs	CMCs
Casting	Gravity	2	1	2	2	2	0	0	0	1	0	0	0	0	0	0
	Low pressure	2	0	2	2	2	0	0	0	2	0	0	0	0	1	0
	High pressure	1	0	2	2	2	0	0	0	1	0	0	0	0	2	0
	Investment	2	2	2	2	2	0	0	0	0	0	0	0	0	0	0
Moulding	Injection	0	0	2	0	0	0	0	0	2	2	2	2	2	0	0
	Compress	0	0	2	0	0	0	0	0	2	2	2	2	2	1	0
	Blow	0	0	0	0	0	0	0	0	2	0	2	2	0	0	0
	Foam	0	0	0	0	0	0	0	0	0	2	2	2	0	0	0
Deformation	Cold	2	0	2	2	2	0	0	0	0	0	0	0	0	0	0
	Warm	2	0	2	2	2	0	0	0	0	0	0	0	0	0	0
	Hot	2	2	2	2	2	0	0	0	2	0	0	0	0	0	0
	Sheet	2	1	2	2	2	0	0	0	0	2	2	2	0	1	0
Machining	Turn	2	2	2	2	2	0	1	0	0	2	2	0	2	2	0
	Mill	2	2	2	2	2	0	1	0	0	2	2	0	2	2	0
	Grind	2	2	1	2	2	0	2	2	2	0	0	0	0	2	2
	Polish	2	2	2	2	2	0	2	2	2	0	0	0	0	1	2
Powder Methods	Sinter/HIP	2	2	2	0	0	0	2	2	1	2	2	0	0	2	2
	Slip cast	0	0	0	0	0	0	2	2	1	0	0	0	0	0	1
	Spray forming	2	2	2	2	2	0	2	2	2	2	2	0	2	0	0
	Hydration	0	0	0	0	0	2	0	0	0	0	0	0	0	0	0
Composite Forming	Lay-up	0	0	0	0	0	0	0	0	0	0	2	0	2	0	2
	Mould	0	0	0	0	0	0	0	0	0	2	2	2	0	0	0
	Squeeze-cast	1	0	2	2	2	0	0	0	0	0	0	0	2	2	0
	Filament wind	0	0	0	0	0	0	0	0	0	0	0	0	2	0	0
Molecular Methods	PVD	0	2	2	2	0	0	2	2	0	0	0	0	0	1	0
	CVD	0	2	2	2	0	0	2	2	0	0	0	0	0	1	2
	Sputtering	2	2	2	2	2	0	0	0	0	0	0	0	0	0	0
	Electroforming	1	0	2	2	0	0	0	0	0	0	0	0	0	0	0
Special Methods	Electrochemical	2	2	2	2	2	0	0	0	0	0	0	0	0	2	0
	Ultrasonic	1	2	0	0	0	0	2	2	2	0	0	0	0	0	2
	Chemical	2	2	2	2	2	0	2	2	2	0	0	0	0	0	0
	Thermal Beam	2	2	2	2	2	0	2	2	2	2	2	2	2	2	2
Fabrication	Weld/braze	2	2	2	2	2	0	0	0	0	0	2	0	0	0	0
	Adhesive	2	2	2	2	2	2	2	2	2	2	2	2	2	2	2
	Fasten	2	2	2	2	2	2	2	2	2	2	2	2	2	2	2
	Microfabrication	2	2	2	2	2	0	2	2	2	2	2	2	2	2	2

Material Class

Metals — Ferrous, Refractory, Precious, Heavy, Light

Ceramics & Glasses — Cementitous, Vitreous, Fine, Glasses

Polymers & Elastomers — Thermosets, Thermoplastics, Elastomers

Composites — PMCs, MMCs, CMCs

Process Class

Chart P2: Hardness, *H*, against melting temperature, *T_m*

The match between process and material is established by the link to material class of Chart P1 and by the use of the melting point–hardness chart shown here. The melting point imposes limits on the processing of materials by conventional casting methods. Low melting point metals can be cast by any one of many techniques. For those which melt above 2000 K, conventional casting methods are no longer viable, and special techniques such as electron-beam melting must be used. Similarly, the yield strength or hardness of a material imposes limitations on the choice of deformation and machining processes. Forging and rolling pressures are proportional to the flow strength, and the heat generated during machining, which limits tool life, also scales with the ultimate strength or hardness. Generally speaking, deformation processing is limited to materials with hardness values below 3 GPa. Other manufacturing methods exist which are not limited either by melting point or by hardness. Examples are: powder methods, CVD and evaporation techniques, and electro-forming.

The chart presents this information in graphical form. In reality, only part of the space covered by the axes is accessible: it is the region between the two heavy lines. The hardness and melting point of materials are not independent properties: low melting point materials tend to be soft; high melting point materials are generally hard. This information is captured by the equation

$$0.03 < \frac{H\Omega}{kT_m} < 20$$

where Ω is the atomic or molecular volume and k is Boltzmann's Constant (1.38×10^{-26} J/K). It is this equation which defines the two bold lines.

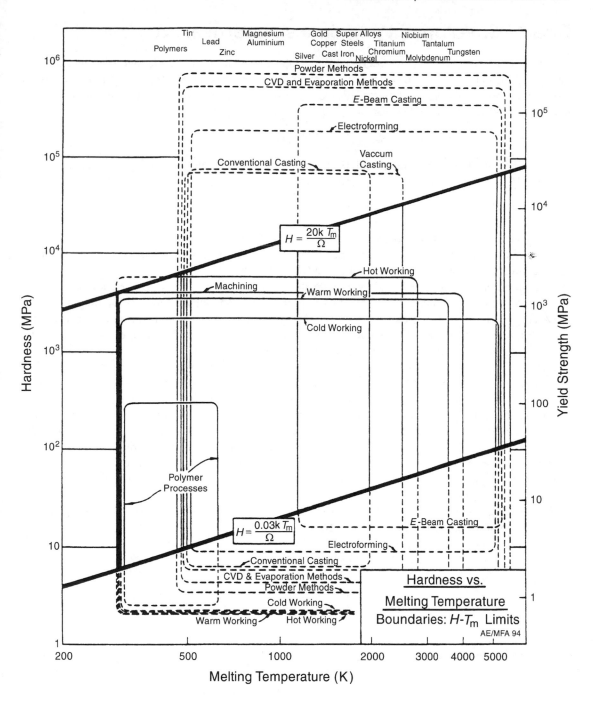

Tin Magnesium Gold Super Alloys Niobium
Polymers Lead Aluminium Copper Steels Titanium Tantalum
Zinc Silver Cast Iron Chromium Molybdenum Tungsten
Nickel

Powder Methods
CVD and Evaporation Methods
E-Beam Casting
Electroforming
Conventional Casting
Vaccum Casting

$$H = \frac{20k\ T_m}{\Omega}$$

Hot Working
Machining
Warm Working
Cold Working

Polymer Processes

$$H = \frac{0.03k\ T_m}{\Omega}$$

E-Beam Casting
Electroforming
Conventional Casting
CVD & Evaporation Methods
Powder Methods
Cold Working
Warm Working Hot Working

Hardness (MPa)

Yield Strength (MPa)

Melting Temperature (K)

Hardness vs.
Melting Temperature
Boundaries: *H*-*T*m Limits
AE/MFA 94

Chart P3: Volume, *V*, against slenderness, *S*

Manufacturing processes vary widely in their capacity to make thin, slender sections. For our purposes, slenderness, S, is measured by the ratio t/ℓ where t is the minimum section and ℓ is the large dimension of the shape: for flat shapes, ℓ is about equal to $\sqrt{A}$ where A is the projected area normal to t. Thus

$$S = \frac{t}{\sqrt{A}}$$

Size is defined by the minimum and maximum volumes of which the process is capable. The volume, V, for uniform sections is, within a factor of 2, given by

$$V = At$$

Volume can be converted approximately to weight by using an 'average' material density of $5000\,\text{kg/m}^3$; most engineering materials have densities within a factor of 2 of this value. Polymers are the exception: their densities are all around $1000\,\text{kg/m}^3$.

The size-slenderness chart is shown opposite. The horizontal axis is the slenderness, S; the vertical axis is the volume, V. Contours of A and t are shown as families of diagonal lines. *Casting processes* occupy a characteristic field of this space. Surface tension and heat-flow limit the minimum section and the slenderness of gravity cast shapes. The range can be extended by applying a pressure, as in centrifugal casting and pressure die casting, or by preheating the mould. *Deformation processes* — cold, warm and hot — cover a wider range. Limits on forging-pressures set a lower limit on thickness and slenderness, but it is not nearly as severe as in casting. *Machining* creates slender shapes by removing unwanted material. *Powder-forming* methods occupy a smaller field, one already covered by casting and deformation shaping methods, but they can be used for ceramics and very hard metals which cannot be shaped in other ways. *Polymer-forming methods* — injection moulding, pressing, blow-moulding, etc. — share this regime. *Special techniques*, which include electro-forming, plasma-spraying, and various vapour-deposition methods, allow very slender shapes. *Joining* extends the range further: fabrication allows almost unlimited size and complexity.

A real design demands certain specific values of S and V, or A and t. Given this information, a subset of possible processes can be read off.

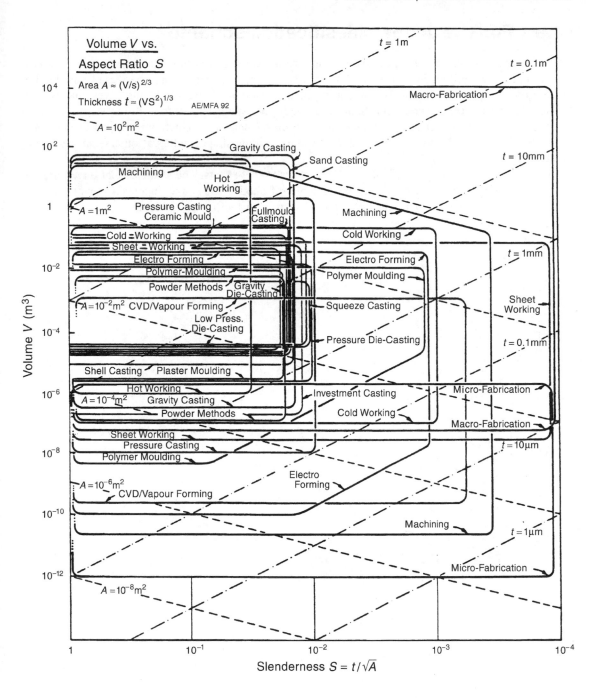

Chart P4: The shape classification scheme

Certain processes are well adapted to the manufacture of shapes with certain symmetries. Lathes, for example, are well adapted to the making of axisymmetric shapes; other shapes are possible but more difficult. Extrusion, drawing and rolling make prismatic shapes. Indexing gives shapes with translational or rotational symmetry. Shapes can be further subdivided into uniform, stepped, angled or dished. Uniform shapes are obviously easier to make than ones which are stepped or have side branches. Some processes are only capable of making hollow shapes, whereas others can make only solid ones.

The chart gives a shape classification which relates to these facts. The shapes are arranged in such a way that complexity, defined here as the difficulty of making the shape, increases downwards and to the right. Examples of each shape are given in order to facilitate the identification of the shape category which describes the desired design.

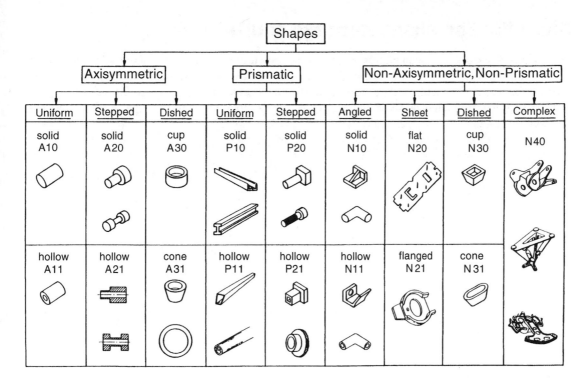

Chart P5: The shape–process matrix

The shape capabilities of manufacturing processes are summarized in this chart. It uses the same format and classifications as the material–process matrix. Processes are listed vertically and the various shapes (referred to by their codes) are listed horizontally. The capability of a process to make a shape is indicated by the number 0 or 1: the number 1 means that the process can make the shape; the number 0 that it cannot.

Process Class		A10	A11	A20	A21	A30	A31	P10	P11	P20	P21	N10	N11	N20	N21	N30	N31	N40
Casting	Gravity	1	1	1	1	1	1	1	1	1	1	1	1	1	1	1	1	1
	Low pressure	1	1	1	1	1	1	1	1	1	1	1	0	1	1	1	1	1
	High pressure	1	1	1	1	1	1	1	1	1	1	1	0	1	1	1	1	1
	Investment	1	1	1	1	1	1	1	1	1	1	1	1	1	1	1	1	1
Moulding	Injection	1	1	1	1	1	1	1	1	1	1	1	1	1	1	1	1	1
	Compress	0	1	0	1	1	1	1	1	0	0	0	0	0	0	1	1	0
	Blow	0	1	0	1	1	1	0	0	0	0	0	0	0	0	1	1	0
	Foam	1	1	1	1	1	1	1	1	0	1	1	0	1	1	1	1	0
Deformation	Cold	1	1	1	1	1	1	1	1	1	1	1	1	1	1	1	1	1
	Warm	1	1	1	1	1	1	1	1	1	1	1	0	1	1	1	1	1
	Hot	1	1	1	1	1	1	1	1	1	1	1	0	1	1	1	1	1
	Sheet	0	1	0	1	1	1	0	1	0	1	1	1	1	1	1	1	0
Machining	Turn	0	1	1	1	0	0	1	1	1	1	0	0	0	1	0	0	0
	Mill	0	0	0	0	0	1	1	0	1	1	1	0	1	1	1	1	1
	Grind	1	1	0	0	1	1	1	1	1	1	1	1	1	1	1	1	1
	Polish	1	1	1	1	1	1	1	1	1	1	1	1	1	1	1	1	1
Powder Methods	Sinter/HIP	1	1	1	1	0	0	1	1	1	1	0	0	0	1	0	0	0
	Slip cast	0	0	0	0	1	1	0	0	0	0	0	0	0	0	0	1	0
	Spray forming	1	1	1	0	1	1	0	1	1	1	1	1	1	1	1	1	0
	Hydration	1	0	0	0	0	0	0	0	0	0	0	0	0	0	0	0	0
Composite Forming	Lay-up	0	1	0	1	0	1	0	0	0	1	0	1	1	1	0	0	1
	Mould	0	1	0	1	1	1	0	0	0	1	0	1	0	1	0	0	1
	Squeeze-cast	1	1	1	1	1	1	1	1	1	1	1	0	1	1	1	1	1
	Filament wind	0	1	0	0	1	1	0	0	0	0	0	0	0	0	0	0	0
Molecular Methods	PVD	1	1	1	1	1	1	1	1	1	1	1	1	1	1	1	1	0
	CVD	1	1	1	1	1	1	1	1	1	1	1	1	1	1	1	1	1
	Sputtering	1	1	1	1	1	1	1	1	1	1	1	1	1	1	1	1	1
	Electroforming	1	1	1	1	1	1	1	1	1	1	1	1	1	1	1	1	1
Special Methods	Electrochemical	1	1	1	1	1	1	1	1	1	1	1	1	1	1	1	1	1
	Ultrasonic	1	1	1	1	1	1	1	1	1	1	1	1	1	1	1	1	1
	Chemical	1	1	1	1	1	1	1	1	1	1	1	1	1	1	1	1	1
	Thermal beaming	1	1	1	1	1	1	1	1	1	1	1	1	1	1	1	1	1
Fabrication	Weld/braze	1	1	1	1	1	1	1	1	1	1	1	1	1	1	1	1	1
	Adhesive	1	1	1	1	1	1	1	1	1	1	1	1	1	1	1	1	1
	Fasten	1	1	1	1	1	1	1	1	1	1	1	1	1	1	1	1	1
	Microfabrication	1	1	1	1	1	1	1	1	1	1	1	1	1	1	1	1	1

Shapes: Axisymmetric (A10, A11, A20, A21, A30, A31); Prismatic (P10, P11, P20, P21); Non-Axisymmetric, Non-Prismatic (N10, N11, N20, N21, N30, N31, N40)

Chart P6: Complexity against volume, *V*

Complexity is here defined as *the presence of features such as holes, threads, undercuts, bosses, re-entrant shapes, etc., which cause manufacturing difficulty or require additional operations.* For purposes of comparison, a scale of 1 to 5 is used with 1 indicating the simplest shapes and 5 the most complicated. Each process is given a rating for the maximum complexity of which it is capable corresponding to its proximity to the top left or bottom right shapes in Chart P4.

This information is plotted on the complexity level-size chart shown here. *Deformation processes* give shapes of limited complexity. *Powder routes* and *composite forming methods* are also limited compared with other methods. *Polymer moulding* does better. *Casting processes* offer the greatest complexity of all: a cast automobile cylinder block, for instance, is an extremely complicated object. *Machining* processes increase complexity by adding new features to a component. *Fabrication* extends the range of complexity to the highest level.

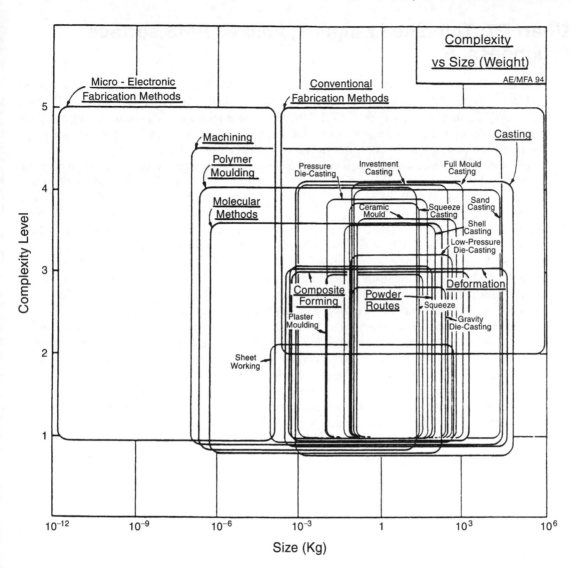

Chart P7: Tolerance range, *T*, against RMS surface roughness, *R*

No process can shape a part *exactly* to a specified dimension. Some deviation Δx from a desired dimension x is permitted; it is referred to as the *tolerance*, T, and is specified as $x = 100 \pm 0.1$ mm, or as $x = 50^{+0.01}_{-0.001}$ mm. Closely related to this is the *surface roughness* R, measured by the root-mean-square amplitude of the irregularities on the surface. It is specified as $R < 100\,\mu$m (the rough surface of a sand casting) or $R < 0.01\,\mu$m (a lapped surface; Table 11.2).

Manufacturing processes vary in the levels of tolerance and roughness they can achieve economically. Achievable tolerances and roughnesses are shown in this chart. The tolerance is obviously greater than $2R$ (shaded band): indeed, since R is the root-mean-square roughness, the peak roughness is more like $5R$. Real processes give tolerances which range from about $10R$ to $1000R$. Sand casting gives rough surfaces; casting into metal dies gives a better finish. *Moulded polymer* inherit the finish of the moulds and thus can be very smooth, but tolerances better than ± 0.2 mm are seldom possible because of internal stresses left by moulding and because polymers creep in service. Machining, capable of high dimensional accuracy and smooth surface finish, is commonly used after casting or deformation processing to bring the tolerance or finish to the desired level. Metals and ceramics can be *surface-ground* and *lapped* to a high tolerance and smoothness.

Precision and high finish are expensive: processing costs increase almost exponentially as the requirements for tolerance and surface roughness are made more severe. The chart shows contours of relative cost: an increase in precision corresponding to the separation of two neighbouring contours gives an increase in cost, for a given process, of a factor of two.

Achievable tolerances depend, of course, on dimensions (those given here apply to a 25 mm dimension) and on material. However, for our purposes, typical ranges of tolerance and surface finish are sufficient and discriminate clearly between various processes.

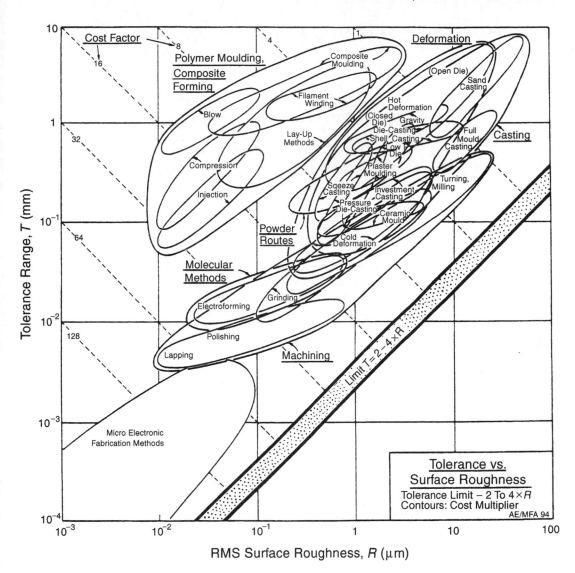

Appendix D

Problems

D1 Introduction to the problems

The problems posed in this section are designed to develop facility in selecting material, shape and process, and in locating data. The first in each section are very easy; some of those which come later are more difficult. Difficulty, when it arises, is not caused by mathematical complexity — the maths involved is simple throughout; it arises from the need to think clearly about the objectives, the constraints, and the free variables. The level of this kind of difficulty is indicated by the number of daggers: † means easy, ††††† means hard. Many of the problems can be tackled by drawing simple bounds (vertical and horizontal lines) onto the charts or by using the material indices listed in Tables 5.7 on page 78 and in Table B1–B7, pages 408–412. Others require the derivation of new material indices; here the catalogue of Appendix A will help. The problems of Section D2 introduce the use of the charts. Sections D3, D4 and D5 explore the way in which indices, some including shape, are used to optimize the selection. Most of the problems use the hand-drawn charts which come with this book. A few require charts which are not part of the hand-drawn set. For these the appropriate output of the *CMS* software (which allows charts with almost any axes to be created at will) is given.

Process selection problems are given in Section D6, which requires the use of the Process Charts of Chapter 11. Section D7 contains data-search problems. Ideally, the use of handbooks should be combined here with the use of computer databases (for information on these, see Chapter 13, Section 13.8). The *CMS* database, particularly, is helpful here. The final problems of Section D8 illustrate the interaction between material properties and scale, and the optimization of properties for a given application.

And a final remark: any one of the Case studies of Chapters 6, 8, 10, 12 or 14 can be recast as a problem, either by giving the design requirements and appropriate material limits and indices, and asking for a selection to be made, or by asking that the design requirements be formulated and limits and indices derived. The Case studies themselves then provide worked solutions.

The best way to use the charts which are a feature of the book is to have a clean copy on which you can draw, try out alternative selection criteria, write comments, and so forth; and presenting the conclusion of a selection exercise is, often, most easily done in the same way. Although the book itself is copyrighted, the reader is authorized to make copies of the charts, and to reproduce these, with proper reference to their source, as he or she wishes.

D2 Use of materials selection charts

D2.1   A component is at present made from brass (a copper alloy). Use Chart 1 to suggest two other metals which, in the same shape, would be stiffer. (†)

D2.2   Use Chart 1 to find the material with modulus $E > 200\,\text{GPa}$ and density $\rho < 2\,\text{Mg/m}^3$. (†)

D2.3 Use the Modulus–Density Chart (Chart 1) to identify the subset of materials with both modulus $E > 100\,\text{GPa}$ and the material index

$$M = E^{1/3}/\rho > 2.15(\text{GPa})^{1/3}/(\text{Mg/m}^3).$$

where ρ is the density. (Remember that, on taking logs, this equation becomes

$$\log(E) = 3\log(\rho) + 3\log(M)$$

and that this plots as a line of slope 3 on the $\log(E)$ vs. $\log(\rho)$ chart, passing through the point $\rho = 1/2.15 = 0.46$ at $E = 1$ in the units of Chart 1.) (††)

D2.4 Which have the higher specific strength, σ_f/ρ: titanium alloys or tungsten alloys? Use Chart 2 to decide. (†)

D2.5 The bubble labelled 'WOOD PRODUCTS' on the Charts 1 and 2 refers to plywood, fibre-board and chipboard. Do these materials have a higher or a lower specific stiffness, E/ρ, than nylons? (†)

D2.6 Are the fracture toughnesses, K_{Ic}, of common engineering polymers like PMMA (perspex) higher or lower than those of engineering ceramics like alumina? Chart 6 will help. (†)

D2.7 The elastic deflection at fracture when an elastic-brittle solid is loaded is related to the strain-at-failure by

$$\varepsilon_{fr} = \frac{\sigma_{fr}}{E}$$

where E is Young's modulus and σ_{fr} is the stress which causes a crack to propagate:

$$\sigma_{fr} \approx \frac{K_{Ic}}{\sqrt{\pi c}}$$

Here K_{Ic} is the fracture toughness and c the length of the longest crack the material may contain. Thus

$$\varepsilon_{fr} = \frac{1}{\sqrt{\pi c}}\left(\frac{K_{Ic}}{E}\right)$$

The materials which, for a given crack-length c show the largest deflection at fracture are those with the greatest value of the material index

$$M = \frac{K_{Ic}}{E} \tag{D1}$$

Use Chart 6 to identify three 'brittle materials' with exceptionally large fracture strains. (††)

D2.8 One criterion for design of a safe pressure vessel is that it should leak before it breaks: the leak can be detected and the pressure released. This is achieved by designing the vessel to tolerate a crack of length equal to the thickness t of the pressure vessel wall, without failing by fast fracture. The pressure p given by this design criterion is

$$p \le \left(\frac{K_{Ic}^2}{\sigma_y}\right)\left(\frac{1}{2RS}\right)$$

where σ_y is the yield strength (for metals, the same as σ_f), K_{Ic} is the fracture toughness, R is the vessel radius and D is a safety factor. The pressure is maximized by choosing the material with the greatest value of

$$M = \frac{K_{Ic}^2}{\sigma_y} \tag{D2}$$

Use Chart 7 to identify three alloys which have particularly high values of M. Comment on their relative materials. (††)

D2.9 An engine test-frame requires a material which is both stiff (modulus $E > 40\,\text{GPa}$) and has a high damping. Damping is the ability of a material to dissipate elastic energy: vibration-deadening materials have high damping. It is measured by the loss coefficient, η. Use Chart 8 to identify a subset of 4 possible materials for the engine test-frame. Comment on their suitability. (†)

D2.10 Use Chart 8 to identify material which should make good bells. (†)

D2.11 Use Chart 9 to identify a small subset of materials with the lowest thermal conductivity (best for long-term insulation), and to identify a small subset with the lowest thermal diffusivity (best for short-term insulation). (†)

D2.12 Use Chart 9 to find two materials which conduct heat better than copper. (†)

D2.13 The window through which the beam emerges from a high-powered laser must obviously be transparent to light. Even then, some of the energy of the beam is absorbed in the window and can cause it to heat and crack. This problem is minimized by choosing a window material with a high thermal conductivity λ (to conduct the heat away) and a low expansion coefficient α (to reduce thermal strains), that is, by seeking a window material with a high value of

$$M = \lambda/\alpha \tag{C3}$$

Use Chart 10 to identify the best material for an ultra-high powered laser window. (Don't be surprised at the outcome — lasers really are made with such windows.) (††)

D2.14 Use Chart 12 to decide whether steels are more or less resistant than cast irons to thermal shock. (†)

D2.15 Table 5.7 tells us that the cheapest material for a column which will not buckle under a given axial load is that with the greatest value of the material index

$$M = \frac{E^{1/2}}{C_m \rho} \tag{C4}$$

where E is the modulus, ρ the density and C_m the cost per kilogram of the material. Use Chart 14 to identify a subset of six materials that perform best by this criterion. How do these compare with the materials used in the construction of buildings? (Remember that, on taking logs, equation (C4) becomes

$$\log(E) = 2\log(C_m \rho) + 2\log(M)$$

and that this plots as a line of slope 2 on the $\log(E)$ vs $\log(C_m \rho)$ chart.) (††)

D2.16 Use Chart 16 to decide whether engineering ceramics are more or less wear-resistant than metals. (Remember that wear-rate, at a given bearing pressure, is measured by the wear-rate constant, k_a; low k_a means low wear rate.) (†)

D2.17 Use an index selected from Table B1, page 408, together with Chart 17, to determine whether the energy-content of a reinforced-concrete panel of prescribed stiffness is greater or less than that of a wood panel of the same stiffness. Treat the panel as a plate loaded in bending. (††)

D2.18 A beam-like component of specified section shape, designed to carry a prescribed bending load without failing, could be moulded from nylon or die-cast from a zinc alloy. Select the appropriate index from Appendix B, Table B1.1 and use Chart 18 to decide which has the lower energy content. (††)

D2.19 A material is required for the blade of a rotary lawn-mower. Cost is a consideration. For safety reasons, the designer specified a minimum fracture toughness for the blade: it is $K_{Ic} > 30\,\text{MPa}\,\text{m}^{1/2}$. The other mechanical requirement is for high hardness, H, to minimize blade wear. Hardness, in applications like this one, is related to strength:

$$H \propto 3\sigma_f$$

where σ_f is the strength (Chapter 4 gives a fuller definition). Use Chart 7 to identify three materials which have $K_{Ic} > 30\,\text{MPa}\,\text{m}^{1/2}$ and the highest possible strength. To do this, position a 'K_{Ic}' selection line at $30\,\text{MPa}\,\text{m}^{1/2}$ and then adjust a 'strength' selection line such that it just admits three candidates. Transfer this strength-limit to Chart 15, and use it to rank your selection by material cost; hence make a final selection. (††)

D3 Deriving and using material indices

D3.1 Material indices for elastic beams in bending

Start each of the four parts of this problem by listing the function, the objective and the constraints.

(a) Show that the best material for a cantilever beam of given length ℓ and given (i.e. fixed) square cross-section $(t \times t)$ which will deflect least under a given end load F (Appendix A, Section A3) is that with the largest value of the index $M = E$, where E is Young's modulus (neglect self-weight).

(b) Show that the best material choice for a cantilever beam of given length ℓ and with a given section $(t \times t)$ which will deflect least under its own self-weight is that with the largest value of $M = E/\rho$, where ρ is the density.

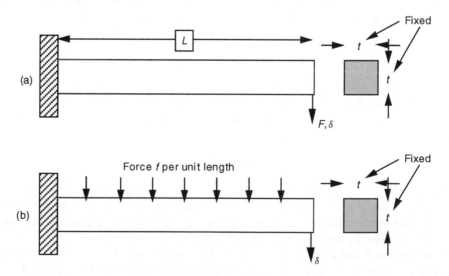

Fig. D3.1 Beams loaded in bending, with alternative constraints.

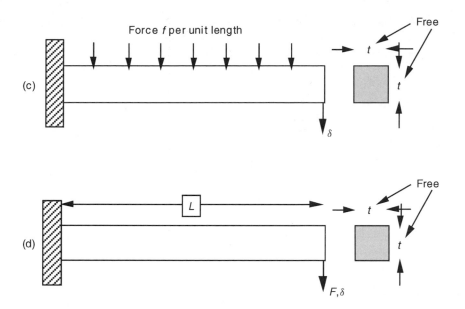

Fig. D3.1 (*continued*)

(c) Show that the material index for the lightest cantilever beam of length ℓ and square section (not given, i.e. free) which will not deflect by more than δ under its own weight is $M = E/\rho^2$.

(d) Show that the lightest cantilever beam of length ℓ and square section (free) which will not deflect by more than δ under an end load F is that made of the material with the largest value of $M = E^{1/2}/\rho$ (neglect self weight). (†††)

D3.2 Effect of geometric constraints on material indices

Derive

(a) equation (5.12) of the text, and
(b) equation (5.13) of the text, using the method in the section which precedes the two equations. (†††)

D3.3 The health service crutch

You are asked to re-design the Health Service Crutch. It is designed for a person with only one serviceable leg, and is, in mechanical terms, a slender column with a padded top (which goes under the armpit) loaded in compression by a known maximum load F. The crutch should be as light as possible. It would seem desirable that it should not fail by elastic buckling. Derive a material index for the lightest column of solid circular section, which will not buckle elastically under an end-load, F.

Write, first, an equation for the weight of the crutch; then one for failure by elastic buckling — Euler's law (Appendix A, Section A5) — and use it to eliminate the free variable — the radius — from the first equation.

Use the result to select a subset of five candidate materials for the crutch, using the appropriate Materials Selection Chart. (†††)

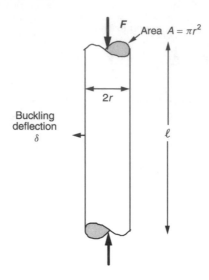

Fig. D3.2 The crutch.

D3.4 The daisy-wheel print head

A daisy-wheel print head (Figure D3.3) is a set of elastic fingers, each carrying a character-face at its tip. The tip is struck from behind by a hammer which compresses it against the paper. Unlike the golf ball of Case study 6.8, only a tiny part of the wheel is loaded (and that in compression), allowing lighter construction: a typical golf-ball weighs 12 g; a typical daisy-wheel, only 8. But the material requirements are different. A golf-ball is positively positioned, lifted from the paper by a spring; a daisy-wheel leaves the paper only because of the elastic recovery of the finger carrying the type. This recovery time is related to the stiffness of the finger, which we model as a cantilever (Appendix A, Section A3) of length ℓ, width b, and thickness t. We wish to minimize the mass m of the finger (to allow rapid repositioning) where

$$m = b\ell t\rho$$

subject to the constraint that the finger is adequately stiff.

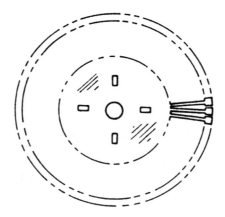

Fig. D3.3 A daisy-wheel print head.

Treat ℓ and b as fixed by the design (they are certainly severely constrained by it), leaving t as the free variable. Derive a material index for selecting materials which meet this stiffness constraint at minimum mass. Use it to identify five polymers which might be suitable for making daisy-wheel print heads. (†††)

D3.5 Materials for vaulting poles

You are hired by a pole vault enthusiast who wishes to equip himself with the very best of vaulting poles. International standards set the length and maximum section of the pole, which must be cylindrical. In use, the pole bends elastically, storing energy which is released at the top of the flight path, projecting the jumper over the bar (Figure D3.4). Assume that the best pole is that made of the material which stores (and then releases) the most elastic energy per unit volume without failing.

Derive a material index and use it to identify a subset of materials which should make good vaulting poles. Are the results consistent with what you know about real vaulting poles? How does the selection change if the criterion is that of storing the most elastic energy per unit weight, instead of volume? (†††)

D3.6 Springs for trucks

In vehicle suspension design is it desirable to minimize the mass of all components. You have been asked to select a material and dimensions for a light-weight spring to replace the steel leaf-spring of an existing truck suspension.

The existing leaf-spring is a beam, shown schematically in Figure D3.5. The new spring must have the same length L and stiffness S as the existing one, and must deflect through a maximum safe displacement δ_{max} without failure. The width b and thickness t are not constrained.

Derive a material index for the selection of a material for this application. Note that this is a problem with two free variables: b and t; and there are two constraints, one on safe deflection δ_{max} and the other on stiffness S. Use the two constraints to fix the two free variables. (†††)

Fig. D3.4 A pole-vaulter.

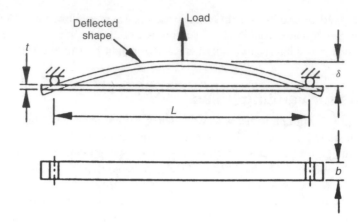

Fig. D3.5 A truck spring. It must have a given stiffness, S, and be capable of deflection through δ_{max} without failure. The objective is to minimize the mass m.

D3.7 Elastic con-rods

It has been suggested that the bearing and gudgeon pin which form the little-end bearing of the con-rod/piston assembly of a reciprocating engine or pump could be replaced by an elastic ligament or hinge, as shown in Figure D3.6.

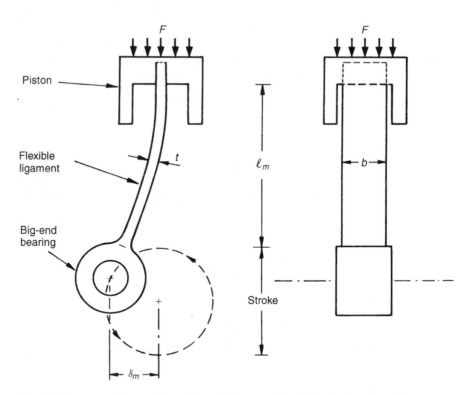

Fig. D3.6 The elastic connecting rod. The flexible ligament is rigidly fixed to the piston.

(a) Discuss the mechanics aspects of this suggestion, assuming that the objective is to maximize the allowable thrust, F. Formulate the constraints which must be met by such a hinge.
(b) Assume the width, b, of the hinge is set by clearances within the device (i.e. it is fixed) and that its length ℓ_m and the maximum lateral deflection δ_{max} (equal to half the stroke) are both specified. Derive a material index for the hinge and use it to identify materials which allow large F without causing the elastic ligament or buckle or fail plastically.

Proceed as follows

(i) Write down an expression for the buckling load of the ligament (see Appendix A, Section A5 for the relevant equations). Use this to identify the minimum thickness t of the ligament which will not buckle under an axial load F.
(ii) Write down an equation for the maximum stress σ in the bending ligament. It is the sum of two contributions, one from the axial load F, the other from the elastic bending (Appendix A, Section A4) which is greatest when the deflection is δ_{max}. The design goal is to maximize F without causing the ligament to fail, which it will of the stress in it anywhere exceeds its failure stress, σ_f. (Fatigue can be allowed for by applying a sufficiently large safety factor, S).
(iii) Substitute for t from (i) into this result, giving a (quadratic) equation for F.
(iv) Now identify the combination of material properties which maximize F. Examine the combination of material properties that appear in these limits: they are the relevant material indices. Use them to identify a small subset of materials which allow high F_{max}. Use common sense to reject any which, for other reasons, are obviously inappropriate, identifying the constraint which causes them to be so. (†††††)

D3.8 The pipeline inspection gadget

A PIG, to someone in the oil and gas business, is not an animal; it is a Pipeline Inspection Gadget (Figure D3.7). PIGs are pulled through pipelines to detect and remove obstructions. Intelligent PIGs do more: they probe the condition of the pipe wall and its surroundings, by vibrational methods, looking for defects. The body of the PIG is a cylindrical shell with an outer diameter fixed by the pipe through which it must pass. The body should be as light as possible, but have natural resonance frequencies which are high so that they do not interfere with the low frequency vibrations used to probe the condition of the pipe itself. Identify a subset of materials from which the PIG body might be made which minimize weight while keeping all natural frequencies greater than a given value ω^*.

Proceed by writing an equation for the mass of the cylinder of length ℓ and wall-thickness t (ignore the end caps). The natural frequencies of the longitudinal flexural modes (Appendix A, Section A12) of the shell are

$$\omega_j = \left(\frac{EI}{m_o}\right)^{1/2}\left(\frac{j\pi}{\ell}\right)^2$$

where $m_o = \rho A$ is the mass per unit length and $j(= 1, 2, 3\ldots)$ is the mode number. The second moment of area for this mode is

$$I_1 = \pi R^3 t$$

and the appropriate area A_1 is

$$A_1 = 2\pi R t$$

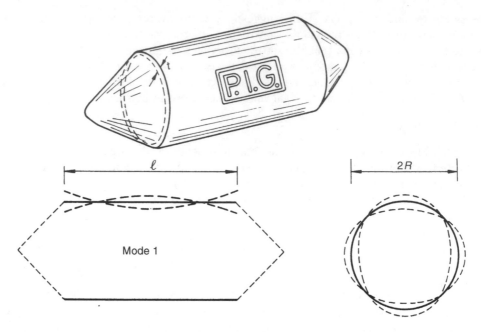

Fig. D3.7 A pipeline inspection gadget. The lowest natural frequencies of its shell must be kept high while at the same time minimizing its mass.

where t is the wall thickness of the shell. The natural frequencies of the circumferential modes (Appendix A, Section A12) are proportional to ω_j again, but with

$$I_2 = \frac{\ell t^3}{12}$$

and

$$A_2 = \ell t$$

Use these to find the lowest natural frequency and use the result to eliminate the free variable, t. Hence derive a material index and use it to make a selection. (†††)

D3.9 The tape-deck opening spring

A spring-loaded device is required to open the tape-deck cover of a compact tape-player when the catch is released. Describe the steps you would follow in proceeding from this 'market need' to the final production of the device, with particular reference to the source and types of material data you would require at each stage of the design process. Your description should include an analysis of the requirement, and some possible mechanical solutions.

A proposed design for the device is sketched in Figure D3.8. It comprises a torsion bar with a rigid actuator arm, and the following features of the design are specified:

   (i) The length of the arm, H;
  (ii) The range of movement of the end of the arm, δ_{max};
 (iii) The opening force F when the lid is shut.

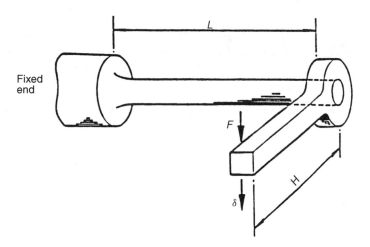

Fig. D3.8 A torsion spring to open a tape-deck lid.

It is required to choose a suitable material for the torsion bar in order to minimize (a) its length (b) its radius and (c) its volume. Find the appropriate groupings of material properties, and draw lines on the relevant Materials Selector Chart to indicate the likely candidate materials. Eliminating materials which have any obvious drawbacks, select the most appropriate material, and write down expressions for the dimensions of the bar in terms of the design specifications. (†††)

D3.10 Material to resist thermal shock

When the temperature of the surroundings of an engineering component changes suddenly from T_o to a lower one T_1, the temperature gradient within it generates stresses, tensile at the surface, which can be damaging. If they exceed the brittle fracture strength σ_f, cracks are nucleated and chunks may spall off. If, instead, they exceed the yield strength, plastic flow takes place and — if repeated — may cause fatigue failure.

Consider a material which is suddenly cooled from T_o to T_1. A thin surface skin, now at T_1, tends to contract but is strongly bonded to the bulk of the component, which is still at T_o. The thin skin is constrained by the bulk, and is therefore elastically strained by an amount which is equal and opposite to the thermal strain. The elastic strain gives (via Hooke's law) a stress. Use this model as a basis for deriving an equation for the thermal stress and equate this to the failure stress to give an equation for the maximum temperature interval which the component can sustain without failure. Read off the combination of properties (the material index) which maximizes this interval. (††)

D3.11 Bearings for bicycles

You are employed by a company which makes bicycles. To counter an influx of cheap imported bicycles from some remote country where the climate is warm and wages are low, the company decides to redesign its standard bike to make it as cheap as possible. You are asked to identify the cheapest material for the bearings. Ball bearings are out; what is wanted is the cheapest material which will give adequate wear resistance when used as a sleeve bearing.

Suggest materials for cheap bicycle bearings. The simplest shape is a sleeve bearing: a thin-walled cylinder of length ℓ, radius r, and wall-thickness t all three of which are fixed by the need to match

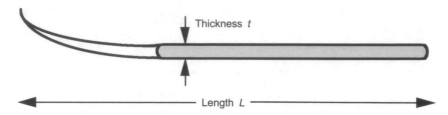

Fig. D3.9 Schematic of a disposable fork.

standard shaft and pedal dimensions. Use Chart 16 to identify possible candidates for the bearing and rank them by cost, using approximate data from Chart 14. Comment on the selection you make. Need for lubrication? Corrosion problems which might greatly increase the wear rate? Does your final conclusion match with your experience of cheap bearings? (††)

D3.12 Disposable knives and forks

Disposable knives and forks are ordered by an environmentally-conscious pizza-house. The shape of each (and thus the length, width and profile) are fixed, but the thickness is free: it is chosen to give enough bending-stiffness to cut and impale the pizza without excessive flexure. The pizzeria-proprietor wishes to enhance the greenness of his image by minimizing the energy-content of his throw-away tableware, which could be moulded from polystyrene (PS) or stamped from aluminium sheet.

Establish an appropriate material index for selecting materials for energy-economic cutlery. Model the eating-implement (Figure D3.9) as a beam of fixed length L and width w, but with a thickness t which is free, loaded in bending. The objective-function is the energy content: the volume times the energy content, q, per unit volume. The limit on flexure imposes a stiffness constraint (Appendix A Appendix A, Section A3). Use this information to develop the index, and use it, with Chart 17, to decide between polystyrene and aluminium.

Flexure, in cutlery, is an inconvenience. Failure — whether by plastic deformation or by fracture — is more serious: it causes loss-of-function; it might even cause hunger. Repeat the analysis, using a strength constraint (Appendix A, Appendix A, Section 4). Use Chart 18 to decide if your initial choice of material is still valid. (†††)

D4 Selection with multiple constraints

The four problems of this section illustrate the application of multiple criteria. All four are solved by using the methods of Chapter 9.

D4.1 A light, stiff, strong tie

A tie, of length ℓ, loaded in tension, is to support a load F, at minimum weight, without failing or extending elastically by more than δ (two constraints). Follow the method of Chapter 9 to establish two performance equations for the mass, one for each constraint, from which two material indices and one coupling equation which links them are derived.

Use these and Chart 5 to identify candidate materials for the tie (a) when $\ell/\delta = 100$ and (b) when $\ell/\delta = 10^3$. Ignore ceramics in your selection. They have great strength in compression but are brittle and, if flawed, have low strength in tension. (††††)

D4.2 A cheap column that must not buckle or crush

The best choice of material for a light strong column depends on its aspect ratio: the ratio of its height h to its diameter d. This is because short, fat columns fail by crushing; tall slender columns buckle instead. Derive two performance equations for the material cost of a column of solid circular section, designed to support a load F, one using the constraints that the column must not crush, the other that it must not buckle.

Some possible candidates for the column are listed in Table D4.1. Use these to identify candidate materials (a) when $F = 10^5$ N and $h = 2$ m; and (b) when $F = 10^3$ N and $h = 20$ m. Ceramics are admissible here, because they have high strength in compression. Reject any candidates which are expensive. (††††)

D4.3 A light, safe, pressure vessel

When the pressure vessel has to be mobile, its weight becomes important. Aircraft bodies, rocket casings and liquid-natural gas containers are examples; they must be light, and at the same time they must be safe. What are the best materials for their construction?

Write, first, a performance equation for the mass m of the pressure vessel. Assume, for simplicity, that it is spherical, of specified radius R, and that the wall thickness, t (the free variable) is small compared with R. The pressure difference, p, across this wall is fixed by the design. The first constraint is that the vessel should not yield — that is, that the tensile stress in the wall should not exceed σ_f. The second is that it should not fail by fast fracture; this requires that the wall-stress be less than $K_{Ic}\sqrt{\pi c}$, where c is the length of the longest crack that the wall might contain. Use each of these in turn to eliminate t in the objective equation; use the results to identify two material indices and a coupling relation between them: it contains the crack length, c.

Figure D4.1 shows the output of the CMS software with the two material indices as axes. Plot the coupling equation onto this figure for two values of c: one of 6 mm, the other of 1 μm. Identify candidate materials for the vessel for each case. (††††)

D4.4 The dare-devil plank

You are seized by the noble notion of raising money for charity by traversing a narrow street on a plank resting on the roofs of the houses on either side, a distance ℓ apart. You feel some concern, first, that the plank is strong enough to support you without breaking (a *strength* constraint), second,

Table D4.1 Data for candidate materials for the column

Material	Density ρ (kg/m³)	Cost/kg C_m (£/kg)	Modulus E (GPa)	Compression strength σ_f (MPa)
Wood (spruce)	490	0.4	15	45
Brick (common, hard)	2100	0.35	25	120
Sandstone	2400	0.4	50	130
Granite	2600	0.6	80	450
Poured concrete	2300	15	15	0.1
CFRP (laminate)	1600	35	100	450
GFRP (laminate)	1780	4,5	28	300
Structural steel	7850	0.4	210	1100
Al-alloy 6061	2700	1.2	69	130

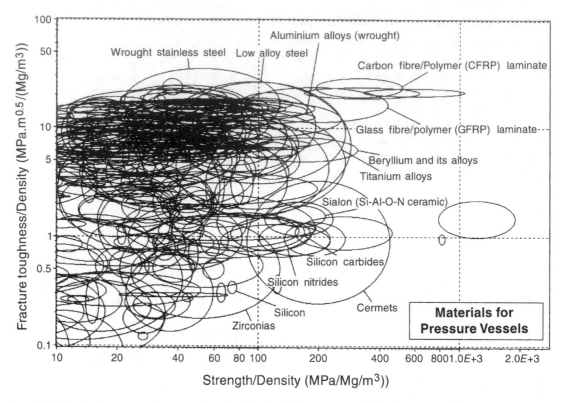

Fig. D4.1 Part of a computer-generated Selection Chart for specific fracture toughness, K_{Ic}/ρ and specific strength, σ_f/ρ.

that it will not sag too much when you are in the middle (a *stiffness* constraint), and, finally, that it be as light as possible so that you can get it up on to the roof by yourself (*objective function* for mass, to be minimized). The plank has a solid rectangular section of thickness t (free) and width b (fixed by your wish for a good footing). Derive two material indices for material selection based on the specification given above. This is a problem with one free variable (t) and two constraints (strength and stiffness). Proceed as follows.

(a) Write down an expression for the mass m of the plank; it is the objective function.

(b) Write down an expression for the failure load F of the plank in terms of its moment of area I, thickness t and its yield or fracture strength σ_f (Appendix A, Section A4); it must not be less than F^* (the force caused by your weight). Solve for t and substitute into the objective function, giving the first mass equation, m_1.

(c) Write down an expression for the deflection δ produced by a load F^* in terms of I and the modulus E (Appendix A, Section A3); it must not be more than δ^* (the maximum acceptable deflection). Again solve for t and substitute into the objective function, giving the second mass equation, m_2.

Given that $\ell = 6$ m, $b = 100$ mm, your weight (times a prudent safety factor) is 100 kg and that a deflection exceeding $\delta^* = 100$ mm would make life difficult, make a selection of material for the beam form the candidates listed in Table D4.2, below. (††††)

Table D4.2 Potential candidate materials for the plank

Material	Density ρ (kg/m^3)	Modulus E (GPa)	Strength σ_f (MPa)
Wood (spruce)	490	15	45
CFRP (laminate)	1600	100	450
GFRP (laminate)	1780	28	300
High-strength steel	7850	210	1100
Al-alloy 6061	2700	69	130

D5 Selecting material and shape

The problems of the section involve the use of shape factors for their solution.

D5.1 Improvised ski-poles

You wish to improvise a ski-pole using standard stock. The pole must meet a stiffness specification and be as light as possible. Current stock includes:

(a) Wood poles in a range of solid circular sections.
(b) Aluminium tubing in a range of diameters with the ratio (tube radius/wall thickness) up to 15.
(c) Steel tubing in a range of diameters with the ratio (tube radius/wall thickness) up to 18.

Derive (or retrieve from Table 8.1, p. 195) a material index for selecting materials for a light stiff beam, allowing for shape. Compare the three candidate stock-materials using this index, tabulating the results. Plot the stock materials onto Chart 1, and draw any conclusions you can about other promising material-shape combinations for the ski-pole. Material properties are listed below.

Repeat the calculation, using the bending strength of the pole as the design constraint rather than its stiffness. (†††)

D5.2 Light tubular display stands

A concept for a lightweight display stand is shown in Figure D5.1. The frame must support a mass 100 kg (to be placed on its upper surface) at a height $h = 1$ m without failing by elastic buckling. It is to be made of stock tubing and must be as light as possible. Derive (or retrieve from Table 8.1, page 195) a material index for the tubular material of the stand which meets these requirements, and which includes the shape of the section, described by the shape factor

$$\phi_B^e = \frac{4\pi I}{A^2}$$

where I is the second moment of area and A is the section area. Tubing is available from stock in the following materials and sizes. Use this information and the material index to identify the best

Table D5.1 Materials for a light ski-pole

Material	Density ρ (kg/m^3)	Modulus E (GPa)	Strength σ_f (MPa)
Wood (spruce)	490	15	45
Cold drawn mild steel tube	7850	210	600
Cold drawn Al tube	2700	69	130

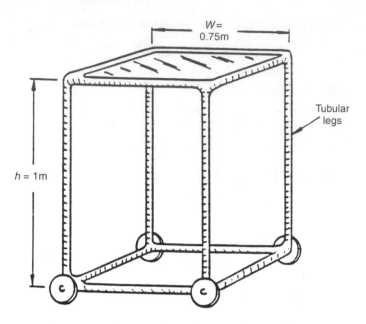

Fig. D5.1 A concept for a tubular display stand.

Table D5.2 Materials for the display stand

Material	Tube radius	Wall thickness/ tube radius
Aluminium alloys	All radii up to 25 mm	0.07 to 0.25
Steel	All radii up to 30 mm	0.045 to 0.1
Copper alloys	All radii up to 20 mm	0.075 to 0.1
Polycarbonate (PC)	All radii up to 10 mm	0.15 to 0.3
Various woods	All radii up to 40 mm	Solid circular sections only

stock material for the stand. Check that a stand made of your selection will actually support the design load. (††††)

D5.3 Energy-efficient floor joists

Section 8.4, page 200, of the text compared solid wood beams with shaped steel I-beams for use as floor joists in buildings on the basis of weight. When the design is stiffness-limited, the wooden joist with $\phi_B^e = 2$ and the shaped steel one with $\phi_B^e = 25$ performed equally well — that is, they had the same weight. But is the steel beam more energy-intensive than the wood?

(a) Locate from Table 8.1, page 195, (or derive if you prefer) the material index for shaped beams loaded in bending. Select that for stiffness-limited design at minimum energy content. Extract data from Chart 17 (p. 370) for modulus E and energy content per unit volume $q\rho$. Hence compare the value of the material index for the two beams.

(b) Answer the problem in a second way. Construct a line of slope 1, passing through 'steels' on Chart 17, and move steels down both axes by the factor 25, as in Figure 8.6, page 202. The shaped steel can be compared with other materials — wood, say — using the ordinary guide-lines which are shown on the figure. Is the conclusion the same?

(c) Repeat the two operations for strength-limited design for the two beams.
(d) What do you conclude about the relative energy-penalty of design with wood and with steel? (††††)

D5.4 Cheap floor joists

The Problem D5.3 required a comparison of wood as a solid beam with steel as an I-beam, on an energy-content basis. Compare the two, instead, on a material-cost basis, using Charts 14 and 15. (††††)

D5.5 An energy-absorbing bumper

The bumper for a military vehicle is to be designed on the basis of its ability to absorb energy in a mild collision without permanent deformation. For the purpose of analysis, the bumper is idealized as a thin-walled tube attached to the vehicle at its ends by rigid supports. The length and outer diameter of the bumper are fixed. Ignore contact stresses and any tendency to buckle at the point of impact.

(a) If the wall thickness of the tube is held constant, identify the combination of materials properties which determine the maximum energy the bumper can store without permanent damage. Use the appropriate design chart to select candidate materials for the bumper.
(b) If the wall thickness of the tube, though thin, can be varied, identify the combination of material properties which, for a given stored energy, will minimise the weight of the bumper. By combining information from two design charts, identify two candidate materials which would be sensible choices for the bumper.
(c) Discuss production methods appropriate to your selection, with particular reference to batch size and to the method of joining the bumper to the vehicle. Comment on how these consideration might influence your final selection for cases (a) and (b) above. (††††)

D5.6 Determining shape factors

A shape factor measures the gain in stiffness or in strength, relative to a solid circular cylinder, by shaping the material. Thus the shape factor for elastic bending if a beam is the ration of the bending stiffness of the shaped beam to that of a solid cylinder of the same length and mass. Shape factors ϕ can be determined analytically by calculating I and A for the section and using the equation

$$\phi_B^e = \frac{4\pi I}{A^2}$$

and the others like it. Alternatively, shape factors can be found by experiment, by measuring the stiffness S and mass m of a beam of length ℓ, made of a material with modulus E and density ρ, by inverting equation (7.26) of the text, (and the others like it) to give

$$\phi_B^e = \left(\frac{4\pi \ell^5 S_B}{C_1 m^2}\right)\left(\frac{\rho^2}{E}\right) \tag{D5.1}$$

(a) Calculate analytically the shape factor for a thin-walled box girder with square section 40 mm × 40 mm and wall thickness 2 mm.
   (Answer: $\phi_B^e = 10$.)
(b) Calculate the shape factor ϕ_B^e of an I-section beam with a bending stiffness 10^8 N/m if the bending stiffness of a solid cylinder of the same material, weight and length is 2×10^6 N/m.

(c) Calculate the shape factor ϕ_B^e from the following experimental data, measured on an aluminium alloy beam loaded in 3-point bending (for which $C_1 = 48$ — see Appendix A, Section A3) with:

Stiffness	$S_B = 7.2 \times 10^5 \, \text{N/m}$
Mass	$m = 1 \, \text{kg}$
Length	$\ell = 1 \, \text{m}$
Density	$\rho = 2670 \, \text{kg/m}^3$
Modulus	$E = 69 \, \text{GPa}$

(Answer $\phi_B^e = 20$.) (†††)

D5.7 Umbrella ribs

Umbrella design has a long history: sophisticated models can be seen in Chinese prints of the 9th century and before. The ribs of an umbrella (Figure D5.2) are loaded in bending; the struts are loaded in compression. The ribs must deflect elastically to take up the curved shape of the umbrella, exerting a specified restoring moment M on the fabric when the umbrella is fully open (thus the bending stiffness is specified). They must do this without failing by plastic collapse or fracture when this moment (or some multiple of it — to allow for mishandling) is applied. The best rib is that with the highest failure moment, M_f, which also meets the stiffness constraint.

(a) Derive a material index for selecting materials for ribs of solid circular section of an umbrella which meets these conditions. Note that the specified working moment M is a constraint; the objective function describes the failure moment, M_f, which is to be maximized. Use the appropriate Materials Selection Chart to identify a subset of materials which would perform well in this application. Comment on any problems of production, joining, cost, etc, associated with each choice.

(b) Umbrellas from the Far East have solid bamboo-wood ribs (they are cut from the solid wall of the hollow bamboo stalk). These ribs have a near-circular section. Bamboo wood has a modulus of 10.5 GPa and a failure stress of 127 MPa. Add this solid material to a copy of the Chart and hence make a judgement about its suitability for umbrella ribs.

(c) Calculate the weight-saving if the ribs were made out of hollow tubes of bamboo wood with a ratio of wall-thickness to diameter of 4:1. (††††)

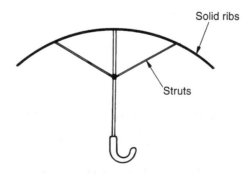

Fig. D5.2 An umbrella or contre-soleil, depending on the season.

D5.8 Bamboo scaffolding

Bamboo scaffolding has been used in China for building construction for over 2000 years. The bamboo lengths are tied together with soft rope. You have been commissioned to investigate and report on this with regard to modern technology, and to propose a concept for a modern scaffolding system to replace the indigenous one.

Figure D5.3 shows a schematic of the bamboo scaffolding, giving the basic dimensions. Note that there are vertical, short and long horizontal, and cross members. The figure also shows a

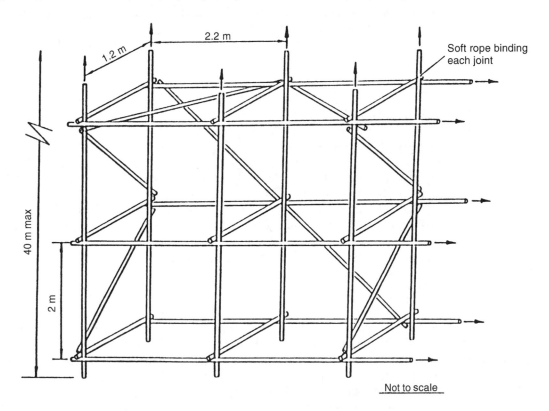

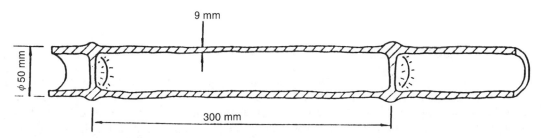

Fig. D5.3 Bamboo scaffolding (above). Bamboo is light, stiff and strong, partly because of its intrinsic structure and partly because of the shape of its section.

cross-section through a short length of typical scaffold bamboo. Bamboo scaffolding is used to a maximum height of 40 meters.

(a) Discuss the problems associated with each member type, describe its potential failure mode, and any problems associated with the fixing points.
(b) For the design of a new proposed system, outline the main design and materials selection steps, showing the feedback between the material choice, the section shape and the modes of loading.
(c) Select the best candidate material for each member, with respect to performance at minimum weight, then with respect to performance at minimum material cost, and finally, with respect to material energy content. Explain your choices.
(d) Discuss the problem of joining scaffolding members, recalling that rapid deployment and removal are essential. (†††)

D6 Selecting processes

D6.1 Making a shaker table

The shaker-table of Figure 6.31 is to be made of a magnesium alloy (melting point $= 905$ K; hardness $= 800$ MPa; density $= 1.78$ Mg/m^3). The diameter of the table is 2 m; the thickness of the table and its webs is 100 mm. The top surface and hub of the table are to be finished to a tolerance, T, of ± 0.07 mm and a RMS roughness, R, of 5 μm. The finish of the remaining surfaces is not critical. Suggest possible process routes. (Use the Size/Slenderness chart, Figure 11.31, and the Tolerance/Roughness chart, Figure 11.33. or the *CPS* software if available.) (†††)

D6.2 Forming a fan

A small fan mixes fuel–air mixture as it enters the burner-can of a high performance aircraft gas turbine. The fan runs at 650°C and is at present made of a stainless steel (density $= 7.9$ Mg/m^3). It is proposed to reduce the weight of the fan (to allow it to accelerate more quickly) by material substitution. In particular, the fan (it is proposed) might be made of boron carbide (melting point $= 2620$ K; hardness $= 24$ GPa; density $= 2.5$ Mg/m^3), without change of shape.

The fan has a simple shape: a disk, 100 mm in diameter and 5 mm thick, carries 8 radial fins attached to one of its faces; the fins are 5 mm thick and project 15 mm normal to the face. The tolerance on disk and fins is ± 0.3 mm, the RMS surface roughness should be less than 3 μm. Use the process charts (or the *CPS* software if available) to identify possible ways of shaping the fan. (†††)

D6.3 A computer case

A case is required for a notebook computer. The sales department insists on an A4 footprint, and a thickness no greater than that of a paperback novel. Translated into more rational units, the outer dimensions of the case are $280 \times 220 \times 20$ mm, with a wall thickness not exceeding 2 mm. It is to be made in two pieces (a base and a lid, each about the same size) from a tough thermoplastic. The tolerance T on the larger dimensions is specified as ± 0.5 mm; the RMS roughness R must not exceed 0.1 μm. (Use the Size/Slenderness chart, Figure 11.31, and the Tolerance/Roughness chart, Figure 11.33. or the *CPS* software if available). (†††)

D6.4 Electron-microscope grids

Small grids of copper (melting point = 1360 K; hardness = 160 MPa; density = 8.96 Mg/m^3) are required to support samples for electron microscopy. The grids are circular disks, 5 mm in diameter and 0.2 mm thick, pierced by a grid of small apertures. A tolerance better than ±0.03 mm, and a surface roughness less than 0.5 μm is required. Use the Size/Slenderness chart, Figure 11.31, and the Tolerance/Roughness chart, Figure 11.33. or the *CPS* software if available. (†††)

D6.5 A granite laser bench

A granite slab is to be used as an optical bench for laser experiments. The top surface is required to be flat to within 0.01 mm, with an RMS roughness of less than 0.2 μm. Can this be achieved by cutting and grinding? (†††)

D6.6 Choosing the cheapest process

A component can be manufactured by machining from the solid, cold forging and cold extrusion. Approximate cost data, in units of the material cost of the component, are given in the table. (Assume that the capital cost of equipment and the cost of power, space and research and development have been absorbed in the overhead rate). Advise on the cheapest process for a batch size of (a) 1, (b) 100 and (c) 10,000. (†††)

Process	Machine	Forge	Extrude
Material cost, C_m	1	1	1
Overhead rate, C_L (hr^{-1})	150	150	150
Tooling cost, C_t	30	3000	9000
Production rate, $\dot{n}$ (hr^{-1})	1	3	15

D6.7 Casting costs

In an analysis of the cost of casting of a small aluminium alloy component, costs were assigned to tooling, overhead and materials in the way shown in the table. The costs are in units of the material cost for the component — in this example, this was £0.04. Identify the cheapest process for a batch size of (a) 100 units, and (b) 10^6 units. (†††)

Process	Sand casting	Investment casting	Pressure die	Gravity die
Material cost, C_m	1	1	1	1
Overhead rate, C_L (hr^{-1})	500	500	500	500
Tooling cost, C_t	50	11 500	25 000	7500
Production rate, $\dot{n}$ (hr^{-1})	20	10	100	40

D7 Use of data sources*

The problems of this section involve data retrieval.

D7.1 Data for cast iron

Use standard data books to retrieve the density, modulus, strength and fracture toughness of plain cast iron. (Primary references: *The ASM Metals Handbook, Smithells* and the *Chapman and Hall Materials Selector*). Then do the same, using the *CMS* software, if available. (†††)

D7.2 Data for magnesium alloys

The candidate materials for a shaker table, identified in Section 6.16, p. 137, included magnesium alloys. Use standard handbooks, first, to locate a castable magnesium alloy with high damping capacity (above 0.5%), and then to retrieve data for the design-limiting properties: modulus and density (for vibration frequency), loss coefficient, and 0.2% proof strength. (Primary references: *The ASM Metals Handbook, Smithells* and the *Chapman and Hall Materials Selector, CMS* software if available.) (†††)

(DTD 5005 or ASTM HZ32A, C80-63)

D7.3 Data for PTFE and leather

Candidate materials for elastic hinges (Section 6.10, p. 116) include PTFE (polytetrafluorethylene, Teflon) and leather. Use standard data books to locate information — as far as you can — for the density, modulus, strength, cost of these. (Primary references: *Materials Engineering Materials Selector, Chapman and Hall Materials Selector*, and the *Handbook of Polymers and Elastomers*.) Then do the same, using the PLASCAMS or the *CMS* software, if available. (†††)

D7.4 Thermal properties of oak

Use standard sources to locate the thermal properties of oak: the thermal conductivity, specific heat, density (and hence the thermal diffusivity) and expansion coefficient. Start by reading approximate values from the Charts. Then try the standard handbooks. Finally, turn to the *CMS* software, if available. (†††)

D7.5 Thermal properties of granite

Granite is used for optical benches and as a stable support for precision metrology. One measure of the suitability of a material for precision instrumentation was (Section 6.20, p. 151) the index

$$M = \lambda/\alpha$$

where λ is the thermal conductivity and α the thermal expansion coefficient. Locate data for these two quantities, for granite (Handbooks first, *CMS* software second). (†††)

---

* All references to handbooks and databases are detailed in Chapter 13, Appendix 13A.

D7.6 Data for silicon carbide

Weight is important in choosing materials for the rotor of a turbocharger: a light rotor speeds up faster, making the charger more responsive to the driver's needs. It is suggested that weight could be saved by making the rotor out of silicon carbide instead of steel. Write a brief report (along the lines of the Case Studies of Chapter 14) detailing the mechanical, thermal, wear and processing properties of silicon carbide, drawing on standard data sources. Comment on the reliability of the data. (Primary sources: Morrell: *Handbook of Properties of Technical and Engineering Ceramics, ASM Engineered Materials Reference Book*, and the *Handbook of Ceramics and Composites; CMS* software if available.) (†††)

D8 Material optimization and scale

The problems of this section illustrate selection when a length-scale is involved.

D8.1 Scaling law for nails

Nails come in a range of sizes as shown in Figure D8.1. When their diameter is plotted against their length on log scales, as shown in the lower part of the figure, the points fall on a line of slope 2/3. This means that nails do not scale isometrically: long nails are relatively thinner than short ones.

(a) Develop a model for the design of a nail, assuming that the length, for a given diameter, is limited by elastic buckling, and that the force required to drive a nail into wood is proportional to the diameter of the nail — this is found, by experiment, to be so. Use your model to explain the scaling law revealed by Figure D8.1. Use data from the figure to determine the value of any unknown constants in your model.

(b) Use your results to explain why copper nails are fatter than steel ones of the same length. Explore whether the model will account accurately for the separation of the lines describing copper and steel nails.

(c) It is required to produce a range of aluminium alloy nails. Using your model, calculate the appropriate diameter for an aluminium nail of length 50 mm.

(d) Comment on the general problem of the way in which material choice affects proportion, and vice versa.

(Young's modulus, E, for steel, is 210 GPa; for copper it is 124 GPa; for aluminium and its alloys it is 71 GPa.) (†††††)

D8.2 Optimizing heat treatment of steel

The drum of a high-speed centrifuge consists of a thin-walled cylinder of radius 1 m which spins about the cylinder axis. The principal loading is that due to centrifugal forces acting on the material of the drum, which may contain internal crack-like defects with a size of up to 2 mm (the resolution limit of the inspection procedures). It is required to maximise the angular velocity of the centrifuge, but cost must also be taken into consideration.

(i) Outline briefly the steps involved in selecting a material for the drum.

(ii) Given a free choice, identify potential candidate-materials, using the Materials Selection Charts as appropriate.

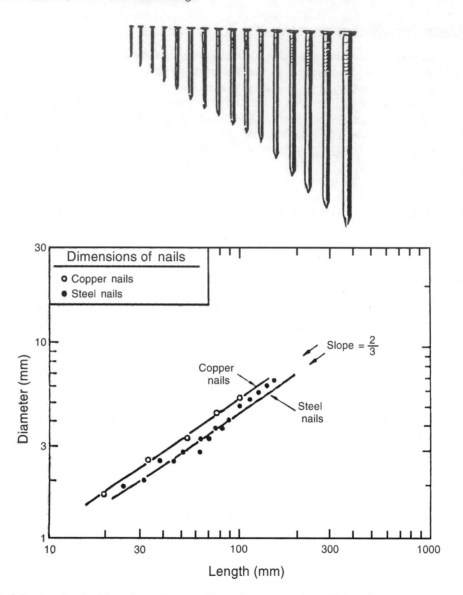

Fig. D8.1 A family of nails (above) and the way their diameter varies with length.

(iii) A heat-treatable steel (density $7.9\,\mathrm{Mg/m^3}$) is selected as a candidate for further investigation. Heat treatment changes both the yield strength and the fracture toughness; the two are related by the equation

$$K_{Ic} = \frac{60\,000}{300 + \sigma_y}$$

where σ_y is in units of MPa and K_{Ic} is in units of $\mathrm{MPa\,m^{1/2}}$. Assuming that the steel contains no defect larger than 2 mm in diameter, calculate the optimum yield strength to be obtained by heat treatment of the steel and hence find the maximum safe angular velocity. (†††††)

D8.3 Optimizing fibre fraction in composites

You have been asked to evaluate the design of the main cross-beam which supports the mast of a new class of ocean racing catamaran. It must be as light as possible.

(a) The manufacturers have suggested using a glass fibre reinforced plastic (GFRP) beam of length, L, equal to the overall beam of the catamaran, and width, w. They have considered it to be simply supported at each end and loaded by the mast at the centre. The beam must have a stiffness greater than or equal to S. The volume fraction, f, of glass fibres (laid unidirectionally) can be varied between 0 and 0.6, and the depth, d, of the beam varied to produce the best combination of properties. The Young's modulus and the density of the composite depend linearly on the volume fraction of the glass fibres. Ignoring self-weight, show that the optimum fraction of glass in the beam for minimum weight is:

$$f = \frac{\rho_m}{2(\rho_g - \rho_m)} - \frac{3E_m}{2(E_g - E_m)}$$

where ρ_m, E_m are the density and modulus of the epoxy matrix polymer, and ρ_g, E_g are those of the glass. Can the optimum for this shape be fabricated?

(b) Briefly, discuss the considerations that influence the choice of production method for this beam, with reference to precision, tolerance, surface finish, batch size, and joining. (†††††)

Index